Rivers of Eden

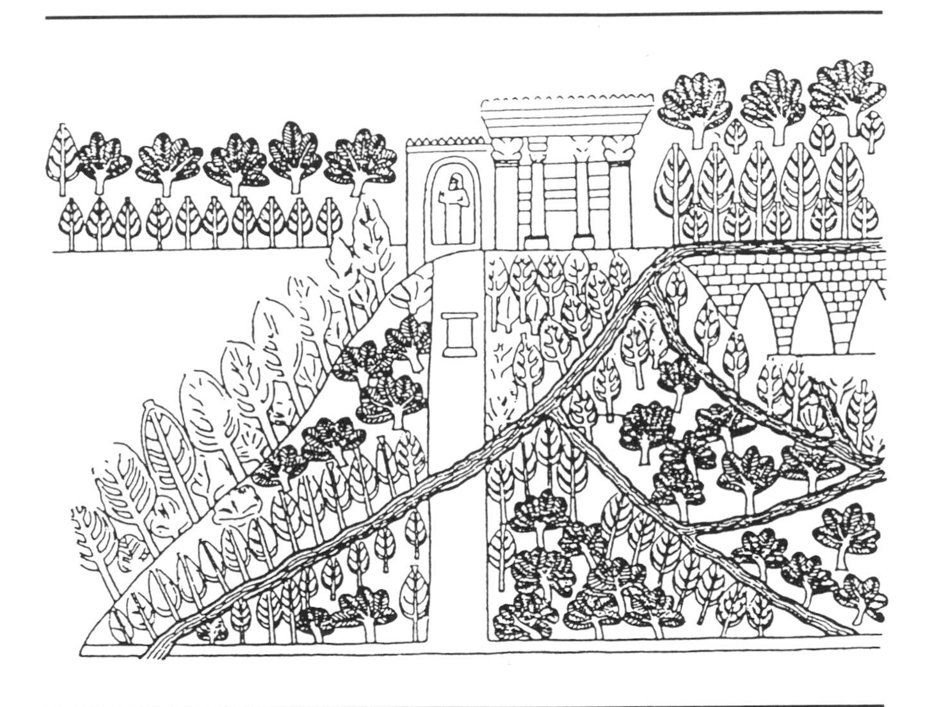

A canal-watered garden in ancient Assyria.

Rivers of Eden

The Struggle for Water and the Quest for Peace in the Middle East

Daniel Hillel

New York *Oxford*

Oxford University Press
1994

Oxford University Press

Oxford New York Toronto
Delhi Bombay Calcutta Madras Karachi
Kuala Lumpur Singapore Hong Kong Tokyo
Nairobi Dar es Salaam Cape Town
Melbourne Auckland Madrid

and associated companies in
Berlin Ibadan

Published by Oxford University Press, Inc.,
200 Madison Avenue, New York, New York 10016

Library of Congress Cataloging-in-Publication Data
Hillel, Daniel.
Rivers of Eden : the struggle for water and the quest for peace in the Middle East / Daniel Hillel.
p. cm. Includes bibliographical references and index.
ISBN 0-19-508068-8
1. Water resources development—Middle East.
2. Water-supply—Political aspects—Middle East.
3. Riparian rights—Middle East.
I. Title. HD1698.M53H55 1994 333.91'00956—dc20
94-19092

Illustration Credits
U.N. Photo 153777/Jeffrey Foxx, p. 9; U.N. Photo 132536/Gamma, p. 11; U.N. Photo 80769, p. 12; U.N. Photo 103913, p. 116; U.N. Photo 156041, p. 117; U.N. Photo, 107130, p. 179; U.N. Photo 111971, p. 195; U.N. Photo 131336, p. 214; Courtesy Netafim Co., Israel, p. 226; Courtesy Netafim Co, Israel, p. 227

1 3 5 7 9 8 6 4 2

Printed in the United States of America
on acid-free paper

Dedicated to all who advance the cause of

PEACE שלום أَلسَّلام

Preface

Behold, I extend peace like a river, and the honor of nations like a mighty stream.
Isaiah 66:12

They will not hear any vain discourse, but only salutations of peace.
Koran XIX:62

If the mythical Eden ever had a real geographical location, it must have been somewhere in the Middle East. Humanity's first abode, as described in the Book of Genesis, was, literally, a Garden of Delight. In it grew "every tree that is pleasant to the sight and good for food," all watered by a primordial river and its distributaries. Adam and Eve, whose names meant "earth and life," were put in the garden for a purpose—"to serve and preserve it."

Alas, as the biblical story unfolds, the first humans soon abused God's trust, yielding to temptation and consuming beyond their needs. In so doing, they banished themselves from the garden by despoiling it. They and their descendants were thenceforth condemned to a life of toil: "Cursed is the earth for thy sake. . . . With the sweat of thy brow shalt thou eat bread till thou return to the earth." One of their sons learned to domesticate animals and became a herder, while the other learned to till the soil and grow crops.

The part of the Middle East where humans settled some 10 millennia ago is known as the Fertile Crescent. It was a hospitable region, blessed with a benign climate and abundant biological resources. Now this same region is a largely degraded landscape in which most of the people are impoverished and insecure.

The problems besetting the Middle East today are multifaceted and exceedingly complex. Ethnic, national, religious, economic, social, and historical conflicts combine to make the present scene particularly contentious. The issues are highly emotive and so do not lend themselves readily to dispassionate analysis and rational solution. Some people in the region do not believe a peaceful solution is possible at all, and thus are driven by desperation and bitterness to acts of violence. The vast majority of the

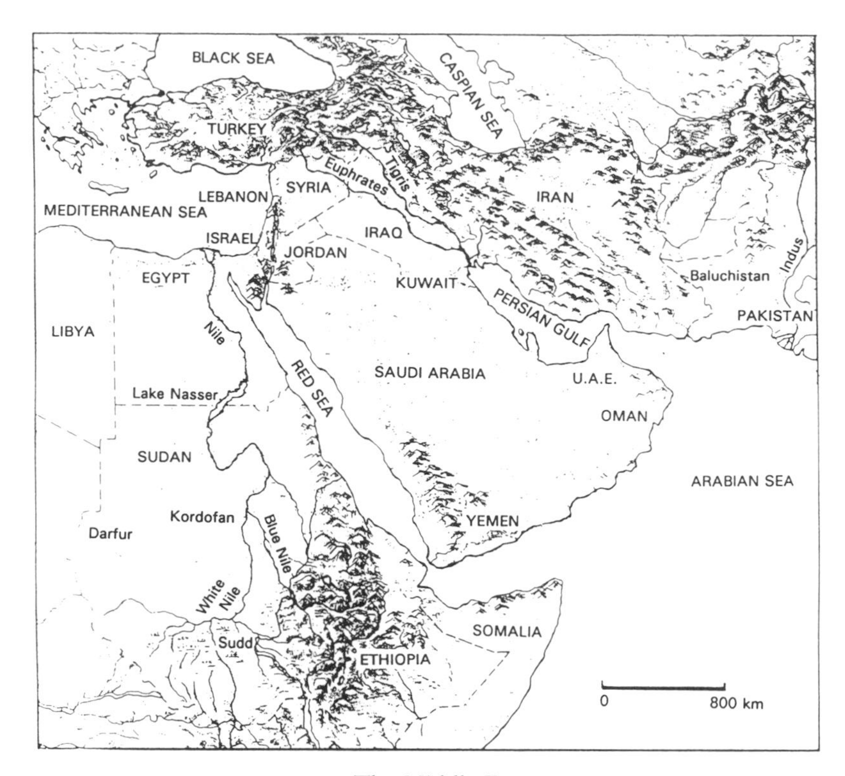

The Middle East.

people, however, are weary of conflict and yearn for a just peace and an opportunity to turn their faith and energy toward progress and development.

The good of the region and all its people demands a new approach, cognizant of the past but not bound by it, aware of the deprivations and grievances of the present but not obsessed by them. Needed above all is a forward-looking vision of the region's potential strengths and positive destiny, attainable in a context of international cooperation.

During the greater part of the twentieth century, the nations of the Middle East have been embroiled in seemingly endless turmoil. Now, in the last decade of this tumultuous century, prospects have arisen for a momentous change. Political leaders have at last begun to respond to the real needs of their people and to pursue peaceful compromises rather than violent confrontations among their nations. However, no mere political formula will heal the region's maladies unless it redresses their root causes.

The general aim of this book is to reveal an important dimension of the Middle East's fundamental predicament, one that is still insufficiently understood and therefore typically ignored by politicians and by the observers who too often report only superficially on the scene. That dimension is the severely wounded environment. Its manifestations include widespread destruction of vegetation and natural habitats, erosion of uplands and watersheds, waterlogging and salinization of valleys, desertification of semiarid areas, and—in particular—depletion and pollution of water resources. Combined with an uncontrolled explosion of population, these processes undermine the economic and social welfare of entire societies and cause the dislocation of numerous people. Environmental degradation thus contributes to despair and extremism.

The more specific message of this book is that the acute shortage of water, if left unresolved, is bound to further exacerbate tensions and conflict. No country in the region can resolve its water problem independently without encroaching upon the resources of its neighbors. Hence no comprehensive water development can take place without peace, and—conversely—no peace is possible or sustainable without such development. The very problem that engenders rivalry and threatens to instigate war can and must be turned into an opportunity and a powerful inducement to promote peace.

The book describes the background and nature of the water problem and offers some ideas toward its peaceful resolution. In so doing, it necessarily explores numerous related issues in the context of the region's history and contemporary circumstances. The ultimate aim is to bring the positive prospects of the Middle East more clearly to the attention of policymakers and opinion makers, who might henceforth act more effectively to rehabilitate the ravaged lands and waters of the region and thereby alleviate the plight of the people dependent on those resources. Only thus can the Middle East be restored to the Eden and center of culture that it once was.

As author, I have striven to present the contentious problems with fairness to all sides, while avoiding inflamed political polemics. Of course, no human being is omniscient or able completely to transcend private perceptions. Yet my conscious intention has been to analyze the sensitive issues as honestly and impartially as I could. The region is now poised on a fateful watershed divide between a dangerous *casus belli* and a promising *casus pacis.* My fervent wish is that the latter prevail over the former, and that a hopeful future will indeed overcome an unhappy present.

Several colleagues read all or parts of the manuscript and offered constructive comments. Among them were Dr. David Hopper, former Senior Vice President of the World Bank; Dr. Shawki Barghouti, Chief of the Irrigation Division and later of the India Division of the World Bank; Mr. Shaul Arlosoroff, then head of the World Bank-UNDP Water and Sanitation Program; Professors Muhammad Jiyad (Near Eastern Studies), David Alexander (Geography), and Haim Gunner (Environmental Sciences) at the University of Massachusetts; Mr. Hassan Khalilieh of the Near Eastern Studies Program at Princeton University; Mr. Joseph Spieler and Ms. Lisa Ross, my literary agents; Ms. Joyce Berry, Senior Editor, Ms. Susan Hannan, Development Editor, and Ms. Laura Calderone, Assistant Editor, at

Oxford University Press; and Dr. Michal Artzy, of Haifa University's Department of Archaeology, to whom I am particularly grateful for apprising me of historical sources pertaining to water in the ancient Middle East.

During the preparation of the book, I have held discussions with numerous additional colleagues and experts. Among them were Professor Dan Zaslavsky and Mr. Menachem Kantor, each a former Water Commissioner of Israel; Mr. Joshua Schwartz of Tahal, Water Planning Engineers, and Mr. Elisha Kally, formerly of the same group; Professor Hillel Shuval of the Hebrew University; Dr. Munther Haddadin, chief Jordanian water negotiator at the Arab-Israeli peace talks; as well Dr. Maher Abu-Taleb of Amman; Professor Said Assaf of the Arab Institute for Research and Transfer of Technology; Professor Muhammad El-Raey, Vice Dean of Environmental Studies at Alexandria University; Professor Abdallah Bazaraa and Dr. Magdy Saleh of the engineering faculty at Cairo University; Dr. Dia El-Quosy, Director of the Water Distribution and Irrigation System Research Institute of Egypt; Messrs. John Hayward and Ulrich Kuffner of the World Bank task force on the development of water resources in the Middle East; Dr. David Brooks of the International Development Research Centre of Canada; Professors Yvonne Haddad and Adnan Haydar of the University of Massachusetts; and others too numerous to list. Each of them contributed valuable insights and thus helped to improve the content of this book. None of them, though, is to be held accountable for the book's shortcomings, for which I alone bear responsibility.

A special acknowledgment is due to the John Simon Guggenheim Foundation for awarding me a Fellowship and lending support to the research that has culminated in this volume.

After many years of studying water in the Middle East, I have come to realize that it is an issue so fluid and pervasive, and so affected by rapidly evolving circumstances, as to defy complete containment or closure. I therefore offer this book to prospective readers in the hope that it might at least whet, if not entirely quench, their thirst for knowledge of the vital subject addressed herein.

Amherst, Mass. D. H.
January 1994

Contents

1. **An Overview**, 3
   A transregional journey, and a realization
2. **Waters of Life**, 20
   The planet's, and the region's, most precious fluid
3. **Ancient Civilizations**, 41
   The rise and demise of societies in an arid environment
4. **Modern States**, 74
   How the contemporary map of the region was drawn
5. **The Twin Rivers**, 92
   New rivalries over the ancient waters of Babylon
6. **The Mighty Nile**, 111
   The mother river and her many thirsty children
7. **The River Jordan**, 143
   Dividing the earth's most storied river
8. **The Flowing Streams of Lebanon**, 177
   Disputed waters in a fractious state
9. **Fountains of the Deep**, 190
   Tapping the underground pools, rechargeable or fossil
10. **The River of Waste**, 210
    An unrealized resource: conservation, efficiency, reuse
11. **Augmenting Supplies**, 232
    Inducing rain and runoff, trading waters, desalinization
12. **Criteria for Sharing International Waters**, 264
    Ethical, legal, and pragmatic considerations
13. **Water for Peace**, 279
    Calming turbulent waters

Notes, 299

Bibliography, 319

Glossary, 331

Index, 343

"In the beginning, when God created the heaven and the earth, the earth was without form and void . . . and the spirit of God hovered over the face of the waters. Then God said: 'Let there be light!' " (Genesis 1:1,2) and the act of Creation began.

Thus was water specified as the only substance to preexist Creation. Reflected in this depiction is an early intuition that the presence of water constituted the precondition for the initiation of life, and later for the advent of civilization. Today more than ever, water is the essential requirement of human welfare everywhere, particularly in that ancient Cradle of Civilization, site of the original Garden of Eden, the arid region known as the Middle East.

Rivers of Eden

1

An Overview

A river rose out of Eden to water the garden, and from thence it was parted and became four heads. Genesis 2:10

I begin with a personal testament. The topic of this book has long been of intense interest to me. Born in a man-made oasis in the semidesert of southern California, I was taken at an early age to Palestine, then in the first stages of reclamation from centuries of desolation. I spent part of my childhood in a pioneering settlement in the Jezreel Valley where, in biblical times, Gideon drove off the pastoral nomads who would encroach periodically on the laboriously cultivated fields of settled farmers.

Here, the ancient contest for survival and supremacy between the desert and the sown, between the mutually dependent yet ever competitive descendants of Abel and Cain, has been waged since civilization began. And it was here, at the frontier of life, that I was first captivated by the region's environment and its contrasting counterpoints of sky and earth, soil and water, rainfall and drought, native and domesticated plants and animals, wilderness and agriculture.

I remember myself as a child of nine, sloshing barefoot in an irrigation furrow, striving with spade in hand to direct the frothy waters as they slaked the harsh dry clods, marveling at the exuberant growth of tender saplings in the tiny orchard that rose up so defiantly in the midst of the wind-blown dry plain. At the end of our working period, my friends and I would immerse ourselves in the ditch and wallow in the squishy ooze like lazy water buffaloes. And on occasional moonlit nights during the long rainless summer, we would sneak away from the grownups to wade into the dark waters of a hilltop reservoir, there to float quietly on our backs and gaze at the luminous sky.

I learned to swim in the Jordan River. Between the ages of twelve and fourteen, I spent my summers in a kibbutz in the Jordan Valley, just south of the Sea of Galilee. There, a dam had been built to control the flow of the river below its outlet from the lake. When the sluices were shut, that stretch of the river became a deep quiescent pool, with soft muddy banks from which my friends and I could dive or slide into the water and swim across to the thicket of reeds on the opposite bank.

The ultimate in daring was to trek downstream, past the dam, to where the river began its sinuous course through the spectacular Rift Valley toward the distant Dead Sea. In late spring, when the dam was opened and the river was in spate and flowed full force, we would throw ourselves into the gushing current and be swept around the curve of a nearly circular meander, then grab onto the overhanging tamarisk branches to emerge from the water and walk across the narrow neck of the peninsula. There we would reenter the river on the upstream side and repeat the feat again and again. The whirlpools were treacherous on the convex side of the curve, where the swirling water undercut the steep and unstable bank, so the entire deed was a rather foolhardy test of youthful courage. But we did all this with sheer delight and heady abandon, completely mindless of the river's epic past and sacred significance.

Though I was not aware at the time, those experiences were seminal moments in my life; they influenced the entire course of my career. An early fascination with water and land grew over the decades to become both a vocation and an avocation, a professional pursuit and a labor of love.

Early in the fateful year 1914, on the eve of the First World War, two young British archaeologists, C. L. Woolley and T. E. Lawrence, explored the deserts of Negev and Sinai, then frontier provinces of the Ottoman Empire. Ostensibly to study the antiquities of these deserts, the mission's major and clandestine aim was to gather intelligence for the British War Office in anticipation of imminent conflict over the eastern approaches to the Suez Canal. However, archaeology was more than a mere guise or clever ruse by which to allay the suspicions of the Turkish authorities, for the study turned out to have authentic scholarly significance. It was published in 1915 by the Palestine Exploration Fund under the title *The Wilderness of Zin*.

The younger of the two archaeologists, then only 25 years of age, soon went on to become the fabled "Lawrence of Arabia," himself a part of the region's history. *The Wilderness of Zin*, however, remained for many years the definitive work on the ancient civilization of the Negev. Sketchy and superficial though it may seem in the light of present-day knowledge, it was, for its time, a remarkable document, filled with fascinating information and intriguing insights.

Still, when the State of Israel came into being, in 1948, that desert seemed largely to be a tabula rasa, a vast and mysterious terra incognita of rugged mountains and barren valleys in which lay hidden (according to

Woolley and Lawrence) the ruins of no fewer than six ghostlike cities, remnants of long bygone civilizations. What a lure for inquisitive young adventurers!

An inquisitive young adventurer was I in 1951, when, at the age of 20, having just gained a master's degree in the earth sciences from an American university and regarding myself the "compleat" scientist, destined for great discoveries, I joined with a band of equally unrealistic dreamers to embark upon the greatest adventure of our lives. Together, we founded the frontier settlement of Sdeh-Boker ("Herdsman's Field," to signify the hoped-for linking of pasturing and farming) in the very heart of the Negev highlands, where the depth of a year's total rainfall seldom exceeds the hollow of a man's cupped hands. In this land of primeval grandeur and pervasive silence, the loudest noise ever heard was the incongruous rumble of a rare torrent hurtling down a ravine after a chance thunderstorm over distant mountains.

With neither plan nor sponsor, we pitched our tents on the parched plateau near the chasm called the Valley of Zin. Standing on the dizzying edge of that chasm, gazing into the desolate, boulder-strewn ravine and across the cratered moonscape toward the mist-shrouded massif of Edom, in which nests the legendary temple city of Petra, we felt ourselves transposed in time, as if carried back magically into the distant past. In our mind's eye, we could see Ishmael the archer, hiding behind a rock with bow in hand, hunting an ibex; or Moses with his rod leading the Israelites, a band of destitute fugitives, God-intoxicated and inspired by a vision of their own destiny and mission.

All too soon, our fantasy and reverie were shattered by reality. Our first year in the Negev was almost totally rainless. When the fervently anticipated wadi flood finally came, it washed away the dikes we had built so laboriously in crude imitation of the ancient dikes whose remnants we found in the same area. It was then we realized that we must study the records of prior civilizations very thoroughly, in order that we might build more solidly upon, and extend, their experience. The *Wilderness of Zin* and the Bible were the only textbooks available to us at first.

The Bible, incidentally, conveys different meanings to different people. To some, it is a purely religious tract. To others, it is an authentic historical account. In fact, it is a complex, heterogeneous compendium of writings composed over many centuries, embodying the living experience, lore, mythology, and spiritual quest of generation after generation of Israelites, living and struggling in and around the land first called Canaan. The Bible thus describes how a band of pastoral desert nomads settled permanently on the land and learned the ways of soil and water husbandry. As such, it is replete with information that is of interest to environmental scientists, inasmuch as it refers—again and again—to the interaction of a nation with the land and its waters, and with its flora and fauna.

Wishing to learn more about the ecology of the desert, I spent five years studying the water regimes of different habitats in the Negev and

Sinai. I lived for a time with a tribe of Bedouins then leading an extremely austere life grazing their emaciated goats and camels on the sparse desert range. I spent many days with the herders and their flocks, wandering over winding wadis and rocky hillslopes.

One day I sat in a tent with the tribe's children at the feet of their old *mualem* as he was teaching his charges *turuk el adab,* the proper "ways of the world." The man had no formal schooling, only the wisdom of the ages to impart to his pupils. And he did so in the time-honored way of the sages, by metaphor, fable, and analogy. "How much is one and one?" asked the venerable instructor. "Two," ventured the eldest of the children. "Sometimes it is so, but sometimes not," replied the *mualem.* Then he explained: "If you put one she-goat in a pen and then another, surely you will have two. But if you put one he-goat with another, they will fight and one might be killed so that one and one would end up being one; or both might die so that one and one would be none. On the other hand, if you put in one she-goat and one he-goat, you might eventually have three, or four, or more!" So, concluded the worldly-wise old teacher: "How much is one and one? That depends on circumstances and on inclinations. Nothing is certain. And so it is with human beings, whether good or bad. And may Merciful Allah provide for the good."

In the years that followed, I have had frequent occasion to ponder the wisdom of the old Bedouin master. His dictum encapsulated an ecological truth with profound implications. In Euclid's simplistic axiom, the whole is equal to the sum of its parts: one plus one equals two. That is so when the parts are inert, non-interactive, sterile. Not so in real life, where the whole includes not merely the sum of discrete parts but also the gamut of their complex interactions, both synergistic and antagonistic. In a living community (an "ecosystem") of plants, animals, and humans, each entity is defined not alone by its individual characteristics but also by its relationships and interactions with all other entities sharing the same domain.

Regarding the volatile inclinations and circumstances of the Middle East, the wise old teacher certainly had a point. As long as the people of this sensitive region could live in relative stability in their environment and among themselves, they could develop culture and attain prosperity and progress. But whenever that stability was disrupted, strife and destruction resulted. Often, the cause for strife has been rivalry over scarce and abused resources: farming and grazing lands, and—in particular—access to water. So sensitive and jealous are the region's people toward any usurpation of their rights to these resources, that at times strife could be triggered by merely perceived rather than real threats to their integrity.

In the history and lore of the Middle East, examples abound. The region's penchant for futile violence is exemplified by the legendary "War of Basoos." In ancient times, so goes the story, two neighboring tribes had long coexisted in peace. They had shared a single well located on the boundary of their territories. Each tribe had access to the water from its own side of the common trough. One day, a camel from one tribe wandered across the boundary and drank from the wrong side of the trough. Incensed at this breach of his tribe's sacred territorial rights, the son of the sheikh whose realm was violated killed the offender. Alas, that hapless creature just happened to belong to the son of the neighboring sheikh, who then rose in anger to smite the killer of his beloved *naaqa*. There began a blood feud between the tribes that lasted five generations and cost countless lives on both sides. It was known as the "War of Basoos," though no one remembered that Basoos, the cause of the strife, had only been an errant camel.

When the fighting ended, neither tribe had won over the other and both were relieved to reestablish their original sharing arrangement. Even the self-generating hatred born of reciprocal violence could not prevail over the imperative to resolve the dispute and restore an equitable water supply to all.

The region generally termed the Middle East encompasses northeastern Africa and southwestern Asia. It is, to be sure, a geographically imprecise and Eurocentric term, and I use it—albeit reluctantly—simply because it has somehow gained universal acceptance. In the nineteenth century, it was common for European, and particularly British, geographers to divide the "Orient" into three regions. The one nearest Europe, extending along the eastern shores of the Mediterranean Sea and roughly coterminous with the old Ottoman Empire, was called the Near East. The second region, from the Persian Gulf to Southeast Asia, including Persia and India, was termed the Middle East. The third region, farthest from Europe and covering the countries of East Asia facing the Pacific Ocean, was designated the Far East.

The Middle East became a strategic reference for the British during the First World War when the operational theater of their "Mesopotamia Expeditionary Force" came to be distinguished as Middle East from the Near East of Palestine and Syria in which their separate "Egyptian Expeditionary Force" operated. In 1932 the Royal Air Force "Middle Eastern Command," then centered in Iraq, was amalgamated with the "Near Eastern Command" in Egypt, but the new command retained the title "Middle East." Then, at the outbreak of the Second World War, the British Army headquarters in Cairo followed the R. A. F. in calling itself "G. H. Q. Middle East."

Since the entire region was then within the sphere of influence of the British, their frequent use of the designation "Middle East" to describe the countries of Northeast Africa and Southwest Asia in military and news communiques made the term familiar throughout the world. Thus it was that an arbitrary designation employed by and for the convenience of a colonial power that has long since lost its influence became accepted internationally.

There are, however, no clear boundaries around the region thus defined. It is generally assumed to encompass the states or territories of Turkey, Syria, Lebanon, Iraq, Iran, Israel and the rest of Palestine, Jordan, Egypt, Sudan, Libya, and the various states of the Arabian Peninsula (Saudi Arabia, Kuwait, Yemen, Oman, Bahrain, Qatar, and Oman). Over time, some writers have tended to enlarge the region by inclusion of Pakistan, as well as the northwestern African states of Tunisia, Algeria, and Morocco, which are connected in language and culture to the Islamic Arab states. Others, however, have tended to reduce the area by excluding from it any African states other than Egypt.

Confusion over the specific territorial limits of the region appears to stem from a failure to distinguish among cultural, religious, national, political, historical, and geographical criteria. Some would identify the region as the domain of the Arabic people and language, but such identification would fail to account for the various non-Arabic nationalities and cultures in the region, including the Turks, the Iranians, the Kurds, and the Israelis. Similarly, attempts to identify the region with the Muslim world ignore the non-Muslim communities within the region as well as the many Muslim communities outside the region.

Though it could be argued that the changed circumstances since the demise of the British Empire should have rendered the expression "Middle East" obsolete, it has in fact been perpetuated through its continued use by various agencies of the United Nations, in international diplomatic and commercial circles, and—perhaps most influentially—by journalists. Willingly or otherwise, therefore, one must accept common usage.

The Middle East is our ancient cultural and spiritual home, the birthplace of Western civilization. Here, in the wake of the last ice age some ten thousand years ago, our predecessors made the fateful transition from nomadic hunting and gathering to settled farming. Here they first domesticated plants and animals, tilled and irrigated fields and orchards, established villages and cities, built temples and monuments, and organized nation states that grew into empires. Here they invented ceramics, metallurgy, mathematics, and writing. And here they conceived and enunciated universal ethical and religious ideals and codified them into laws.

Seeing the ravaged state of much of the Middle East today, one wonders how all of that could have taken place in this seemingly inhospitable corner of the earth. (The fabulous oil-based wealth of the few only accentuates the deprivation of the many.) The contrast between the brilliant record of the past and the depressed condition of the present demands an explanation. Why does the region that once led the world in innovation and culture now lag behind? How did the famed Fertile Crescent, the legendary site of the Garden of Eden, become so degraded, and its people so embittered?

Those haunting questions lay dormant in my mind for many years and might have continued so indefinitely had not a chance journey raised them into my consciousness and made the quest for answers a compelling task.

In the summer of 1988 I had occasion to fly from Khartoum via Cairo to Zurich, and—only a day later—from Zurich to Karachi. The beginning of my journey was in the arid plains of Sudan, where I was on an advisory mission for the World Bank to help supply water to drought-stricken villages. The end of my journey was in Pakistan, where I took part in the effort to improve irrigation in the desert province of Baluchistan. Those were not the first nor the last of my many assignments in the Middle East, a region in which I have, in fact, spent a major part of my life. In some palpable sense, however, that trip was the most revealing: a graphic transect across the entire region that illuminated its startling contrasts, its vast and largely empty spaces, its pervading aridity and despoiled environment, its uneven distribution of resources, and—paradoxically—its great potential.

In the weeks preceding that flight, I had traversed the plains of Darfur and Kordofan and the Red Sea province in Sudan. These provinces are contiguous with the great Sahel, a belt of land girding the entire continent of Africa and constituting the transition zone between the Sahara in the north and the tropical forest region in the south. It is that very habitat of grasslands and savannas where a group of primates evolved into the human species, and whence our progenitors ventured forth to colonize the entire earth.

Originally verdant with grasses and shrubs, the areas I saw had been parched beyond recognition by the droughts of the 1970s and 1980s, the cruelest in recent memory, whose effects were exacerbated by excessive human exploitation. Traveling cross-country over hundreds of miles, I could see not a single blade of grass or a leafy bush. The land lay forlorn,

Figure 1.1 The African Sahel during a drought.

prey to the sere winds of the desert. Carcasses of camels and cattle were strewn about, half buried in the loose dust. Only the vultures showed signs of life, as they swooped down from their perches on naked trees whose trunks were polished by pelting grains of sand.[1] The villages were practically empty, with but a few frail people, too old or too young to trek across the desiccated plains toward the already overcrowded cities along the Nile.

And yet I knew that Sudan is a land blessed with prodigious amounts of water. The Sudd area of southern Sudan is often described as the world's largest freshwater swamp, covering an expanse larger than the Everglades of Florida and the Okefenokee of Georgia. It is a practically impenetrable morass, a maze of streams and seasonally inundated marshes fed by inflows from the White Nile and its many tributaries. Only the inaccessibility of the area and a persistent civil war have thus far prevented some of the water from being diverted to Sudan's dry areas where it might turn human despair into hope.[2] For irrigation in Sudan has been practiced successfully over many decades. If improved and expanded, it could make the country one of the most productive in the world, thanks to the nearly ideal conditions created by the deep and fertile soils, the warm and clear climate, and the abundant potential supplies of water. Even without the additional water from the Sudd, much could be done to free up water now used wastefully by traditional flood-irrigation, if Sudan were to adopt new and efficient techniques of water conservation.

Such were my thoughts at the time. But for the moment my assignment was ended and I was at the Khartoum airport, there to board an

Figure 1.2 Women and children collecting firewood in the drought-denuded Sudan.

Figure 1.3 Shallow wells devoid of water during a drought in the African Sahel.

aircraft that would take me away from that drought-afflicted land. Despite the early hour, the heat was already oppressive. As the airplane rose, I saw Sudan's capital lying under a choking pall of reddish dust, an African "dust bowl" even more widespread than the one that scourged America's Great Plains in the 1930s. Arrayed along the narrow streets were gaunt eucalyptus trees, once shade-giving but now bereft of foliage, standing with arms outstretched like beggars in silent supplication. And there, flowing incongruously through the city, were the twin ribbons of water: the churning Blue Nile, dark with mud from the volcanic highlands of Ethiopia, and the sedate White Nile, milky with organic residues from the Central African swamps and rain forests. The two rivers join here to form the "elephant trunk" that gives the city its peculiar zoomorphic name, Khartoum.

The aircraft left the haze-shrouded city behind, and for the next few hours it followed the northward course of the united Nile, a river that slithers like a creeping vine through the wasteland of the eastern Sahara, giving rise to the largest oasis on earth. The Nile is also the world's longest river, its sources so distant from its outlet that one can easily perceive how the ancient Egyptians, living along its lower reaches, could have believed that it sprang by divine miracle out of the very bowels of the desert. The gently curving strip of life watered by the Nile is so incredibly narrow, so tenuous, so anomalous, that one marvels how it exists at all in the face of the desert's merciless sun, thirsty winds, and engulfing sands. But exist it does, with an exuberance of lush vegetation rising defiantly against the grim Nubian desert pressing on both sides, constantly threatening to obliterate it. No place on earth demonstrates more vividly what the presence or absence of water means to life.

For several hours I gazed through the narrow window, transfixed by the sight of the Nile wending its way through the expansive Sahara. Here and again I noticed the cataracts, where the river bursts through natural barriers, its suddenly maddened waters cutting the rocks into jagged teeth that jut out of the swirling current. Along other stretches, the translucent desert air revealed a river in a tranquil mood, its lustrous waters crawling languidly like a lazy snake.

Figure 1.4 Women in the Sudan carrying water from the White Nile.

Then a most improbable sight, a seeming mirage: a glistening, widening pool of absolutely motionless fluid, resting in the valley like a mass of molten silver. I had to blink to reassure myself that I was not suffering eye fatigue from staring too long at the blindingly bright landscape, before realizing with a start that I was seeing Lake Nasser. A phenomenon contrary to nature: a huge elongated lake in the midst of the largest and driest desert in the world, a giant reservoir created by the hubris of engineers who took it upon themselves to plug the ancient river at Aswan. Here the seasonal surge of the mighty Nile is held back, contained, tamed. Henceforth, the fabled floods that had visited Egypt for millennia are no more than a fading memory. Seen from above, however, the Aswan High Dam itself appears so puny that one wonders how so small a plug can hold back so massive a body of water, 10 or more kilometers wide and over 500 kilometers long—one of the world's greatest man-made lakes.

As I scanned the lake that day, I was surprised to see how much it had shrunk as a result of the prolonged East African drought: the receding shoreline had left behind a wide fringe of exposed lake bottom, dark with dry and cracked alluvium. That was the year that the water in Lake Nasser had fallen to such a low level that Egypt was in danger of having to curtail its production of food and electricity. As it happened, that threat was averted by the eleventh-hour appearance (just a few weeks after my journey) of freakish floods, which, capriciously, devastated Khartoum with a sudden surfeit of water even as they replenished Lake Nasser and restored the dwindling waters—and hopes—of Egypt.

My flight continued past Aswan and its High Dam, along the Nile of Upper Egypt, past the storied ancient sites of Kom Ombo, Idfu, and Luxor—and on to Cairo, at the cusp of the river where it separates into distributaries and fans out like the veins of a palm leaf to form the wide Delta. The divided waters flow listlessly here, as if seeping through a giant sponge.

Egypt's capital lay below us, half smothered by a turbid cushion of smog. As the plane descended into the haze, I saw the swarming streets of that overcrowded city with its 12 million or more people and its haphazard traffic of every description. The plane landed, and soon took off again, passing over the fruitful fields of the spreading Delta with its intricate network of irrigation and drainage canals.

The entire Delta is covered by a rich dark alluvium, which is the reluctant gift of distant Ethiopia. Annually, the volcanic slopes of that mountainous country are raked by the torrential monsoons that swell the Blue Nile and send it forth gushing with water and silt. The ancient Egyptians, noticing that the Delta was being augmented by the Nile, actually believed that the river's water itself was transmuted into solid earth. Because that dark loamy soil contrasted so starkly with the light-hued sands of the surrounding desert, the Egyptians named their country for it: *khami,* meaning black. And, since that fertile soil was considered the mother lode of all substances, the Greeks borrowed its name for the term *khemia,* from which was derived our term "chemistry," the science of matter in general.

Within the Delta, another anomaly: splotches of sterile land, glistening with a fine powder of crystallized salt. Soil salinity is a worsening malady in this ancient granary. For thousands of years, while salinization plagued Mesopotamia and other irrigated river valleys, it was a relatively minor problem in Egypt. Its recent exacerbation here is one of the unforeseen effects of damming the river.

Before it was dammed, the Nile's annual pulsation and the associated fluctuation of the water-table created an automatically self-leaching cycle in which the salts were flushed away by the Nile itself. Now, however, the water in the river is always kept at a high level, which inhibits the natural drainage that had previously kept the soils free of salts. Thus it is that Egypt had begun lately to suffer an eleventh plague to which it had seemed immune for so many centuries.

I observed this malady on the ground during my several working assignments in Egypt. While there, I also noticed that some farmers were carrying donkeyloads of sand from the adjacent desert and spreading it on their fields in a last-ditch effort to keep the surface above the rising groundwater, thus hoping to stave off the waterlogging of their soils.

As we flew along the edge of the Delta I glimpsed some of the "New Lands": desert tracts on the fringes of the floodplain, reclaimed from the Sahara in recent years. The diversion of water to these tracts is part of Egypt's effort to expand farming to keep pace with its swelling population, which has grown 20-fold in less than two centuries. But the effort is only partially successful. During a prior assignment in Egypt, I had seen huge center-pivot irrigation rigs, installed in hopes of achieving instant modernization, standing motionless and rusting under the sun. At home in the American West, these supermachines are thoroughly alien to Egypt. Since their parts are all made abroad, a single broken cog suffices to paralyze one of these behemoths and turn it from a lifegiver into a grotesque relic. Such are the perils of centrally planned technology transfer. Had the same investment been applied to the development of smaller units made locally and adapted to the scale and character of irrigation in Egypt, and had the people operating them been given direct ownership or effective incentives to maintain them, the scheme might have worked much better.

The view of Egypt's New Lands brought to mind my earlier experience in the Negev, where I took part in developing irrigation methods suitable for desert areas. I remember one day in particular, when a long-awaited pipeline had been completed to an outpost settlement and a sprinkler system installed in a tract of dry land that was thenceforth to become an irrigated field. Among the guests invited to mark the joyous occasion was a Bedouin sheikh of the neighboring tribe. At the appointed moment, the valve was turned and all at once the sprinklers began to spew glistening jets into the clear desert air and onto the parched soil. The sheikh was suddenly caught in the swirl, and there he stood, his beard and black robe dripping, momentarily dazed by the splattering waters—a nature-defying sight neither he nor his ancestors had ever dreamed of seeing in their desert abode. The rest of us felt a twinge of embarrassment, fearing his dignity might be offended, but the kindly sheikh soon broke into a joyful smile. "By Allah," he interjected, "an upside down rain is rising from the earth!" We all laughed together with him then, sharing his sense of wonderment.

Old memories gave way to new sights as my journey continued over the old town of Alexandria, Egypt's gateway to Europe. Here I could discern the receding shoreline of the Delta, which, no longer fed by the Nile silt now held back behind the Aswan High Dam, is being scoured by the waves and currents of the Mediterranean. Then the aircraft flew across the azure sea and on to Europe—a close neighbor of Africa, yet a world apart.

For anyone coming from Africa or the Middle East, landing in Zurich, an exemplar of Western modernity, cannot be but an eerie experience. Seen from above, the pale green misty hills, the neatly arranged houses and fields, and the orchestrated traffic flowing on geometrically designed roadways—the entire ensemble seems unreal, like a mechanical toy. Then, on the ground, the coolness of the weather and of the people, their fastidious sense of order, their self-absorption and—above all—their conspicuous affluence (based in part, ironically, on the profligate use of Middle Eastern oil), create so stark a contrast with the "less developed" nations as to make it evident that something is very wrong in our world. Observing this, I felt as though I were experiencing an out-of-body vision, detached yet aware that I, too, am part of this smug culture, so oblivious to the poverty that prevails elsewhere.

The following day's flight from Zurich to Karachi was equally revealing. Over the Balkans into Asia Minor, across the ancient lands of Anatolia, Phoenicia, Mesopotamia, and Persia, passing the mountain ranges with their humid west-facing slopes and their starkly arid easterly rain-shadows, seeing the Fertile Crescent as it is, imagining how it must have been. How indeed could civilization begin and flourish in such an austere landscape?

The degraded condition of the Fertile Crescent today may have something to do with a changed climate or with the devastations of repeated wars. The main cause, however, seems to be the prolonged exploitation of a fragile environment by successive generations of forest cutters and burners, grazers, and cultivators, all diligent and intelligent but cumulatively destructive nonetheless.

Visible from the air, the ancient cities of southern Mesopotamia are now mounds of hardened earth, mute time capsules in which the remains of successive civilizations are entombed. In their heyday, these cities drew their sustenance from the fertile plain, the veritable Garden of Eden fructified by the brimming Tigris and Euphrates. Now, from the vantage of an airplane hovering 10,000 meters above ground, I observed wide stretches of barren, salt-encrusted terrain, alternating with saline marshes, the entire plain crisscrossed with remnants of abandoned irrigation canals. Long ago, these were tended by enterprising irrigators whose early success eventually doomed their own land.

Equally haunting is the sight of the sloping highlands of the Middle

East, which have borne the brunt of rainfed farming and relentless overgrazing longer than any other area in the world. Here, the land has been stripped of its natural vegetation (including the fabled cedars of Lebanon), with only sparse groves of the original forest cover surviving. The natural mantle of fertile soil, perhaps a meter thick, has been raked off the hillsides by the rains and laid down in the valleys. Now one sees skeletal hills exposing jagged rocks, with soil remaining only in shallow discontinuous pockets.

Flying over the scraggy hills of southern Iran, I glimpsed many of the ancient "chain-well" systems, dating back to Persian times. Called *karez* in Persian or *qanat* in Arabic, they are ingenious water supply systems, consisting of one or more mother wells drained laterally through a tunnel that emerges at ground surface some distance—often many kilometers—away from the source. The path of the tunnel is marked on the surface by a series of mounds surrounding vertical shafts that once provided access to the horizontal tunnel. From above, the qanats resemble strings of pearls dangling from the foothills to the valley floor. For many centuries, these remarkably durable systems served as the principal means for supplying water to towns, villages, and orchards (called *pairi-daeza* in ancient Persian, whence our word "paradise").

When I first visited rural Iran in the 1950s, chain-well systems operated everywhere, and I had the chance to observe the qanat diggers (called *mughani*) at work. They were an exclusive ancient caste proudly practicing a hazardous craft that had been passed from father to son for centuries. Into the secret passageways they would burrow, molelike, with nothing more than a rope, a basket, a hand spade, a candle, and the fatalistic courage and skill of ages to serve them inside the lonely depths. Now many of these systems have been abandoned, victims of the modern motorized pump. The allure of cheap petroleum fuel has induced the drilling of deep tubewells, which all too quickly can drain an entire aquifer dry. And what has happened, I wondered, to the once indispensable qanat diggers? True, their job was extremely hazardous and arduous, but it was honorable and important. Now deprived of their traditional livelihood and mission, these artisans of the land and providers of water have been dispossessed. Sadly, they are just one example of how traditional village life is dying everywhere in the Middle East, with nothing but the squalor of overcrowded cities to absorb its displaced people.

Abutting southeastern Iran is Pakistan's desert region of Baluchistan, the destination of my journey. On my first tour of duty there, in 1982, I was shown the traditional farming practices of the district. Somewhere along a narrow winding road southwest of Quetta, I noticed an old farmer irrigating a young apple orchard. The saplings were planted in a basin surrounded by an earthen dike, and the farmer was flooding the basin with an enormous volume of water pumped out of his new motorized well.

Surprised to see such profligate waste of water in so dry a region, I

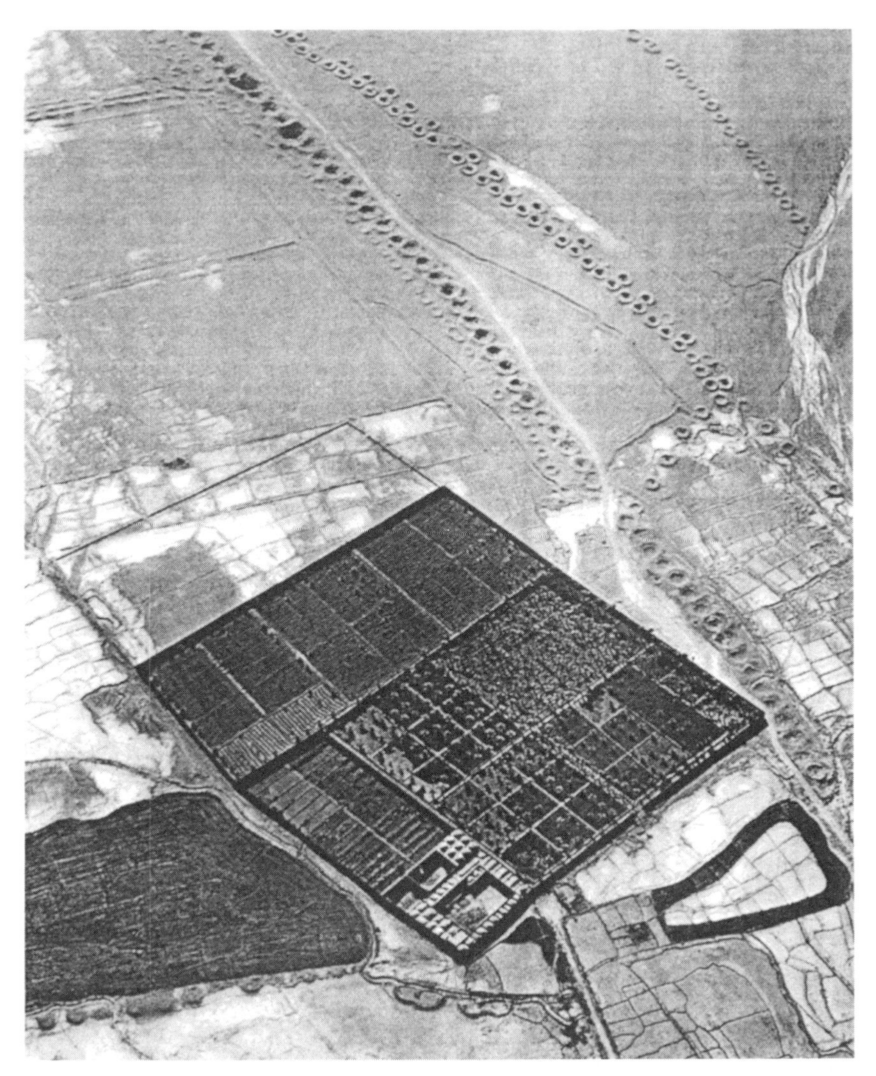

Figure 1.5 Aerial view of chain-well systems supplying water to orchards and fields in Iran.

had my interpreter ask the farmer if he believed this much water was needed for his small trees. His response was unhesitating: to make the trees big, one needs to give much water. Might it be possible, I inquired, to grow big trees without flooding, but instead by supplying them with less water, delivered slowly one drop at a time? The old farmer evidently thought the question quite ridiculous, for he shook his head at my ignorance and winked at the interpreter. No, he replied indulgently, it is not possible. And is it possible, I persisted, to put several hundred people into a container and lift them into the air and then carry them around the world? Of course, replied the farmer, that's an airplane, that's science, and science can do anything! So why can't science grow trees by watering them one drop at a time, I asked. The old farmer shook his head once more and repeated: big trees need big amounts of water. The man had evidently learned the art of irrigation in the Punjab, an area with plenty of water, where irrigation by flooding had long been the only method known.

Five years later, I visited Baluchistan for the fourth time, after helping to introduce modern drip irrigation there. Again, I traveled along the same route and passed by the same orchard. This time, the old farmer was sitting in the shade of his grown trees, drinking tea, while his young son was tending the narrow tubes that irrigated them drop by drop. The old farmer recognized me and remembered my question. "By God," he muttered philosophically, "science really can do anything!"

My flight came to an end at Karachi. Yet for me, it was not so much the end of a journey as the beginning of a new quest; not merely a traversal of space but indeed an awakening, an epiphany, a perception of the thread connecting the present with the past and imparting to the region's diversity an overarching commonality of environment and fate. The literally heightened perspective from an overflying aircraft allowed me to see the Middle East whole, undivided by ethnic differences or arbitrary political boundaries.

The view of a region from above reveals much about its environment. But a true understanding also requires a view from the ground: an insight into the character and condition of the people who live and interact with that environment. The Middle East is a particularly problematic region, not only because of the austerity and fragility of its environment but also because of its location at the juncture of three continents and the variegated cultures of its peoples. It is the home of differing religious, ethnic, and linguistic groups, each with its own lore and tradition. These groups are engaged in intricate interrelationships shaped by past as well as present rivalries. Yet all share a common environment and are similarly subject to its constraints. All depend for their future on the prospects for the region as a whole. The ethnic and geographical variety that breeds conflict can also spur cultural and material interchange and development.

The abused state of the Middle East's environment is a primary cause of the people's frustration. Impoverished and displaced from the land, many yearn for a bygone era, an idealized golden age, a redeeming return to a purer and simpler life. Some are so obsessed with the bitter grievances of the past and present that they seek redress through violence against those they hold to blame. Both responses divert attention from the region's real ills and are barriers against effective action to alleviate them. What the region needs is a rechanneling of the fervor of its people from sterile conflict toward a bold program of economic, social, and environmental rehabilitation, including the coordinated development and efficient use of the region's precious water resources. Rather than remain imprisoned by the past, the people of the region need to set their sights on a

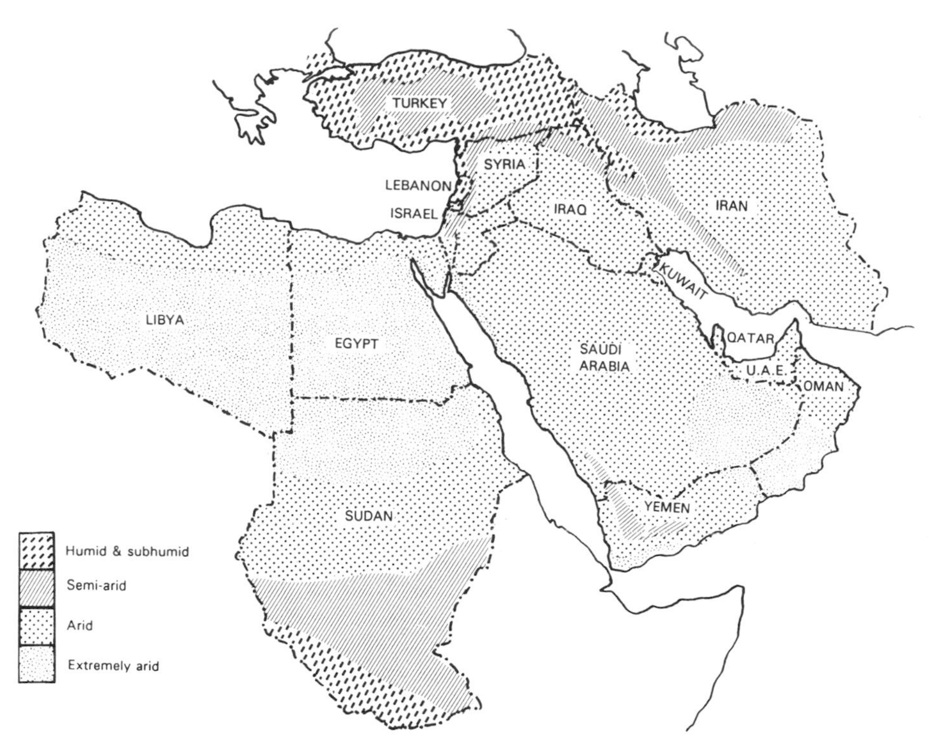

Figure 1.6 Climatological zones in the Middle East.

more promising future. The international community can help fulfill that promise.

The fundamental premise of this book is that there exists an important and insufficiently appreciated environmental dimension to the overall problem of the Middle East and that at its core lies the issue of water, the lifeblood of the region. By its nature, this issue cannot be resolved by each state separately but only by neighborly coordination. The current peace process promises to create a climate within which such coordination will be possible, bilaterally as well as regionally. However, the complex environmental and hydrological issues cannot be resolved by political formulas negotiated by diplomats alone. The tasks of environmental rehabilitation and particularly of water resource development and efficient utilization require expertise, insight, bold vision, and a considerable investment of effort and capital. Whatever the cost, however, it is certain to be much less than that of the only possible alternative, which is the failure of peace and the renewal—even escalation—of conflict.

2

Waters of Life

From water we made all living things.
Koran XXI:30

Our planet is the planet of life, and it is so because it alone among the nine planets of the solar system is blessed with the precise ranges of temperature and pressure that make possible the existence in a liquid state of a singular substance called water.

Water is so ubiquitous on our globe, covering nearly three-quarters of its surface, that the entire planet really ought to be called "Water" rather than "Earth." Alas, as Coleridge's Ancient Mariner complained, most of the water everywhere is unfit to drink. Less than 1 percent of the water on earth is "fresh" water, and that amount is unevenly distributed. Humid regions are endowed with an abundance of it, even with a surfeit, so that frequently the problem is how to dispose of excess water. Arid and semiarid regions, on the other hand, are afflicted with a chronic shortage of it. In some extremely arid regions, fresh water is so scarce that terrestrial life is barely possible.

Life as we know it began in an aquatic medium, and water is still the principal constituent of all living organisms. It is, literally, the essence of life. As Vladimir Vernadsky wrote some 100 years ago: "Life is animated water." Though we appear to be solid, we are really liquid bodies, similar in a way to gelatin, which also seems solid but is in fact largely water, "gelled" by the presence of organic material. The analogous material in our bodies is protoplasm. Actively growing herbaceous plants are some 90 percent water by mass. So are we at the time of birth. And although we tend to dry up a bit as we grow older, we still remain mostly water (about 65 percent)—over 40 liters of it, encapsulated in trillions of cells. Our brain is, incidentally, one of the more watery (75 percent) of our tissues.

As water is lost from the body continuously by perspiration, it must be replaced frequently. People living and working in a desert climate perspire as much as 10 liters per day.[1] Without dependable access to a spring, stream, well, cistern, pipeline, or some other source of water, human activity in and arid region is impossible, and life itself is in danger.

Even more vulnerable to water deficits are plants growing in a dry climate. Only a small fraction of the water normally absorbed by plant roots from the soil is retained. Most, often more than 99 percent, is lost as vapor in the process known as transpiration, which is analogous to the perspiration of animals. Transpiration is an inevitable consequence of the exposure to the dry atmosphere of a large area of foliage, necessary to facilitate absorption of sunlight and of carbon dioxide. Most crop plants are extremely sensitive to lack of sufficient water to replace the amount lost in transpiration. Water deficiencies impair plant growth and, if

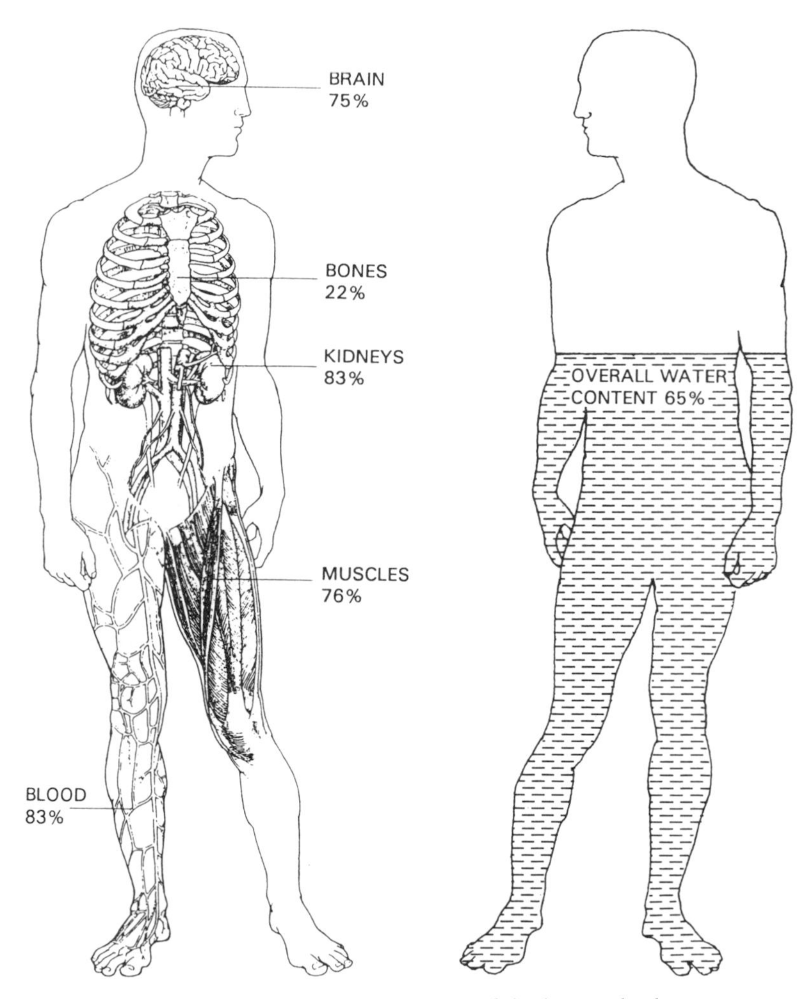

Figure 2.1 The water content of the human body.

extended in duration, can be fatal. A plant not able to draw water from the soil at a rate commensurate with the evaporative demand of the atmosphere will soon begin to dehydrate, wilt, and eventually desiccate. The danger of desiccation in dry regions is particularly acute because the evaporative demand of the unquenchably thirsty atmosphere there is practically continuous, whereas rainfall occurs only sporadically.[2]

To survive during the long dry spells between rains, plants must rely on the diminishing reserves of moisture contained in the pores of the soil, which itself loses water by evaporation and downward percolation. So tenuous and delicate is the water economy of most crop plants in arid regions that even short-term deprivation can cause sufficient stress to reduce yields or even result in total crop failure. Hence the importance of artificial irrigation for the reliable production of crops in such environments. Irrigation not only increases the yield of individual crops but also extends the effective growing season so as to permit multiple cropping, that is to say the successive production of several crops each year.

The Middle East is a transitional zone between maritime and desert climates. Only a small fraction of the region receives sufficient precipitation to sustain rainfed agriculture, whereas most of the region is semiarid to arid. Fortuitously, there are rivers and aquifers that convey water from zones of surplus precipitation to zones of deficit, and so it happened that farming, which was first developed in this region's more humid zones, soon moved to the river valleys in the arid zone, where crops could be grown under irrigation. Here the development of civilization depended absolutely on the availability of water and on the fertility of the soil.

Water as the prime requisite has decisively affected the course of human history ever since. Its abundance has enabled societies to flourish,

Figure 2.2 Location and extent of the world's arid and semiarid regions and the major rivers associated with them. Arid and semiarid regions are shown stippled.

its scarcity has caused them to wither. No consideration of the fate of societies, past or present, can ignore its role. Prime examples are the civilizations of the Tigris-Euphrates, the Nile, and the Indus rivers, called "hydraulic civilizations" by the anthropologist Karl Wittfogel.[3] Their irrigation-based economies produced a sufficient surplus to allow the development of the first cities in history, where artisans, traders, priests, scribes, administrators, and kings resided; and where art and science could flourish.

Significantly, the word for water in classical Persian (an Indo-European language) and, appropriately, the first word in the Persian dictionary, is *ab.* From it were derived the words *abad,* meaning abode, and the word *abadan,* meaning civilized. (The Arabic word for civilization, *tamadun,* is related to *madina,* meaning city. Similarly, our modern word "civilization" is related to the Latin *civitas,* or city.)

Our ancient forebears revered water. To them, a spring of fresh water bubbling up from the ground seemed indeed to be alive. Hence it inspired animistic and divine associations. It was "living waters" to the Hebrews, "running water" to the Arabs. The Egyptian priests posited that the earth itself was created out of the primordial waters of Nun, and that such waters still lay everywhere below the ground. Noticing that the Delta was being augmented by the Nile, the Egyptians could easily come to believe that their land was being produced by the river's water, transmuted into solid earth.

A similar notion was prevalent among the Mesopotamians. According to the Akkadian (and Babylonian) creation myth, *Enuma Elish,* all the world was originally sea, until Marduk smote the evil goddess Tiamat, splitting her body apart like a shellfish. One part he set above as firmament, to keep the upper waters from flooding the earth; and the other he laid down as a rocky foundation for earth and sea. He then bound a rush mat upon the face of the lower waters and piled mud on it to create dry land. This is reminiscent of the Hebrew Bible's depiction of the initial state of chaos, when "the earth was unformed and void, and darkness was upon the face of the deep, and the spirit of God hovered over the face of the waters." (The word for "the deep" in Hebrew is *Tehom,* likely derived from the Mesopotamian Tiamat.) In Ugaritic mythology, the god Baal vanquished and tamed the primeval waters, transforming them into beneficent rain.

To signify water, the ancient Egyptians drew a wavy line 𓈖 reminiscent of a wriggling snake. The symbol is preserved in the Hebrew and Phoenician letter *mem* (representing *mayim,* water), which in turn became the Latin letter *M.* The Sumerian word for water was *a,* which also signified sperm, or generative power, the masculine element that fructifies mother earth. Among the rivers and springs that are held sacred in many countries, most notable are the Nile, Ganges, and Jordan rivers, and the Zamzam spring of Mecca. In the Judeo-Christian and Muslim traditions, as well as in many of the Eastern religions, water is regarded not only as a physical cleansing agent but also as a source of spiritual purification and renewal. To the ancients, who did not perceive that rain is recycled water from the earth, it represented the material manifestation of God's grace from heaven. Hence the word for rain in Hebrew is *geshem,* connoting "substantiation." The Hebrew word for heaven, *shamayim,* can be separated into *sham-mayim,* which suggests "source of water."

Water permeates the lore of the Middle East, as manifested in innumerable legends, parables, proverbs, prayers, and poems. It figures prominently in the customs and codes of the region. In ancient Babylonia, water was used as a means to determine truth and to enforce justice: suspects whose purported crimes could not be proven otherwise were thrown into the river on the assumption that if they were innocent the divine water would deliver them to safety. The Hebrew Bible also alludes to the use of water to divine the future (Genesis 44:5).

Springs and wells have always been the focal points and centers of social life in both nomadic and settled societies. Thus it was by the "well of water" near Haran that Eliezer, Abraham's servant, met Rebekah, "the damsel very fair to look at." She then drew water from the fountain to serve him and his thirsty camels (Genesis 24), thereby proving to be kind as well as pretty. Consequently, she became the wife of Isaac "and he loved her." (That was, incidentally, the first biblical mention of a man's love for a woman.)

Even more romantic is the description of how, a generation later, Jacob (Isaac's younger son) first met his beloved Rachel at the well where she came to water her father's flock (Genesis 29). Chivalrous Jacob singlehandedly rolled away the heavy stone that covered the well, watered Rachel's sheep,[4] then kissed the maiden (his cousin and future wife), who was not only "fair to look at" but also "beautiful of form."

A similar event involved Moses (Exodus 2). After escaping the wrath of Pharaoh, he wandered to the Land of Midian (perhaps in northern Arabia). It was there at the well that he met the daughters of Reuel, the Priest of Midian. After helping them to water their flock by driving off hostile shepherds, the noble stranger was invited to the home of their father. The grateful Priest then gave Moses his daughter Zipporah for a wife.

The Bible provides vivid descriptions of the ageless conflict over water between farmers and pastoralists living along the tenuous boundary between the sown land and the desert. One of the earliest references to a dispute over water is given in Genesis 21, wherein Abraham reproved Abimelech (king of the Philistines) because of the well of water that Abimelech's servants had wrested from Abraham's servants. To settle the dispute, "Abraham set seven ewe-lambs. . . . And he said: 'Verily, these seven ewe-lambs shalt thou take of my hand that it may be a witness that I have digged this well.' Wherefore that place was called Beer-Sheba [*beer* = well; *sheba* = oath, or seven]; because there they . . . made a covenant." That place-name has been retained to this day and is now a large city at the northern edge of the Negev.

Another example is the dispute over Isaac's wells, described in Genesis 26: "Now all the wells which his father's servants had digged in the days of Abraham . . . the Philistines had filled them with earth. . . . So Isaac departed there and encamped in the valley of Gerar. . . . And Isaac's servants digged in the valley and found living water. But the herdsmen of Gerar quarreled with them, saying 'the water is ours!' So Isaac called the well *Esek* [contest]. . . . And he digged another well, but they contested over that one also. So he called it *Sitnah* [hatred]. And he moved, and digged yet another and they contested not. So he named it *Rehoboth* [spaciousness, implying generosity]." Rehoboth, too, has been retained as a place-name in Israel.

Still more poignant is the story in Genesis 21 of the banishment of Hagar and Ishmael, traditionally considered to be the progenitor of the Arabs: "And Hagar strayed in the wilderness, and the water in her container was spent, so she cast the child under a shrub . . . and she wept. Then God opened her eyes and she saw a well of water and she filled the container and gave the lad to drink. . . . The lad grew and dwelt in the wilderness, and he became an archer." Was it inevitable then, and is it inevitable now, that a young man in this environment must become an archer—a warrior?

In the Book of Deuteronomy (11), the Promised Land is described as a land that "drinketh water as the rain of heaven cometh down." But the divine promise to the Israelites is conditional: "If ye shall hearken diligently unto My commandments . . . I will give the rain of your land in its season." That promise is followed by the most dire of all possible threats to a people living on the edge of the desert, the threat that they may be deprived of water: "Take heed . . . lest ye turn aside and serve other gods . . . the Lord shall shut up the heaven so that there be no rain and the land shall not yield her fruit and ye shall perish quickly from off the good land which the Lord giveth you."

Water figures prominently in the early Israelite settlement of Canaan. The outcome of one epic battle against the Canaanites that took place in the Jezreel Valley was determined by a flash flood. The event was described dramatically in the epic poem of the Prophetess Deborah (Judges 5): "The streams trembled, the heavens also dripped, yea the clouds dropped water, the mountains oozed at the presence of the Lord . . . from the heaven they fought. . . . The brook Kishon swept them away, that ancient brook, the brook of Kishon." A similarly decisive battle was fought in the eastern part of the same valley by Gideon (Judges 6), who drove off the nomadic Midianites, the "sons of the east." Before entering into the battle, Gideon consulted the pattern of dew to ascertain God's will, and he relied on the manner of drinking water from a spring to select his best Israelite fighters.

Wells were also sites of battles. One example among many is the cruel story (II Samuel 2:13–17) of how the young men commanded by Abner (serving the son of Saul) dueled with the young men of Joab (serving David) by the pool of Gibeon. That well, incidentally, can still be visited today.

In David's lamentation over the death of Saul and the beloved Jonathan, he inveighed against Mount Gilboa for being the site of the Philistine victory over the Israelites. His indignation could find no stronger expression than: "Ye mountains of Gilboa, let there be no dew nor rain upon thee, neither fields of choice fruits, for there was defiled the shield of the mighty."

Significantly, the coronation of King David's son Solomon, the first Jewish king to be crowned in Jerusalem, took place at the city's only perennial water source—the spring of Gihon.

Altogether, the Hebrew Bible mentions water (*mayim*) directly no fewer than 580 times and indirectly many more times as it alludes to rain and dew and rivers and wells. But perhaps the ultimate evocation of the spiritual quality of water is contained in Psalm 23: "He leadeth me beside still waters, He restoreth my soul."

Arabic culture also arose out of life in the desert, where the supply of water has always been tenuous. Hence water, often associated with the myth of the amniotic fluid that nurtures the life of infants and conveys them to birth, has been held to be sacred and its waste considered sacrilege. Water still suffuses the imagery and symbolism of the Arabic language.

Water is a major theme in the Koran. Though the Koran is much shorter in length than the Bible, the word "water" (*ma'*) occurs in it 60 times, rivers over 50, and the sea (*bahr,* which also signifies lakes or large rivers) over 40 times. In addition, there are numerous indirect references to water in words signifying fountains, springs, rain, and clouds.

"With water we made all living things," states the Koran repeatedly, thus drawing attention to and emphasizing the vital importance of water. Allah's own throne is on the waters (sura XI:9) in the highest world. "It is Allah who drives the winds that raise the clouds and spreads them along the sky as He pleases and causes them to break up so that you can see the rain issuing out from the midst of them" (XXX:48). "It is He who . . . sends down pure water from the sky" (XXV:48), "He drives the rain to the barren land to bring forth crops" (XXXII:27), and "He leads it through springs in the earth" (XXXIX:21).

Water is a gift to humans and a sign of Allah's mercy, hence it must be regarded with respect and gratitude. "We send the water from the sky and give it to you to drink" (XV:22). "Consider the water you drink: If We had pleased, We could make the rain bitter: Why then do you not give thanks?" (LVI:68–70). "We send down pure water from the sky to give life to a dead land and provide drink for what We have created" (XXV:48–49). "And you see the earth barren and lifeless but when We send down water upon it, it thrills and swells and puts forth every joyous kind of growth" (XXII:5), and "with water We bring forth the buds of every plant . . . the thick-clustered grain, palm trees laden with dates . . . vineyards and olive groves and pomegranates which are alike and different" (VI:99). "Let man reflect on the food he eats, how We pour down the rain in torrents; We open the soil for the seed to grow; how We bring forth the corn" (LXXX:24–32). These and many other such statements serve as reminders that fresh water originates with God and not with man.

One chapter (sura LV) of the Koran refers to four gardens in which are "two springs flowing free" and "two springs pouring forth water in continuous abundance." Another sura (XLVII) refers to the four rivers of Paradise, describing them as "rivers of water incorruptible, rivers of milk of which the taste never changes, rivers of wine, a joy to those who drink, rivers of honey pure and clear." *Tasnim* is the sacred fountain that signifies the highest level of spiritual ecstacy. Near a river or pool in Paradise called *Kauthar,* the Prophet will stand to greet the faithful.

The Koran also recognizes differing qualities of water: "Watered with one water, yet We make some excel others in taste" (XIII:4), "one palatable and sweet, the other salt and bitter" (XXV:53 and XXXV:12).

The provision of fresh water for ceremonial ablution and purification prior to prayer is essential to Muslim religious practice. "When you rise to pray, wash. . . . If you are polluted, cleanse yourselves" (V:6), "He sends down water from the sky to cleanse you" (VIII:11).

Moreover, the faithful were promised eternal life in a paradise, described as "gardens beneath whose trees rivers flow" (V:119). To inhabitants of a hot and dry climate, such statements evoke in the mind's eye the ideal of serenity and well-being. The sudden appearance of an oasis in the desert, the miraculous gift of upwelling water, represents nothing less than paradise on earth. The feeling is conveyed in the classical Arab adage listing the three manifestations that most gladden the heart: *al-ma', al-khadra', wa'l-wajh al-hasan* ("water, greenery, and a beautiful face").

Countless Arabic poets have extolled the marvels of water in all its manifestation—as rain, dew, spring, stream, lake, or sea. Beneficent rain ("the gentle weeping of the clouds that causes the buds to smile and the flowers to laugh") is popularly referred to as *rahmat* ("mercy"), representing Allah's grace upon Earth. This usage alludes to such passages in the Koran as "He sends water so that you may taste His mercy"[5] (XXX:46) and "He sends down saving rain for them after they have lost all hope and spreads abroad His mercy" (XLII:28). The mutual attraction between humanity and God was expressed by the Muslim poet Maulana Rumi in the verse: "When the thirsty seek the water in the world, the water too is seeking the thirsty in the world."

The quest for the Water of Life, capable of restoring youth and bestowing immortality, is another recurrent poetic theme. That quest is associated with Khidr, said to be the companion of Moses (Koran, XX). He had discovered the Fountain of Life hidden in the darkness and partaken of its water, thereby becoming immortal. Khidr is the patron saint of the wayfarers, appearing whenever a pious person is in need. In some respects, he is the Islamic equivalent to the immortal Elijah, the messenger of God in the Jewish tradition.

Water can symbolize the dual aspects of the divine: the wrathful (manifested in violent storms and torrents) and the merciful. In the words of the Koran: water shall provide "life for the just and faithful, and death for the iniquitous and unfaithful."

Just as gardens and springs are bestowed upon the righteous, so they are withheld from the wicked. The latter are inevitably ruined: their gardens desiccated, their wells depleted, and their dams breached. "How many were the gardens and springs they left behind," when through arrogance and greed they mistook the gardens of this world for their lasting entitlements (XLIV:25, LXVIII:17–33). Such evildoers are condemned to drink the boiling waters of hell.

In Islamic architecture, the various waterworks (such as barrages and canals for diverting river water, wells for tapping groundwater and qanats for transporting it, and finally the jetting spouts and reflecting pools built to enliven enclosed gardens and palaces) were much more than merely functional devices. They served as sources of spiritual solace and aesthetic delight and as celebrations of the blessed mysteries and the profound meanings of water. The very term "Islamic Garden" evokes the imagery of lush shrubbery and fruit-laden trees abundantly supplied by streams, pools, and elegant fountains. To people living in a harsh, dry climate, such gardens served to alleviate the oppressive heat, dust, and fatigue of the desert, and thus to soothe the senses and gratify the soul.

Throughout antiquity and into the Middle Ages, the notion prevailed that the fresh water of springs and streams emanated spontaneously from the bowels of the earth.[6] The ancients also had difficulty perceiving that rain and snow originate as water that evaporates from the earth's surface and condenses in the upper atmosphere to form clouds. Water vapor seemed too thin and transparent to be considered substantial. (The classical Hebrew word for "insubstantial" is *hevel,* literally meaning "vapor." Hence the nihilistic opening statement of Ecclesiastes—"vapor of vapors, all is vapor"—is translated as "vanity of vanities, all is vanity").

For a long time, no one imagined that all the water flowing in the innumerable springs and mighty rivers could possibly result from so seemingly insubstantial a source as rain and snow. (The very notion seemed ridiculous, especially in view of the Nile, which appeared to rise out of the rainless desert.) The first person known to have suggested that idea was Leonardo da Vinci, but it was only in the latter part of the seventeenth century that the English astronomer Edmund Halley and, separately, the Frenchman Claude Perrault proved the principle by measurement and calculation based on the catchments of the Thames and Seine, respectively.

Today we know that the ultimate source of all fresh water is the continual natural distillation process by which water vaporized from the earth's surface is condensed in the atmosphere to form assemblages of droplets, which then coalesce and fall back to earth as precipitated rain or snow. The precipitation reaching the ground infiltrates the soil or flows overland as surface runoff. Of the amount infiltrated, some of the water is sucked

Figure 2.3 The importance of water in Islamic architecture: A reflecting pool in the patio of the Alhambra, the 14th-century Moorish palace in Granada, Spain.

up by plant roots while the remainder percolates beyond the root zone to the groundwater beneath.

The earth and the atmosphere are thus engaged in an endless reciprocal passing game whose main article of exchange is water. The forces that impel this exchange are the sun's radiant energy and the earth's gravitational pull. Evaporation draws water to the atmosphere, precipitation returns it to the earth. This cyclic exchange is called the hydrological cycle, and its study is the science of hydrology.

Both the supply of water by precipitation and the need of terrestrial

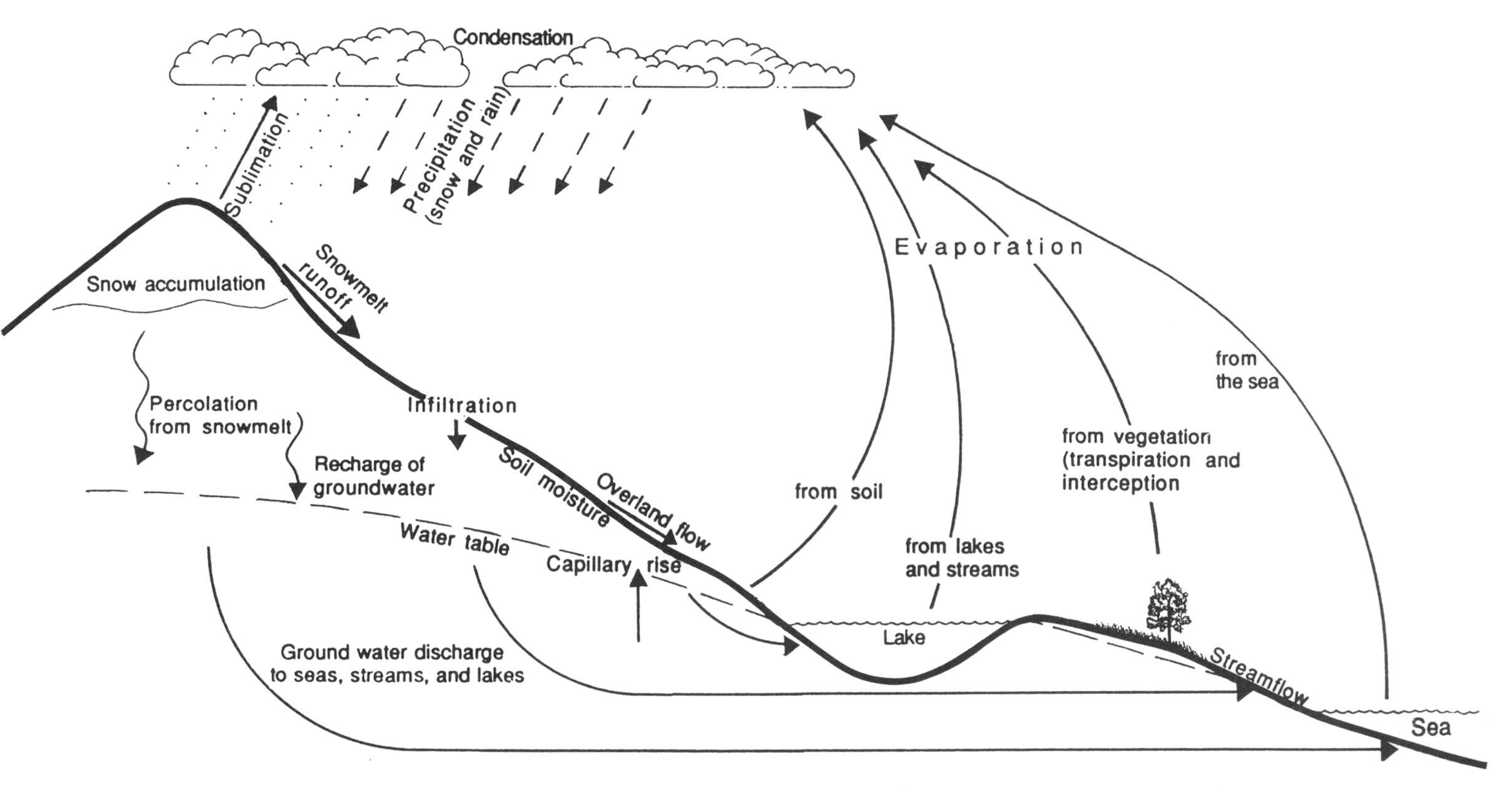

Figure 2.4 The hydrological cycle.

organisms for water (strongly affected by the evaporative demand of the atmosphere) differ from place to place and season to season. The relationship between supply and demand determines the water economy of every region. A humid zone is one in which the natural supply generally equals or exceeds the demand, with the excess water flowing out of the zone via rivers or underground aquifers. A humid zone permits dependable farming without requiring irrigation (though supplementary irrigation is sometimes helpful, as even humid zones are not totally immune to occasional drought). In a subhumid or semiarid zone, rainfall is sufficient in most seasons but droughts occur frequently enough to make the practice of rainfed farming a somewhat hazardous venture. An arid zone, on the other hand, is one in which an imbalance normally exists between the demand for water by crops and its natural supply by rainfall, the latter being too meager to meet the former over most of the area in most seasons. Extensive grazing, rather than continuous cultivation, becomes a major form of land use in such areas, which are often contested between nomadic herdsmen and settled farmers. To survive in an arid zone, farmers try to supplement rainfall with irrigation, wherever water is available from streams or wells.

The situation is basically different in real desert areas, which can be characterized as "extremely arid." Here average annual rainfall is insufficient to sustain agricultural crops, so ordinary rainfed farming is impossible. The biblical definition of the desert is "the land unsown." To subsist in the desert, people must devise ingenious schemes to obtain supplementary supplies of water, either by wresting the precious fluid from underground aquifers, if available, or by collecting it off slopes of barren ground during brief episodes of rainfall, or by conveying it from another region.

By some strange accident of history and geography, almost one-fourth of the world's population is struggling to eke out an existence in the earth's semiarid and arid zones, often at the very edge of the desert. Here, by a cruel stroke of nature, the requirements of crop plants for water are the greatest, whereas the supplies by natural precipitation are the least. In these zones, the scales are weighted at the outset against stable agriculture. The balance must be rectified by augmentation of water supply and by strict water conservation. Where irrigation water is available, however, arid regions can be amazingly productive, thanks to their warmth, abundant year-round sunshine, and—in some places—the presence of fertile soils.

The severe droughts that afflicted sub-Saharan Africa in the 1970s and 1980s have turned the world's attention increasingly to issues of water supply. In some regions, notably in eastern Africa, the paucity of precipitation has resulted in massive famine, coupled with extensive land degradation—a process called *desertification.* A growing shortage of water has occurred even in such wealthy and agriculturally productive regions as the western United States, to the extent that it has cast doubt on the ability of this bountiful country to provide surplus food to the hungry parts of the world in the future as it has in the past.

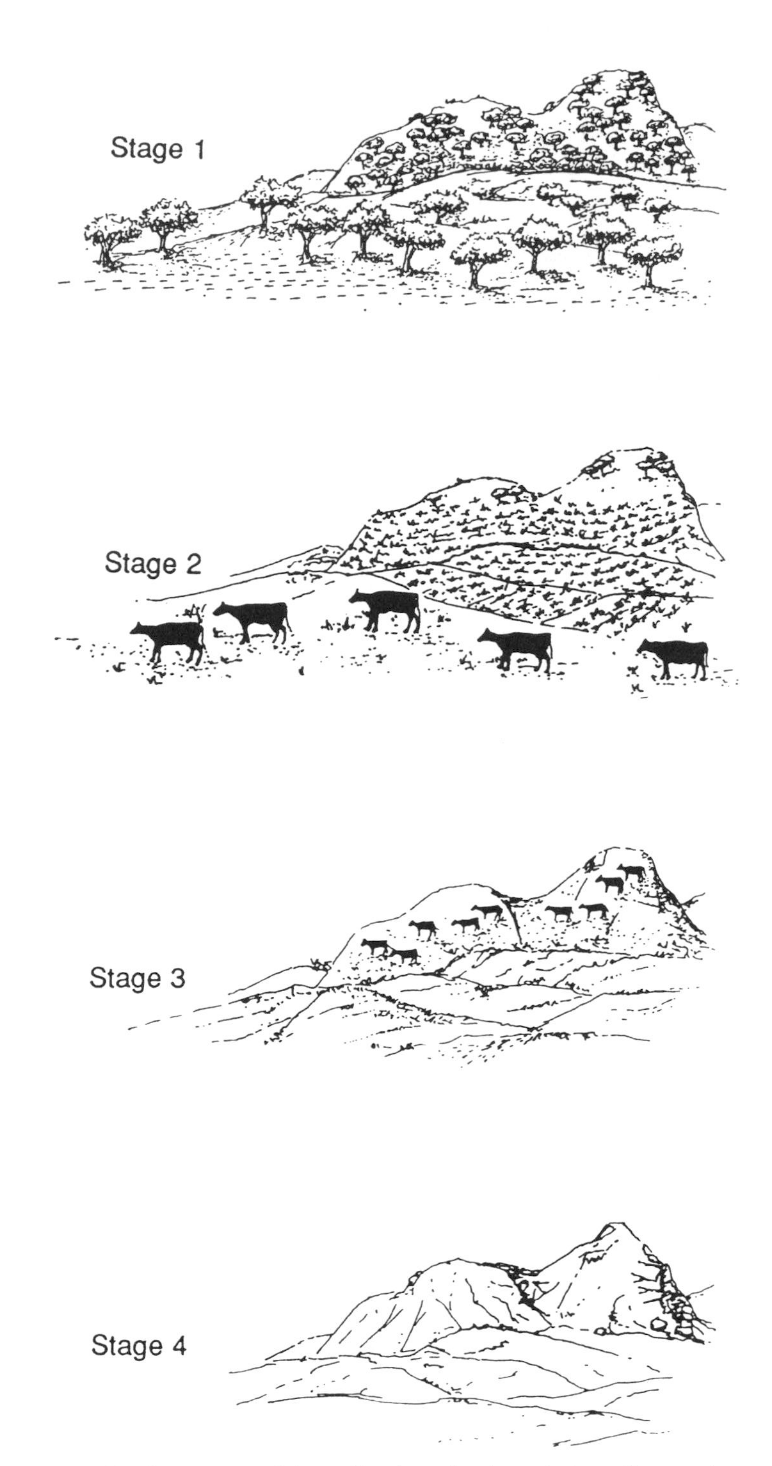

Figure 2.5 The process of desertification: the degradation of vegetation and land in semiarid regions, caused by excessive exploitation.

Figure 2.6 Overgrazed land around a well in the Negev.

People living in arid regions who cannot migrate elsewhere during a drought try to survive by overgrazing or overtilling the land. The denudation of vegetation exposes the soil to erosion by wind and by subsequent rains. The resulting damage to productivity causes a semiarid region to resemble a desert. In the past the indigenous population was much less numerous than it is today, and temporary migration was often possible. Even so, periodic famine was a fact of life in such regions. In our time the situation has worsened. The growth of population and the restriction of traditional tribal patterns of migration and transhumance have increased pressure on land, vegetation, and water resources. Consequently, famine affects more people and the process of desertification has extended to vast areas, particularly in the Sahel belt south of the Sahara and throughout much of the Middle East.

Concern has risen not only over the quantity of fresh water available but also over its quality. Many of the world's rivers, lakes, and aquifers, serving as sources of water, have become so polluted with toxic chemicals or contaminated by pathogens that they are no longer safe for human use.

The problem of providing safe and sufficient water supplies to people everywhere is greatly exacerbated by the rapid growth of population. Not only is the entire world's population growing at an alarming rate, which is serious enough in itself; but—worse yet—the population of the poorest and most arid countries is increasing even faster than the global average. In the Middle East, for example, the growth rate in most countries exceeds 3 percent per year, doubling the population of each succeeding generation.

Ever-expanding populations threaten to overwhelm the capacity of drought-prone countries to feed them.

Earth's population as a whole is expected to double from the present 5 billion to about 10 billion by the middle of the next century. Over 90 percent of that growth is projected to take place in the so-called Third World—otherwise known as the "developing nations" of Africa, Asia, and South America—which may, by the year 2050, constitute as much as 85 percent of the world's population. This staggering rate of population growth is causing the migration of many millions from rural villages to urban centers, and from impoverished regions to wealthier ones. If the situation of people in poor countries continues to deteriorate, the rich countries can expect a practically unstoppable influx of "environmental refugees": desperately poor people compelled to leave the countries that, tragically, can no longer support them.

In the wealthy countries of the "First World" (including the United States and Canada, Western Europe, and Japan) the annual supply of fresh water per capita generally exceeds 1,000 cubic meters per year (CM/Y). China, now rapidly developing its economy and raising its living standards, is basing its water resource planning on renewable fresh water supplies of about 500 CM/Y per person. Many countries, however, will need to make do with much less. Israel affords only some 350 CM/Y at present, an amount that will diminish as the population grows—though it is supplemented with recycled water (wastewater that is recovered, treated, and reused for irrigation). Other countries with even less fresh water per capita include the Kingdom of Jordan, which has only some 275 CM/Y.

In contrast with population, which is progressively increasing, annually renewable fresh water supplies are generally limited. In many cases, they can be augmented only marginally by such methods as cloud seeding and seawater desalinization (which is still too expensive to apply on a large scale). Therefore, the imperative is to conserve supplies and to use water more efficiently. Conservation calls for reducing waste and storing water collected during times of surplus for use during times of shortage. The traditional engineering approach has been to build more and bigger dams, but finding suitable sites for new dams has become more difficult, and—worse yet—existing dams are being clogged by silt. They also lose much water by evaporation. The preferred approach, in many cases, is to increase underground storage by means of the artificial recharge of groundwater reservoirs (aquifers).

Greater populations will require expansion and intensification of agriculture and greater reliance on irrigation. The problem here is that while expanding irrigation will require more water, so will the simultaneous growth of cities and industries. Early in the twentieth century, some 90 percent of global water use was for irrigation. That proportion had diminished to 80 percent by 1960, to about 70 percent by 1990, and is expected to fall further to some 60 percent by the year 2000. The inescapable need is to find ways to grow more food with less water. Fortunately, there is much room for improvement of water use efficiency, albeit not without incurring secondary problems. Among those are the need for greater investment in technology and the danger of environmental pollution with the residues of agro-chemicals, which tend to be used in greater amounts as the intensity of production increases.

The recycling of wastewater is another challenging task. It requires treatment to remove pathogenic microbes and other disease-causing agents. Such treatment can be done in basins or directly in the soil and subsoil during the process of groundwater recharge.

Superposed on all these problems is the worrisome prospect of a global change of climate, resulting from the so-called greenhouse effect. If it occurs, as some scientists have predicted,[7] such a change is likely to disrupt the preexisting patterns of weather, water supply, and agriculture.[8] The adjustments required, including shifts in the mode and geographical distribution of agricultural production, will doubtlessly increase the burden borne by nations located in the arid zone. The result may be a breakdown of social order on a scale even greater than the world has recently witnessed (as, for example, in Ethiopia, Somalia, and Sudan) and a general threat to peaceful progress and to international stability.

In the last few years, the world has become all too painfully aware of the conflict over oil resources in the Middle East. That conflict has obscured the region's older and more acute problem of resource scarcity. Underlying the craving for oil is a growing thirst for water. The two fluids do not mix, and as oil is the lighter of the two and superficially more valuable, it tends to float to the top of the public's attention. But though some of the countries of the Middle East are oil-rich, nearly all are water-poor, and getting poorer. Compounding the water scarcity is the sad state of the region's ecology as a whole, resulting from centuries of deforestation, overgrazing, soil erosion, salinization, and the depletion or contamination of water supplies. The obsession with oil is in part a symptom of the desperation resulting from the loss of productivity and economic security.

To some observers, oil may seem to be the region's major source of wealth, but that is illusory. Great though the oil reserves of the Middle East might seem, they are finite and non-renewable. Sooner or later, they will run out. After that, with the mirage of quick riches gone, the region will necessarily return to a dependence on its basic land, water, and human resources. In the meanwhile, the crucial question is what is being done with the oil-generated wealth to lay the basis for permanent, self-sustaining development. Unfortunately, the answer thus far is bitterly disappointing. Our generation has witnessed the greatest transfer ever of financial resources: trillions of dollars paid by the industrial countries to the oil producers of the Middle East. Tragically, much of that wealth has been squandered in a vain pursuit of military security or supremacy, or in self-indulgence, rather than in long-term development projects that could promote progress for the region as a whole.

Missing in the contentious clamor over the Middle East is a measure of insight into the deprivations that drive the current conflicts by condemning so many of the region's people to wrenching poverty, cruelly accentuated by the fabulous wealth of the few. To understand the causes of that poverty, we must draw back from the immediate conflicts (with their apparent ethnic, ideological, political, and military aspects) and consider their deeper wellsprings.

The Middle East is in the throes of a human-induced environmental crisis that is among the worst in the world. Throughout the region the traditional village life based on farming is deteriorating. In an area perched between sea and desert, where climate is notoriously unstable, traditional farming is being literally undermined not only by the long-term degradation of the soil and water resources but also by the runaway rate of population growth. As a result, millions of impoverished people, no longer able to derive their livelihood from the land, now gather in overgrown cities, without adequate housing, sanitation, or prospects for employment. Extremism feeds on the resulting discontent.

Casual observers, including many reporters who scout the scene for a quick headline, often miss the real story that lies beyond the dramatic pronouncements and acts of violence. The problem of the region is not merely tangled politics; the Middle East is a severely wounded region needing systemic rehabilitation, and the water shortage is at the heart of that malady.

Water scarcity is, to be sure, a prevalent problem in many parts of the world, but nowhere does it seem as likely to engender conflict as in the Middle East. Divided as they are by arbitrary boundaries that cut across natural hydrological and geographical units, the countries of the region find themselves in a state of perpetual tension over common resources to which they lay overlapping claims. Specifically, many countries in the Middle East are linked by common rivers or aquifers subject to overuse or contamination, so actions by one country inevitably affect the quantity and quality of water available to its neighbors.

Nearly all of the region's readily available waters are now being utilized in an uncoordinated, competitive manner. With escalating demands, further development of each country's water supplies will require increasing investments and is likely to stir up ever more bitter rivalries. According to the World Bank, the costs of water supply and sanitation in the Middle East are the highest in the world, being over twice those in North America and five times those in Southeast Asia. Irrationally, many Middle East countries subsidize water used for inefficient irrigation so lavishly as to encourage rather than curb its wasteful use.

Considering the intense rivalry over water, in this region as well as elsewhere, we can readily understand how the very word *rival,* originally a neutral term used in Roman law to designate a neighbor sharing with another the waters of a *rivus* (a stream generally used for irrigation), has acquired the negative connotation of competitor, adversary, even enemy. Sharers of so vital a resource as water may choose either to cooperate or to compete. Somehow, our language has forgotten the first option and has retained only the second. Is that folly inherent in human nature and therefore inevitable, or is it merely an aberration that can be, and ought to be, rectified?

The water problems of the present-day Middle East are epitomized by the dispute over the twin rivers, Tigris and Euphrates, which discharge through Iraq into the Persian Gulf. These, like the Nile, are "exotic" rivers, deriving their waters from outside the arid region into which they flow. Their common origin is the highlands of eastern Anatolia (a part of modern Turkey), which receive copious amounts of rain and snow. From there the Tigris flows to Iraq (only briefly skirting the northeast corner of Syria), while the Euphrates courses through northeastern Syria for some hundreds of kilometers before reaching Iraq. The Euphrates and its tributaries are Syria's only significant rivers, and that country's main hope for expanding its agricultural production.

Both Iraq and Syria stand to lose a good part of their current supply of water from the Euphrates when Turkey completes its Southeast Anatolia Project. That plan calls for the construction of 13 hydroelectric and irrigation projects on the upper reaches of the twin rivers, including the massive Ataturk Dam on the Euphrates. As Turkey extracts more water upstream, less remains for the two countries downstream, and their already fierce rivalries with Turkey and with each other are certain to escalate. Recently completed dams in Turkey now control the flow of water to Assad Dam, a prime source of Syria's electrical power as well as its irrigation water. When the Anatolia project becomes fully operational, Syria stands to lose about 40 percent of its potential supply from the Euphrates. Syria's population, currently about 14 million, is expected to swell to 22 million or more within 20 years. The projected dwindling of the Euphrates' inflow, along with the already worrisome depletion of Syria's groundwater resources, may curtail Syria's agricultural development and cause severe food shortages within the next decade.

Even more sensitive is the situation of Iraq, which has a larger population than Syria's (nearly 20 million) and which lies further downstream. Affected by both Syria's and Turkey's water withdrawals, Iraq may lose as much as 80 percent of its flow from the Euphrates. Particularly vulnerable is the wheat- and rice-producing area of northwestern Iraq. To compensate for the reduced flow of the Euphrates, Iraq should be able to make increased use of the Tigris, though the flow of that river, too, will eventually be reduced by the extensive water development due to take place in Turkey's eastern Anatolia.

The situation in that subregion is further complicated by the Kurdish issue. The Kurdish nation—ancient, proud, and rooted in the mountain district they call Kurdistan (including parts of Iraq, Iran, Turkey, and Syria)—has become increasingly militant in demanding its long-denied rights to independence. In pursuit of those rights, the Kurds in Iraq have been rebelling against the central authority of Baghdad, while the Kurds of southeastern Turkey have been disrupting Turkish development of the area they consider part of their own domain.

An equally vexing problem, though not yet so acute, exists along the Nile River. Egypt, being a desert with practically no effective rainfall, depends entirely on irrigation from the Nile to produce food for its people. Although Egypt has an agreement with Sudan over allocation of Nile waters, other countries that control the river's headwaters are not bound by it. The Nile is, in any case, subject to the variable pattern of the region's climate. The miracle of 1988 (when Egypt was saved from the consequences of a prolonged drought by the sudden and fortuitous resurgence of the Blue Nile that refilled Lake Nasser) may not recur in so timely a fashion in the future. There is also some concern that the predicted global warming[9] may reduce inflows to the Nile.[10]

An ambitious plan by Egypt and Sudan to augment the flow of the White Nile by cutting a canal through the Sudd swamps has been bogged down by the civil war in southern Sudan, where the non-Muslim (Christian and animist) populace has been rebelling against the fundamentalist Muslim regime in Khartoum. There are also serious concerns that the project called the Jonglei Canal, when and if it is implemented, will cause great damage to the indigenous people of the Sudd and to their environment. Another potential problem to Egypt and Sudan is the real possibility that Ethiopia, the source of the Blue Nile, which accounts for some 80 percent of the water flowing into Egypt and most of the water used for irrigation in Sudan, will attempt to utilize more of that water to answer the needs of its own hungry people. Referring to this possibility, Anwar Sadat, former president of Egypt, warned that his country would consider going to war over any curtailment of its water supplies. The population of Egypt, less than 3 million in the early part of the nineteenth century, is now nearing 60 million and increasing by a million and a quarter every year. Already, the country that was the granary of the Mediterranean region in Roman times is forced to import half of its food supplies.

Another focus of rivalry is the Jordan River basin. Syria and the Kingdom of Jordan have agreed on a joint plan to build a large "Unity Dam" on the Yarmouk River, a major tributary to the Jordan River, intended to supply Syria with power and the desert kingdom with water to supplement its insufficient irrigation system. Israel's objecting, as well as the shifting alliances and rivalries among the riparians, may or may not allow the plan to be implemented. Meanwhile, Syria's increasing diversions from the headwaters of the Yarmouk are damaging the eventual prospects for the planned dam, and are thus reducing Jordan's potential supplies. The desert kingdom could face a crisis before the end of this decade that would require transferring still more water from the already deprived farming sector to answer the domestic needs of the country's increasingly urbanized population, now growing at an astonishing annual rate of about 3.5 percent. In addition to its internal population growth, Jordan has had to absorb a sudden influx of some 350,000 Palestinians (10 percent of the state's population), who were displaced from Kuwait and the other Gulf States in the wake of the 1991 Persian Gulf War.

Complicating the issue further is Israel's concern over the residual flow to the lower Yarmouk, from which it has for many years drawn water for its own needs. Israel's occupation (since 1967) of the Golan Heights, long a province of Syria, allows it to oversee diversions from the Yarmouk, as well as to prevent diversions of water from the upper tributaries of the Jordan River itself. Hence Israel is not likely to relinquish its hold over the Golan unless it receives firm assurances that its water supplies will not be threatened.

Meanwhile, Palestinians on the West Bank are vying with Israel for a larger share of common groundwater resources. Israel now draws nearly a third of its fresh water supplies from wells in its own territory that tap into an aquifer that originates in the highlands of the West Bank. Should the Palestinians, after achieving independence, undertake intensive well drilling and pumping in their domain, the amount of water available to Israel could diminish markedly. One reason for Israel's reluctance to withdraw from the West Bank is its fear of losing control over that aquifer.

Other troubled waters can be found throughout the Middle East, notably in the states along the Persian Gulf (Oman and the United Arab Emirates), whose development boom has caused the rapid depletion of the Damman aquifer, their common subterranean reservoir. There are also problems between Iraq and Iran over the tributaries to the Tigris.

One positive trend is the diminishing dependence of national economies on water-intensive agriculture. During the first decade of Israel's existence, for example, some 30 percent of the country's gross national product and exports depended on irrigation. Today, irrigated agriculture accounts for only about 3 percent of GNP, as the economy has shifted into industries and services requiring less water per unit of production value. A similar trend seems destined to take place in the neighboring countries as well, and is already noticeable in Jordan and Egypt. Another hopeful development is the continued improvement of irrigation methods to minimize the appalling waste of water and the salinization of soil in the field. A more far-reaching change would shift some forms of agriculture from open fields, with their inherently high consumption of water, to greenhouses. This could permit greater crop production per unit amount of water consumed, though it would require much greater capital investment and operating costs.

The water shortages besetting the Middle East, though endemic, are not insoluble. They can be alleviated, but only in the context of regional cooperation. Given the will to embark upon coordinated programs, the countries of the region can have many opportunities to develop water resources jointly and to utilize them far more rationally than they have so far. The important principle to realize is that regional cooperation is not a zero-sum game, in which any gain to one side must be balanced by loss to another. Rather, it can produce a net gain for all.

A case in point is the entire set of tangled interrelationships among Israel, Syria, Jordan, and the West Bank. The simplistic formula of "land for peace" ignores the fact that land implies water. The Golan Plateau is not simply a sliver of land between Israel and Syria; it straddles the headwaters of the upper Jordan and the Yarmouk. The West Bank is not simply disputed land; it rests on the catchment of an important aquifer. The security zone in southern Lebanon lies astride the Litani River, whose waters are not yet fully utilized. Disputes over these territories will only be resolved when an agreement regarding those water resources is reached among the neighboring states.

An agreement between Israel and the Kingdom of Jordan, for example, would allow the Yarmouk's excess flood waters to be stored behind a new dam on the Yarmouk, or possibly even in Israel's Sea of Galilee, thus augmenting the perennial storage capacity available to Jordan. Adding water to the Sea of Galilee would, in turn, help to reduce that lake's salinity level. During seasons of high floods the extra water could be transferred to recharge aquifers in both countries and free up additional water for the Palestinians. Moreover, cooperation between Israel and the Kingdom of Jordan may lead to the construction of a canal conveying seawater from the Red Sea to the landlocked Dead Sea. Such a canal could permit desalinization and produce electric power for industries and towns. It would also make possible the development of inland fisheries and recreation facilities in the largely uninhabited Arava Valley. Finally, such a joint project would prevent further shrinkage of the Dead Sea and could induce the two countries to cooperate in the extraction and marketing of the Dead Sea's minerals. There are, however, serious environmental and economic objections to this plan.

All these issues and possibilities will be described more fully in the subsequent chapters of this book. Let the reader be forewarned that the problems are formidable. Our narrative, however, will not be a pessimistic jeremiad. Quite the contrary: great though the problems may be, they are not intractable. The challenge is to go beyond the mere recitation of problems toward defining what realistic prospects exist for resolving them.

An awareness of the options for collaborative development of water resources in the Middle East can serve as an inducement to attaining and sustaining regional peace. Political accommodation and hydrological cooperation are mutually conditional and mutually reinforcing. The hydro-political realities are such that none of the countries involved can solve its own water problems unilaterally. The solution to the problems of the Middle East requires an era of peace so that natural, financial, and human resources can be shifted from futile, wasteful, and self-destructive wars to sustainable development. To the cynics who pride themselves on being "realistic," such a notion may seem utopian. They miss the point: a regional accord must prevail in the Middle East because the only alternative—a perpetuation or exacerbation of conflict—is literally a dead end. The bitterness of the past must be overcome in pursuit of a better future.

3

Ancient Civilizations

With the ancients is wisdom
And in length of days understanding.
Job 12:12

The study of history does not merely resurrect a dead past. In the words of Thucydides: "Knowledge of the past helps to anticipate the future." We might also add that it helps us to understand the present. For if we ignore the past and focus our attention exclusively on the predicaments of the moment, we may find ourselves repeatedly surprised by a host of bewildering problems seeming to come out of nowhere, without antecedents and hence without direction. How did these problems arise? Chances are, the seeds of what we witness today were planted long ago by our predecessors, even as we may be planting the seeds of future problems, perhaps unknowingly, at present.

Those who fail to heed the lessons of the past are admonished in the Koran: "Do they not travel through the earth and see what was the end of those before them? . . . They tilled the earth and populated it in great numbers. . . . There came to them their apostles with clear signs, which they rejected, to their own destruction. It was not Allah who wronged them, but they wronged their own selves" (XXX:9).

There is only one past, but there are many different ways to study and interpret it. Herein we choose to take the environmental point of view. Historians who ignore the role of environmental factors may ascribe the decline of a society to moral decay, lack of military preparedness, superiority of an enemy's technology, or whatever. All these and more can be important, but as often as not it is the degradation of basic resources that seals the fate of a once-vibrant society. Nowhere is this principle more clearly evident than in the region where intensive human exploitation of land and water resources began earliest. We therefore wish to consider how early human civilizations interacted with and within the fragile environment of the Middle East, and how they affected the subsequent fate of the region's peoples.

The sine qua non of civilization in the Middle East is the availability of adequate land and water resources, and the history of the region—boiled down to its essentials—is the adaptation of societies to those fundamental resources and the ways they managed or mismanaged them. The first settled agricultural communities apparently developed in the relatively humid highlands girding the Fertile Crescent on the west, north, and east. Here agriculture could rely on natural rainfall, hence we call the societies and cultures that evolved under these conditions "rainfed civilization." Some of the communities eventually developed into city states, each based on its own agricultural hinterland. Initially confined to intermontane valleys, their agricultural activities gradually encompassed sloping lands as well.

A quite different type of civilization developed later, when agriculture was introduced into the major river valleys of the Middle East. Here the climate was arid, rainfall was deficient, and crops needed to be irrigated with water drawn from the rivers. We therefore designate the civilizations that formed there "riverine civilizations" (or "hydraulic civilizations").

Both the rainfed and riverine civilizations long coexisted in the Middle East with a third form of society, which developed in the semiarid subregions, namely, the pastoral seminomadic or nomadic societies. To be sure, the three forms of civilization were not isolated or mutually independent, as they traded and interacted with one another. Pastoralists were often allowed to graze in the vicinity of agricultural settlements and urban centers, to which they supplied animal products and from which they purchased tools and other supplies. From time to time, however, nomadic tribes were driven by drought or by population pressure to invade, and often to settle among or in place of sedentary societies.

Much of the territory we call the Middle East lies within the great expanse of desert (*sahra* in Arabic) that stretches in a wide and practically continuous belt from the Atlantic coast of Africa to Sind, Thar, and Rajasthan in the Indian subcontinent. Along the western, northern, and eastern fringes of the West Asian part of this desert belt lies an arc of mountain ranges that intercepts seasonal rain-bearing clouds coming from the northwest. It is this arc of mountains and intermontane vales, which—along with the more arid valleys that receive its runoff waters—constitutes the Fertile Crescent. Its complement in Northeast Africa is the valley of the Nile, which, though itself a desert, receives the runoff that is shed from the more humid highlands of Ethiopia and the tropical rainbelt of Central Africa.

The climate of the region was not always what it is today. Remnants of flora and fauna, of ancient streams that now lie buried beneath sand dunes, and even of deep water-filled subterranean aquifers, all attest that some tens of millennia ago the region was much more humid than at present. Fluctuations of climate occurred repeatedly in the past and continue to occur, to varying degrees and on varying time scales. The most momentous climate change took place at the end of the era known as the Pleistocene and the beginning of the Holocene, ten thousand to twelve thousand years ago. The last ice age ended and warming trend prevailed. Northerly areas and highlands that had been cold and forbidding now burst forth with a profusion of plants and animals that responded to the longer and warmer growing season.

Having survived the hardships of the ice age, doubtlessly thanks to their growing ingenuity and acquired skills, stone-age humans now found themselves in a more auspicious ecological situation, in which they could not only survive but even prosper and multiply. In the Middle East, they found a particularly favorable region for subsistence and habitation. On discovering a few favorable locations, clans of humans would naturally tend to prolong their stay there so as to avail themselves of the local advantages. Such advantages may have included an assured supply of water, an abundance of game or edible plant resources, readily available raw materials such as flint or wood, and protection against inclement weather or potential enemies.

Remnants of early permanent habitation, the so-called Natufian culture dating to the 12–14 millennia B.P. (Before Present), have been found by archaeologists in the hills of Israel and Jordan. The Natufians were apparently among the first hunter-gatherers to make the transition to a settled lifestyle, and were the forerunners of the earliest farmers of the Neolithic age. They built mud and stone houses, devised flint-bladed sickles to harvest wild grains, and used mortars and pestles to grind that grain for the preparation of food.

Numerous potential crops grew wild in the rainfed arc of mountains, hills, and valleys fringing the Fertile Crescent, including plants with edible grains, fruits, nuts, stalks, leaves, and bulbous roots. Prominent among the native plant resources of the region were wild species and varieties of the Graminea and Leguminosa families, whose seeds could be collected and stored to provide food for several months. Most native plants scatter their seeds as soon as they mature, and the seeds therefore are difficult to harvest efficiently. A few anomalous plants, however, due to chance mutations, retain their mature seeds. The preferential harvesting of such seeds, and their propagation in favorable plots of land, constituted the real beginnings of agriculture, providing the early farmers with crops that could be harvested more uniformly and dependably than could the wild plants.

The progenitors of the region's crops—namely, wild emmer wheat, wild einkorn wheat, wild barley, and various edible legumes—can still be found growing in the hills of northern Israel, Lebanon, western Syria, southern Turkey, northeastern Iraq, and western Iran. Their persistence in these locations, close to the sites of the earliest discovered farming communities, suggests (along with other evidence) that the climate of the region as a whole has not changed fundamentally since Neolithic times, though it may have fluctuated somewhat in the interim. As settlements and villages acquired permanence, several fruit-bearing trees (which require years to mature) could also be domesticated. These included figs, olives, and dates, as well as grapes, pomegranates, and almonds. The earliest animal domesticates were sheep, goats, pigs, dogs, and cattle.

The biblical story describing the banishment of Adam and Eve from the Garden of Eden may be taken to symbolize humanity's transformation from the relatively carefree "child of nature" hunting-gathering-wandering phase of existence to a life of toil and responsibility as permanently bound tillers of the soil.

The Agricultural Transformation is very likely the most momentous turn in the progress of humankind, and many believe it to be the real beginning of civilization. Often called the Neolithic Revolution, this transformation evidently first took place in the Middle East between 10,000 and 8,000 years ago. The ability to raise crops and livestock resulted in a larger and more secure supply of food. At the same time, it required attachment to controllable sections of land, and hence it brought about the growth of larger coordinated communities. The economic and physical security so gained accelerated the process of population growth, and necessitated further expansion and intensification of production. A self-perpetuating pattern thus developed, so the transition from the nomadic hunter-gatherer mode to the settled farming mode of life became in effect irreversible.

As agriculturists, humans began to affect their environment to a greater degree than ever before. They cleared away the natural flora and fauna from selected tracts and in their place introduced and nurtured the species of plants and animals that people preferred. Thus they modified the natural ecosystems of increasingly large areas, until they eventually altered entire regions. Their success, in terms of population growth, was considerable, but this success sometimes resulted in the practically irreparable degradation of the once-bountiful environment in which agriculture began.

An important factor in the evolution of agriculture in the Middle East was the development of the tools of soil husbandry. Seeds scattered on the ground are often eaten by birds or rodents, or desiccated, so their germination rate tends to be very low. Given a limited seed stock, the early farmers would naturally do whatever they could to promote germination and seedling growth. The best way to accomplish this is to insert the seeds to some shallow depth, under a protective layer of loosened soil, and to eradicate the weeds that might compete with the crop seedlings for water, nutrients, and light.

The simplest tool developed for planting was a paddle-shaped digging stick, by which a farmer could make holes for seeds. Later, the stick was modified to form a spade (first made of wood but eventually of metal), which could not only open the ground for seed insertion but also loosen the soil and eradicate weeds more efficiently. In time, such a spade was modified so it could be pulled by a rope, thus opening a continuous slit (furrow), into which the seeds could be sown. This human pulled spade, or *ard*, gradually metamorphosed into the animal-drawn plow. The first picture of such a plow, dating to 3000 B.C.E., was found in Mesopotamia, and numerous later pictures have been found both there and in Egypt. It was not long before these early plows were fitted with a seed funnel, so the acts of plowing and sowing could be carried out simultaneously. The same ancient implement is still very much in use today throughout the Middle East.

Although the development of the plow represented a huge advance in convenience and efficiency, it had an important side effect. As with many other innovations, the benefits were immediate, but the full range of consequences took several generations to unfold, long after the new practice became entrenched. The major environmental effect was that plowing made the soil surface—now pulverized and cleared of weeds—much more vulnerable to erosion, especially on hillslopes. Over a period of centuries, as cultivation of sloping ground became widespread, the cumulative result of this erosion spelled the nearly complete removal of the thick mantle of soil that had originally covered the region's uplands. In the history of civilization, contrary to the idealistic vision of the prophet Isaiah, the plowshare may have been far more destructive than the sword.

To understand the fate of the lands and peoples of the Middle East, we must focus more closely on the soil-water-climate conditions governing the region's agriculture. Pronounced differences in landscape, climate, and hydrology divide the Middle East into distinct environments. The different zones favor or impose different lifestyles, which have been combined with ethnic and political divisions, and so have had a lasting impact on history. The environment not only conditions civilization—it is also conditioned by it.

The physical conformation of the land has obvious effects: mountain ranges and deserts act as barriers to communication, whereas plains enable it and rivers channel it. Major political units evolve in zones of easy com-

Figure 3.1 Depiction of a seeder-plow on an ancient Mesopotamian seal.

munication, whether in the Mesopotamian Plain (Sumer, Babylonia, Assyria) or on the Iranian or Anatolian plateaus (Elam, the Hittite Empire, Urartu). On the other hand, mountain ridges and intermontane valleys foster local independence and discourage the amalgamation of large and uniform groupings—political, ethnic, and linguistic.

The regional climate is characterized by an annual cycle of a rainy winter and a totally dry summer. The annual rainfall ranges from well under 100 millimeters in the arid southern subregions to as much as 1,000 mm in the semihumid uplands fringing the Fertile Crescent. Although not generally abundant, and not uniformly or ideally distributed in time and space, the rainfall in the uplands is generally sufficient to support a rich natural community of trees, shrubs, and herbs. When that vegetation was initially cleared, generally by fire, the combination of rainfall and fertile loamy soil could support the growth of many crops. The minimal amount of rainfall generally needed for rainfed crops is about 300 mm; less rain would generally result in crop failure, and more would give a proportionately higher yield.

The effectiveness of rainfall in sustaining crops depends on the presence of a receptive and retentive soil. For crops to succeed, the soil must be able to absorb rather than shed the rain, and it must be able to store it in the root zone. The typical soil of the Middle Eastern uplands is a loam formed on limestone, and it is relatively receptive to rain. However, the amount of water any soil can retain depends on its depth, and herein lies the problem. The residual upland soils tend to be rather shallow, covering the hillsides to a depth that seldom exceeds one meter and on steep slopes is much less than that. Only on plateaus and particularly in intermontane valleys, where sediment is deposited by gravity and water, does the soil attain greater depths.

When rain falls on sloping land, part of the rain infiltrates and part runs off, in varying proportions. A soil that is deep, well-structured, and covered by protective vegetation and a mulch of plant residues will normally absorb 90 percent or more of the rainfall. On the other hand, a soil that is denuded of vegetative cover and deprived of a surface mulch may absorb less than 80 percent of the rain, and in extreme cases less than 50 percent, especially if the rainstorms are intense and if the soil is shallow and its surface has been compacted by tillage, by animal trampling, or by the raindrops themselves. As more runoff is induced, accelerated erosion ensues. The water trickling off the slopes collects in the bottomlands, where it causes frequent flooding and a rise of the water-table. The water-borne sediment clogs the natural drainage outlets. Consequently, marshes tend to form in the coastal and intermontane valleys. Prior to the advent of artificial drainage (and later still of insecticides), such marshes were often infested with malaria.

The insidious process of soil erosion is especially severe in the uplands of the Middle East, where the rains do not appear as gentle showers or drizzles, as they do in northwestern Europe, but as violent squalls. Ironically, the beneficent rain that is so desperately needed and eagerly awaited by farmers can become an agent of destruction, a voracious monster gnawing at the soil and wearing away the land. Therefore the problem of rainfed agriculture on sloping ground is how to control the erosive power of rainstorms by promoting the penetration of rainwater into the soil rather than its escape as surface runoff. The onset of erosion sets up a vicious cycle: erosion strips off the relatively permeable topsoil, thereby making the soil shallower and less absorptive. This effect further diminishes infiltration, induces greater runoff, and accelerates erosion.

So rainfed farming, which began in the flat valleys where erosion was not a serious problem, perforce had to devise specialized methods of soil conservation as it gradually expanded, with the increase in population, onto the uplands.

The experiences of the Israelites, as reflected in the Bible, spanned the entire spectrum of the region's variegated physical environments: from Mesopotamia to Egypt to the desert of Sinai and on to the rainfed hills of Canaan. In the course of their history, the Israelites interacted with all the indigenous cultures of those subregions, as well as with the intrusive culture of the Aegean Sea peoples, while evolving into a distinctive nation and developing a unique culture of their own. Out of that unique culture arose three of the world's great religions: Judaism, Christianity, and Islam.

The story of the Israelite settlement in the land of Canaan is particularly instructive. To destitute nomads coming out of the desert (at the beginning of the Iron Age, circa 1200 B.C.E.), the Promised Land must have seemed like a veritable paradise. However, it was a land already populated, in which the most desirable areas were filled to capacity. Finding the valleys occupied, the Israelites had no recourse but to settle in the highlands of Judea, Samaria, and Galilee, as well as in the northern Negev.

Here they faced new and unfamiliar challenges: a variable and capricious pattern of rainfall; few perennial sources of water for domestic needs or for irrigation; shallow, stony, and erodible soils; a rugged terrain with practically no flat land; and a thicket of oaks, pines, and shrubs that the settlers needed to clear away. So they had to learn the ways of water and soil conservation: how to hew out and plaster cisterns to collect and store rainwater for the dry season, how to clear stony ground, and how to carve out arable fields on steep slopes. They did this by collecting the stones off the ground and using them to construct rock walls on the contours, thus dividing the slope into a series of terraces.[1] Furthermore, to restore soil fertility they allowed the land to rest for one year out of seven,[2] a sort of sabbatical known as *shmitah*.

The aims of terracing were simple: to transform a smoothly sloping terrain into a staircase of nearly horizontal arable plots with adequate control of water and minimal erosion. In the process, however, terracing stamped the permanent imprint of humans on the landscape. It transformed the natural slopes, altered the patterns of natural drainage and

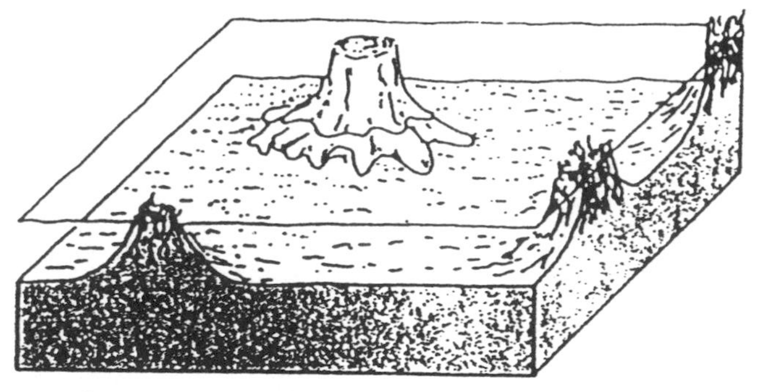

A. Sheet erosion

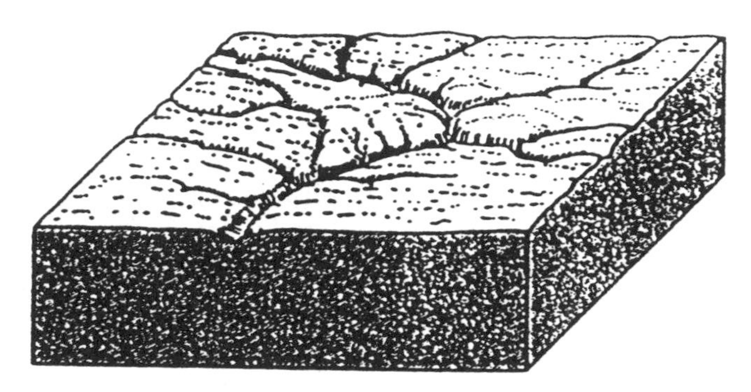

B. Rill erosion

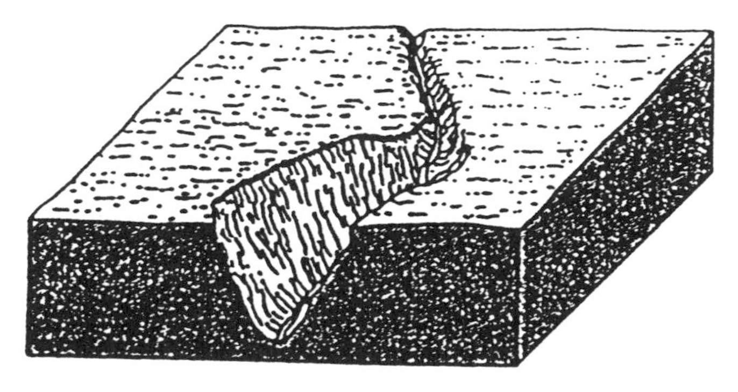

C. Gully erosion

Figure 3.2 Soil erosion by water, following removal of vegetative cover.

erosion, changed the profile and development of soils, and produced culturally controlled flow and sedimentation patterns.

Ironically, the terraces that protected the soil also increased its vulnerability to erosion. Terraces are effective in conserving both soil and water only as long as they are perfectly maintained, otherwise they increase the very hazard they were designed to abate. The moment a terrace wall is breached, whether by the spontaneous collapse of a stone wall under the weight of the soil resting against it, or by grazing goats dislodging the stones, the exposed vertical mass of soil is certain to erode even more rapidly than before the terraces were built. The farmers of ancient times undoubtedly cared for their terraced fields as long as they could live and work peacefully on their land.

Unfortunately, all too frequently in the course of history, wars would intervene. Invading armies would plunder the villages and ravage the land. At times they would even send the population into exile, as did the Assyrians to the Israelites (in the year 722 B.C.E.), the Babylonians to the Judeans (in 587 B.C.E.), and the Romans again to the Judeans in the first and second centuries C.E. In the wake of such disasters, the settled population would be replaced by bands of herding nomads, and their sharp-hoofed goats would trample over the land indiscriminately, devouring the

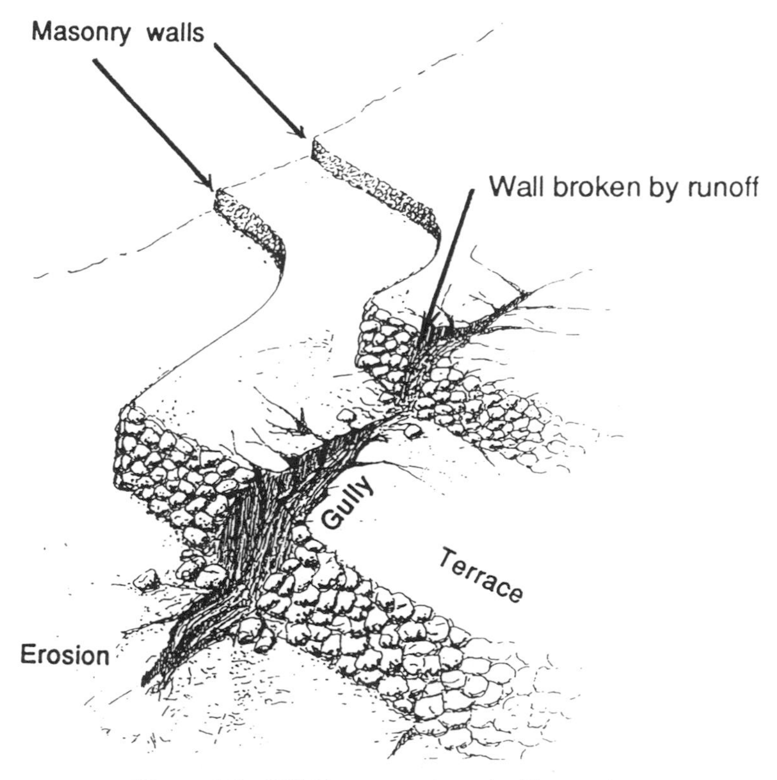

Figure 3.3 Hillside terraces breached by erosion.

vegetation and destroying the terraces. Indeed, the hills of Israel and Lebanon are full of relics of ancient terraces that have fallen into such disrepair that the soil they once held has long since been eroded to bare bedrock. Some of the terraces, however, have somehow remained intact through the ages and are still cultivated and productive, proving that the system is indeed sustainable if properly managed. (At present, unfortunately, many of these terraced plots are considered too small and irregularly shaped to allow efficient farming with modern machinery.)

The history of land husbandry in Lebanon is as instructive as the case of Israel. Lebanon was the land of the Phoenicians, who settled along the coast sometime during the second millennium B.C.E. During the period of 1000 to 500 B.C.E. they became the Mediterranean world's foremost navigators and traders. (As conduits of culture, as well as commercial goods, they transmitted the Semitic alphabet to the Greeks.) Lebanon received enough rain to produce good yields of grain, grapes, olives, and many other crops. But the amount of tillable land along the coast was limited by mountain ranges. These mountains were covered by dense stands of the famed cedars of Lebanon. The wood obtained from them was used for construction of buildings and of ships, and became a prized commodity of international trade. So the Phoenicians carried out systematic logging of the cedars. The wood was sold to treeless Mesopotamia and Egypt. The biblical Book of Kings reports that King Solomon, with the agreement of the Phoenician King Hiram, sent thousands of laborers to bring timber from Lebanon to Jerusalem for the construction of his temple and palaces.

With prosperity came an increase in the population of Phoenicia. Following the deforestation, cultivated fields began to creep up the hillslopes, and erosion ensued. By the ninth century B.C.E. the Phoenicians found that their agriculture and their extensive commerce, based on exporting lumber and various industrial products (including glass and dyes), were inadequate to support their growing population. They then embarked on colonization of other lands, including Carthage on the coast of North Africa, and other colonies in Sardinia, Sicily, and Spain. These colonies supplied food for the home country and accepted its exports. But as the homeland of Phoenicia gradually deteriorated, so did the strength of the nation. Finally, it was defeated by the Greeks, whose King Alexander destroyed the city of Tyre in 332 B.C.E. (Carthage, established on new lands, continued on its own for two centuries more, only to succumb to the Romans in 146 B.C.E.)

The uplands of the eastern Mediterranean littoral were also subject to overgrazing. Though slower acting than land clearing for cultivation, excessive grazing is ultimately no less devastating. Goats can be particularly destructive. They not only browse their favorite shrubs but can climb right up into trees to eat their foliage. They also tend to consume tree seedlings, thus inhibiting forest regeneration.[3] Sheep, too, can do great damage where they overgraze, since they will eat grass down to the roots, and their sharp hooves—like those of goats—tear up the sod and pulverize the soil.

Fires further contributed to erosion. As the wooded vegetation of the Middle East tends to desiccate during the long rainless summer, it becomes extremely flammable. Since early times, the herders of the region took advantage of this by setting repeated fires to suppress the dense woody vegetation (trees and bushes) and thus to encourage the growth of pasturable grasses. In the process, erosion was greatly accelerated.

Altogether, the uplands of the Middle East provide a telling example of how societies have tended to destroy their environment. Many of the areas that once supported a thriving agriculture are now largely unproductive. The face of the land itself is a more eloquent and revealing document than all the written records.

The earliest farmers depended only on seasonal rainfall to water their crops. However, even the relatively humid subregions of the Middle East are subject to a high incidence of drought. In time of need, therefore, it was only logical for farmers to augment the water supply for their crops by artificially diverting water from streams. It was also logical to try to raise crops on riverine floodplains that were naturally irrigated.

At some point, then, farming was extended from the relatively humid areas of its origin toward the region's major river valleys of the Jordan, Tigris-Euphrates, Orontes, Nile, and Indus. As the climate of these river valleys is rather arid, a new type of agriculture based primarily or even entirely on irrigation came into being.

With a practically assured water supply, an abundance of sunshine, a year-round growing season, and deep and fertile soils, irrigated farming became a highly productive enterprise. However, its very success induced an insidious process that could not have been foreseen: land degradation resulting from silting, waterlogging, and salinization.

Nowhere are the consequences of this process more tragically apparent than in southern Mesopotamia (part of present-day Iraq), the domain of the ancient Sumerians, Akkadians, and Babylonians.

In the second half of the fifth millennium B.C.E., when the art and the benefits of farming were already widely established in the rainfed subregions of the Middle East, a group of people of uncertain ethnic and geographic origin, now known as the Sumerians, colonized the lower courses of the Tigris and Euphrates rivers. Their land was a flat plain of brownish alluvium, dusty when dry and miry when wet, and deluged periodically by the sudden overflows of the twin rivers. In Mesopotamia, the myth of the great flood unleashed by God to scourge the sinful—described in the Gilgamesh epic as well as in the biblical story of Noah—arose out of bitter experience.[4]

The Mesopotamian Plain was never a homogeneous environment. Although the plain appears flat to the eye, it is nowhere entirely so. Peri-

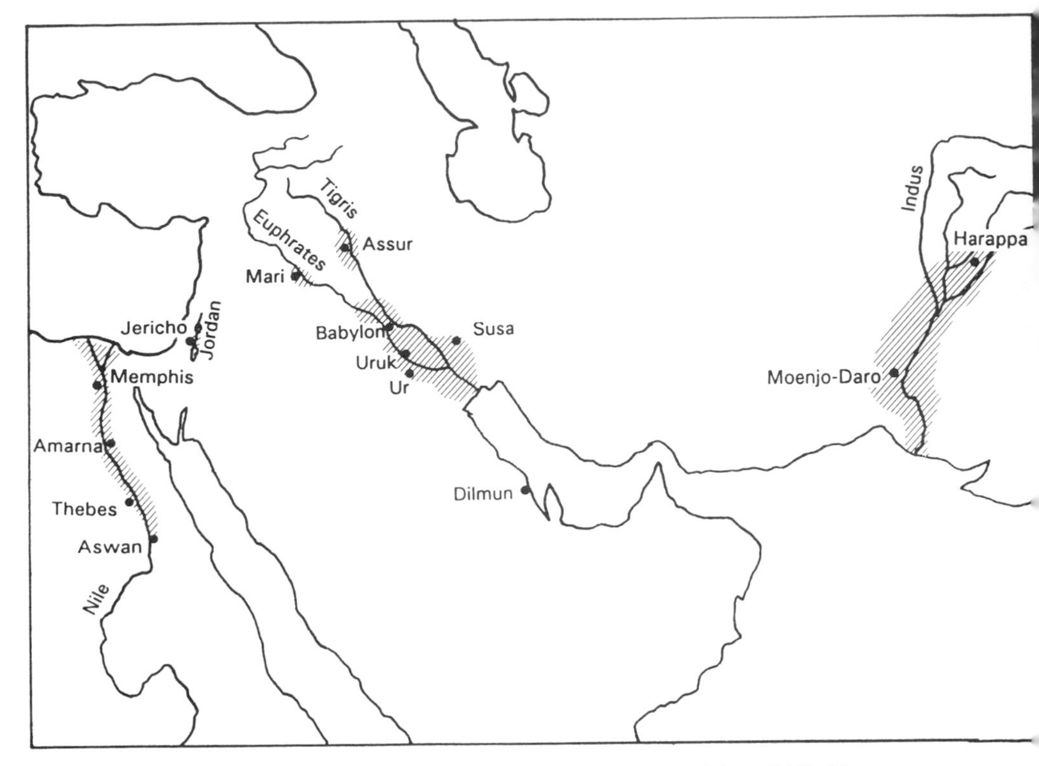

Figure 3.4 The ancient hydraulic civilizations of the Middle East.

odically shifting water courses tend to build up their beds and banks, to the effect that in time the rivers and canals rise above the surrounding land, which is then divided by their banks into basins imperceptible to the eye but critical to the flow of water.

The soils of the alluvial plain are deep, and the relative flatness of the land and the elevation of the rivers permit the diversion of water via canals onto a more or less ordered system of fields. In this manner, an irrigation-based economy was established that depended on the controlled exploitation of the rivers. As the two rivers tend to burst their banks in flood, thus threatening farmers, great efforts were required to contain or tame the floods by building high banks, or levees.

Through diligence and ingenuity, the Sumerians gradually transformed their land from a barren plain with interspersed swamps, periodically too dry and too wet, into a land of extensive grain and forage fields and datepalm plantations. A Sumerian myth refers to the introduction of cereals from the distant highlands. Another Sumerian myth describes farmers as "men of dikes and canals." Those farmers could feed several times as many people as farmers on equal plots without irrigation. (Herodotus, in the fifth century B.C.E., heard that in Mesopotamia grain yielded 200- and 300-fold the amount of seed planted. This is undoubtedly a great exaggeration; even a 30-fold yield[5] is impressive, and much more plausible.) The Sumerians traded the products of their husbandry for the raw materials—building stones, wood, metals, and gems—that were lacking in their land.

The surplus production of their farmers enabled the Sumerians to develop the world's earliest urban society. The first Sumerian cities were small mud settlements. Later, houses were built of sun-dried mudbrick joined by bitumen (an early use of the material oozing out of the ground, now called petroleum, that would acquire great importance in our century). Later still, the Sumerians used burned bricks to build their homes and temples. The latter (called *ziggurratu* in Assyrian, meaning "summit") were towers built on multistoried terraces, referred to in the biblical story of the tower of Babel as challenging the heavens.

We owe much to the Sumerians. They invented writing (which they did by pressing a wedge-tipped reed stylus into tablets of wet clay, to create a script called "cuneiform"). They developed sailboats, wheeled vehicles, the potter's wheel (the first industrial machine with continuous rotary motion), yokes for harnessing animals to carts and plows, weighted levers for lifting water, accounting procedures, literature (including epics and love songs), and lawbooks. They also developed weapons, war machines, and an entrenched bureaucracy (for which we may be less grateful).

Notwithstanding their seminal achievements, the Sumerians did not last. As they began to decline, during the third millennium B.C.E., the Akkadians, who spoke a Semitic language, gradually superseded them. They shifted the center of power northward and extended the domain of their activity from the Persian Gulf toward the Mediterranean littoral. However, after a time Akkad, too, began to decline. They were supplanted by the Babylonians, and eventually by the Assyrians, whose center of power was located far upstream on the Tigris River. An inscription on the tomb of the Assyrian Queen Semiramis, dating to about 2000 B.C.E., proclaimed: "I constrained the mighty river to flow according to my will and led its waters to fertilize lands that before had been barren and without inhabitants".[6]

Thus the center of civilization in Mesopotamia tended in general to shift gradually northward, and the land of Sumer in southern Mesopotamia never recovered its former productivity.

One of the early excavators of the cities of Sumer, C. L. Woolley, puzzled over this demise in his 1936 book, *Ur of the Chaldees:* "Only to those who have seen the Mesopotamian desert will the evocation of the ancient world seem well-nigh incredible, so complete is the contrast between past and present . . . it is yet more difficult to realise, that the blank waste ever blossomed, bore fruit for the sustainance of a busy world. Why, if Ur was an empire's capital, if Sumer was once a vast granary, has the population dwindled to nothing, the very soil lost its virtue?"

The answer is that the Sumerians themselves, unwittingly to be sure, brought about their own decline by causing the degradation of their soils.

The very act of diverting river water onto the valley land, which at first transformed desiccated tracts of desert into a fertile expanse of fields and orchards, gradually created insurmountable problems. Ironically, initial success led to eventual failure.

The first problem was silt. Early in history, the upland watersheds of the twin rivers were deforested and overgrazed. Erosion resulting from seasonal torrential rains proceeded to strip off the soil of those uplands and pour it into the streams, which in turn carried it as suspended sediment hundreds of miles southeastward. As the silt-laden flood waters found their way toward the lower reaches of the valley and their flow became sluggish,[7] more and more of the sediment settled along the bottoms and sides of the rivers, thus raising their beds and their banks above the adjacent plain. Rivers that are elevated above their floodplains are notoriously unstable: during periodic floods they tend to overflow their banks, inundate large tracts of land, and from time to time change course abruptly. The silt also tends to settle in channels and fields, and thus to clog up the irrigation works.

The second and even more severe problem was salt. This problem, along with waterlogging, resulted primarily from the inexorable rise in the water-table—a rise that, in the absence of adequate natural or artificial drainage, inevitably followed the flood-irrigation of low-lying lands.

All waters used in irrigation contain dissolved salts. Since crop roots normally exclude most of the salts while extracting soil moisture, the salts tend to accumulate in the soil and, unless leached out, will in time poison the root zone. In arid regions, rainfall is generally insufficient for leaching, hence irrigation must be applied in excess of crop water uptake so as to remove harmful salts by downward percolation. The excess water and the salts it contains do not disappear, however, but tend to salinize the groundwater and raise the water-table. In low-lying valleys that are poorly drained, continued irrigation raises the water-table toward the surface until it waterlogs the soil.

The rise of the water-table may take place slowly and remain invisible and unnoticed for quite some time, perhaps decades or even generations. Finally, when the water-table comes within a meter or less of the ground surface, a secondary process of capillary rise comes into play. The upward seeping groundwater evaporates at the surface, infusing the topsoil with salt. As the salinization process advances, an irrigator might try to irrigate more in a desperate attempt to flush out the salt with fresh water. But in so doing he is merely accelerating the rise of the water-table and waterlogging the soil.[8] After each irrigation the salt reappears. It oozes from below and blossoms out in mockingly beautiful floral patterns, and more of it each time, until the surface is encrusted with a myriad of glistening crystals and the soil is rendered sterile.

Imagine the helpless consternation of Mesopotamian farmers on first noticing the paradoxical appearance of a white snowlike powder on the ground in the heat of summer, and then realizing that more and more of their fields, previously so productive, were now afflicted with a strange

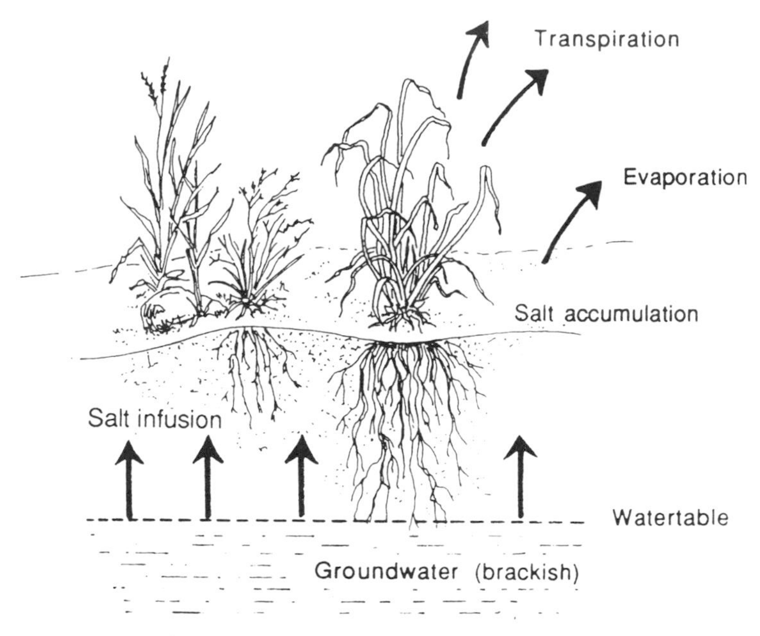

Figure 3.5 The process of soil salinization.

leprosy that kills the soil. Not knowing from where this affliction arrived, those farmers must have thought of it as a curse of the gods.

Working in tandem, silt and salt can destroy an entire region's irrigation-based agriculture. Evidently they did so in ancient southern Mesopotamia, as well as in the Indus Valley. The same processes, incidentally, are at work today no less than in the past and are still the scourge of irrigated agriculture in arid regions. Strong examples can be seen in the former Soviet Union, around the Aral Sea, in the Murray River basin of Australia, in the Indus River basin of Pakistan and India, in Mexico, and—not the least—in the irrigated areas of the southwestern United States.

The Koran (XXV:53) states: "It is He who has released two bodies of flowing water: One palatable and sweet, the other salty and bitter; yet He hath separated between them." A similar statement is made in the form of a rhetorical question in the New Testament (James 3:11): "Doth a fountain send forth at the same place sweet water and bitter?" The answer, unfortunately, is that careless or unwitting irrigators can in fact turn sweet water into salty water, and thereby turn fertile soil into barren ground.

During periods of both scarcity and surplus of water, problems of its equitable distribution must have arisen among users variously located along the river and the diversion canals, whether upstream (closer to the source) or downstream (at the tail end of the delivery system). Regulation of water flow (including construction of canals and their maintenance by frequent dredging of the continuously deposited silt) therefore required coordination, either voluntary or imposed by some central authority.

King Hammurabi of Babylon, who ruled Mesopotamia from 1728 to

1686 B.C.E., was strongly aware of the need for regulating the use of irrigation water from the two rivers. In his famous Code,[9] the laws concerning water management were aimed at preventing an owner of one field from causing damage to the field of a neighbor: "If a seignior [a citizen with full rights] was too lazy to make [the dike of] his field strong and did not make his dike strong and a break has opened up in his dike and he has accordingly let the water ravage the farmland, the seignior in whose dike the break was opened shall make good the grain that he let get destroyed." Documentary evidence indicates that the king directed his provincial governors to dig canals and dredge them regularly, as well as to build flood protection works such as earthen levees. Large-scale flooding was, of course, a constant danger in lower Mesopotamia, especially if the two rivers crested simultaneously. Numerous other documentary remains pertain to water and its management.[10]

The buildup of soil salinity affects the choice of crops to be grown. Although wheat is obviously the preferred food grain, it is more salt-sensitive than barley. Cuneiform sources suggest that the proportion of barley grown in ancient southern Mesopotamia increased progressively in time, along with the spread and increase of soil salinity.[11] In recent times, the relative areas of the two staple crops have ranged from over 90 percent wheat north of Baghdad to more than 90 percent barley in the southern plain, which is more severely affected by salinization despite the attempts by the modern Iraqis to provide artificial drainage there.

At present, the method of preventing or remedying salinization involves lowering the water-table. This is done either by means of lateral subsoil drainage through open ditches or buried perforated tubes, or by means of upward pumping through wells. In ancient times, however, practically the only method for dealing with salinization was a system of alternate-year fallowing. At the end of an irrigation season, the water-table might be within a half meter to a meter of the surface. If the land is kept fallow for a year, it is normally invaded by phreatophytes—native weeds capable of sending down roots to the water-table and drawing water from it. These plants can lower the water-table to a depth of perhaps 2 meters, thus lessening the capillary rise of water and the migration of water-borne salts toward the surface.

But while the ancient practice of alternate-year fallowing could well have retarded salinization, it could not prevent it. At each alternate-year irrigation season, more salt was added to the subsoil and the groundwater. The fallowing system gradually lost its efficacy, and the inevitable result was progressive soil salinization and—finally—abandonment of the land. Left unirrigated, the salinized land might eventually recover its productivity by natural leaching, but that could require decades, or even centuries.

The civilization of Mesopotamia depended on a complex dynamic interplay between the unseen rains falling on distant mountains, the resulting unforeseeable overflowing of the rivers, and the equally mysterious upsurge of the groundwater. Life was thus perceived as a contest between the fresh water from above, represented by the good god Apsu, and the poisonous brine welling up from below, represented by the evil goddess Tiamat. The story of this contest in related in *Enuma Elish*, the Babylonian Genesis. The dual origin of water is echoed in the story of the Deluge in the Hebrew Bible, in which the water engulfing the earth is said to have come from two sources: "the windows of the heavens" and "the fountains of the great deep." In a Sumerian myth, the goddess of love and procreation was envied by her sister and enemy, the goddess of death. The latter meted out her revenge stealthily, in the form of salt-laden water oozing up from below to kill the lifegiving soil.

To the east of Mesopotamia, far across the deserts of southern Persia and of Baluchistan, lies the Indus River Valley. Here, another irrigation-based civilization developed in ancient times, probably under the influence of Mesopotamia. Though the Indus River civilization apparently embraced an area more extensive than that of either the Sumerian or the Egyptian, much less is known about it. No written records have been discovered, hence the language of this civilization has not yet been deciphered and its history and fate cannot be determined with certainty. Some have conjectured that it came to an end in a catastrophic flood of the River Indus. It appears more likely that this civilization, like the Sumerian, succumbed to environmental degradation.

Unlike Mesopotamian cities, which were built of sun-dried mud bricks, the Indus Valley cities were built mostly of baked bricks. The firing of these bricks must have required great quantities of wood. The use of wood as fuel and as building material, along with the grazing of cattle, goats, and sheep, probably resulted in widespread denudation of the watersheds. The resultant erosion, and the increased silting of the river valley, may have worsened the flooding of the cultivated land. Salinization was very probably another acute problem in the ill-drained Indus Valley then, as indeed it is now.

In sharp contrast to the fragility of southern Mesopotamia and the Indus Valley, we have the example of the Nile Valley, which has nourished and sustained more than five millennia of civilization without interruption. That contrast may seem puzzling. The Tigris-Euphrates and the Nile are similar in that they are exotic rivers, drawing their waters from outside the arid regions into which they flow. Mesopotamia and Egypt had similar climates and raised similar crops. Both depended primarily on irrigated agriculture.

Although civilization in Mesopotamia developed somewhat earlier than in Egypt, the two cultural centers were contemporaries for long periods of history, and constituted diametric powers vying for hegemony over the entire Fertile Crescent. Their rivalry went on for many centuries, and was most intense in the first millennium B.C.E., during which great armies (among the first massed armies in the history of warfare) marched back and forth to test which of the two would predominate. Both civilizations had their periods of greater and lesser ascendancy. At various times, both won and lost wars, dominated nations, and were themselves subjugated. Moreover, both suffered occasional episodes of floods and droughts that caused famine and that periodically decimated the population.

Through all these vicissitudes, the civilization of Egypt survived and continued in the same location. In contrast, the nations of Mesopotamia—Sumer, Akkad, Babylonia, and Assyria—each in turn, rose and then declined, as the center of population and power shifted gradually from the lower (southern) to the central and thence to the upper parts of the Tigris-Euphrates valley. What explains the durability of Egypt and the fragility of southern Mesopotamia? The answer might be relevant to the situation of the region in our day, and in the future.

Egypt, in the words of Herodotus, is the gift of the Nile. There were actually two gifts: water and silt. Those, one might think, are the same gifts brought by the Tigris and Euphrates to southern Mesopotamia, but there are important differences. Neither the clogging by silt nor the poisoning by salt were anywhere as significant in Egypt as they were in the Tigris-Euphrates plain, so the land of Egypt could remain productive while the land of Mesopotamia suffered degradation.

The productive land of Egypt is divided into two sections: the long, narrow strip alongside the upstream section of the river is called Upper Egypt. The broad Delta where the river fans out into a series of distributaries is called Lower Egypt, and it includes more than half of Egypt's arable land.

The silt comes mainly from the steep volcanic highlands of Ethiopia, lashed each summer by the monsoonal rains rolling in from the Indian Ocean. The downpours scour the slopes, scraping off their loose mantle of mineral-rich dark soil and splashing it into the annual flood of the Blue Nile. (The good fortune of Egypt, one might say, is derived from the misfortune of Ethiopia.) Added to that silt is the humus contributed by the White Nile from its rain forest and swamp sources.

When the gathering flood reached Egypt proper, it would overflow the river banks and deposit an annual increment of silt estimated to have averaged about 1 millimeter thick on the floodplain. This amount was not so excessive as to choke up the irrigation canals or cover young seedlings, but it was fertile enough to add nutrients to the land and nourish its crops.

Whereas in Mesopotamia the inundation usually comes in the spring, and summer evaporation tends to salinize the soil, in Egypt the Nile begins its rise in the middle of August and attains its maximum height in the beginning of October. Thus, in Egypt the inundation comes at a much more favorable time for the fall planting of winter crops: well after the spring harvest and after the summer heat kills the weeds and aerates the soil.

The narrow floodplain of the Nile (except in the Delta), as well as the deep-cut nature of the river bed and the roughness of the adjacent desert terrain, made it generally impractical to divert and convey water in long canals as was done in Mesopotamia. Hence there was no widespread raising of the water-table. The water-table was controlled by the stage of the river, which, over most of its length, lies below the level of the adjacent land. When the river crested and inundated the land, the seepage naturally raised the water-table. As the river receded and its water level dropped, it pulled the water-table down after it. This all-important annual pulsation of the river and the associated fluctuation of the water-table under a free-draining floodplain with a permeable sandy subsoil created an automatically repeating, self-flushing cycle by which the salts were leached from the irrigated land and carried away by the Nile itself.

The early farmers of Egypt, circa 5000 B.C.E., probably relied on natural irrigation by the unregulated floods to water the banks of the river. As soon as the flood withdrew, they could cast their seeds in the mud. At times, however, the flood did not last long enough to wet the soil thoroughly, and then the crops would fail and famine would ensue. So the Egyptian farmers learned to build dikes around their plots, thus creating basins in which a desired depth of water could be impounded until it soaked into the ground and wetted the soil deeply enough to sustain the roots of the crops throughout the growing season. The diked basins also retained the vital silt and prevented it from running off with the receding floodwaters.

The earliest pictorial record of artificial irrigation is the mace-head of the "Scorpion King" (circa 3100 B.C.E.) that depicts a ceremonial cutting of an irrigation channel. The king is shown holding a hoe, with laborers excavating the channel, and others holding a basket and a broom, all standing alongside the channel. Rectangular irrigation basins can be seen in the background.

The basis of Egypt's productivity was the nearly optimal combination of water and soil nutrients provided by an annual regime that was more dependable and timely than the capricious floods of Mesopotamia. It enabled Egyptian farmers to produce a surplus that fed the artisans, scribes, priests, merchants, noblemen, and—above them all—the Pharaohs who used their coercive power to order the building of self-aggrandizing monuments. Those monuments still stand today, less in testimony to the vainglorious kings who ordered them than to the diligence and organization of a society of labor nurtured by the river and rooted in the land irrigated by it.

Unfortunately for students of Egypt's history, the records of dead kings and noblemen are disproportionate to those relating to the ordinary life of the people. Since Egyptians wrote on sheets made of the pulp of papyrus reeds (our word "paper" comes from the name of this plant), most of their records were destroyed by humidity and time. Enough, however, has survived in papyri and wall paintings to give us a fair picture of ancient Egyptian agriculture. It appears to be remarkably similar to the agriculture practiced by the *fellahin* who live and work along the Nile

Figure 3.6 The Scorpion King inaugurating an irrigation canal, circa. 3100 B.C.E.

today. No nation has lived so long and in such harmony and intimacy with the soil. In the words of an ancient tomb inscription: "I live, I die, I am Osiris. . . . I grow up as grain . . . the earth has concealed me. I live again, I die again, I am barley. I do not pass away."

The political and ecclesiastical structure of Egyptian society was strongly hierarchical. With the nearly absolute authority and proclaimed deity of the Pharaohs, who were presumed to hold sway over the cosmic order and over famine and plenty, ancient Egypt conformed perhaps more closely than Mesopotamia to Wittfogel's concept of a despotic "hydraulic civilization."

As the population of Egypt grew despite periodic setbacks, the necessity arose for intensifying production. Instead of one crop per year, the Nile Valley farmers could grow two, three, or even four, given the year-round warmth and abundant sunshine of the local climate. To do so, they needed to draw water at will from the river or from shallow wells dug to the water-table. They did this at first by manually lifting and carrying buck-

Figure 3.7 Ancient Egyptian depiction of sowing and plowing. Tomb of Nakht, 18th Dynasty.

ets, using shoulder yokes. In time, a new technology was developed: mechanical water-lifting devices. The simplest of these (a concept apparently imported from Mesopotamia) was the *shadouf,* consisting of a long wooden pole used as a lever, with the long arm serving to raise bucketfuls of water and the short arm counterweighted with a heavy stone or a large lump of mud. A more sophisticated device, invented some centuries later and attributed to Archimedes ("Archimedes' screw"), is the *tambour,* which consists of an inclined tube containing a tight-fitting spiral fin. Both of these devices are human-powered. The most elaborate of the ancient mechanical water-lifting devices is the animal-powered *saqia* waterwheel, introduced to Egypt, probably from Persia, during the Persian or Greek (Ptolemaic) occupations. All these devices are still in use today, alongside modern motorized pumps.

The affluence of the river lands, in Egypt as in Mesopotamia, has always attracted the hungry dwellers of the bordering deserts. A document of the nineteenth century B.C.E. mentions nomads begging "to serve the Pharaoh" as "the desert was dying of hunger." Egyptian border garrisons generally warded off such would-be intruders. Occasionally, some were admitted. The Bible relates that Abraham, and later Jacob and his sons, went down to sojourn in Egypt during times of famine in Canaan. Egyptian texts of the twentieth, and again of the thirteenth, centuries B.C.E., state that Asiatic herdsmen were permitted to enter Egypt "as a favor, to keep them and their cattle alive." After the middle of the eighteenth century B.C.E., "wretched Asiatics" invaded and conquered lower Egypt. These intruders, called *Hyksos* (meaning "rulers from foreign lands"), were expelled two centuries later. Egypt finally lost its independence to the Greeks (who built Alexandria) and was later ruled by the Romans, the Byzantines, and a long succession of other foreigners. Yet the agricultural base remained intact. That is, until recently.

When Napoleon invaded Egypt in 1798 (thus highlighting the strategic importance of the Middle East and making it a focus of Great Power struggle in modern times), he brought a contingent of scholars for the

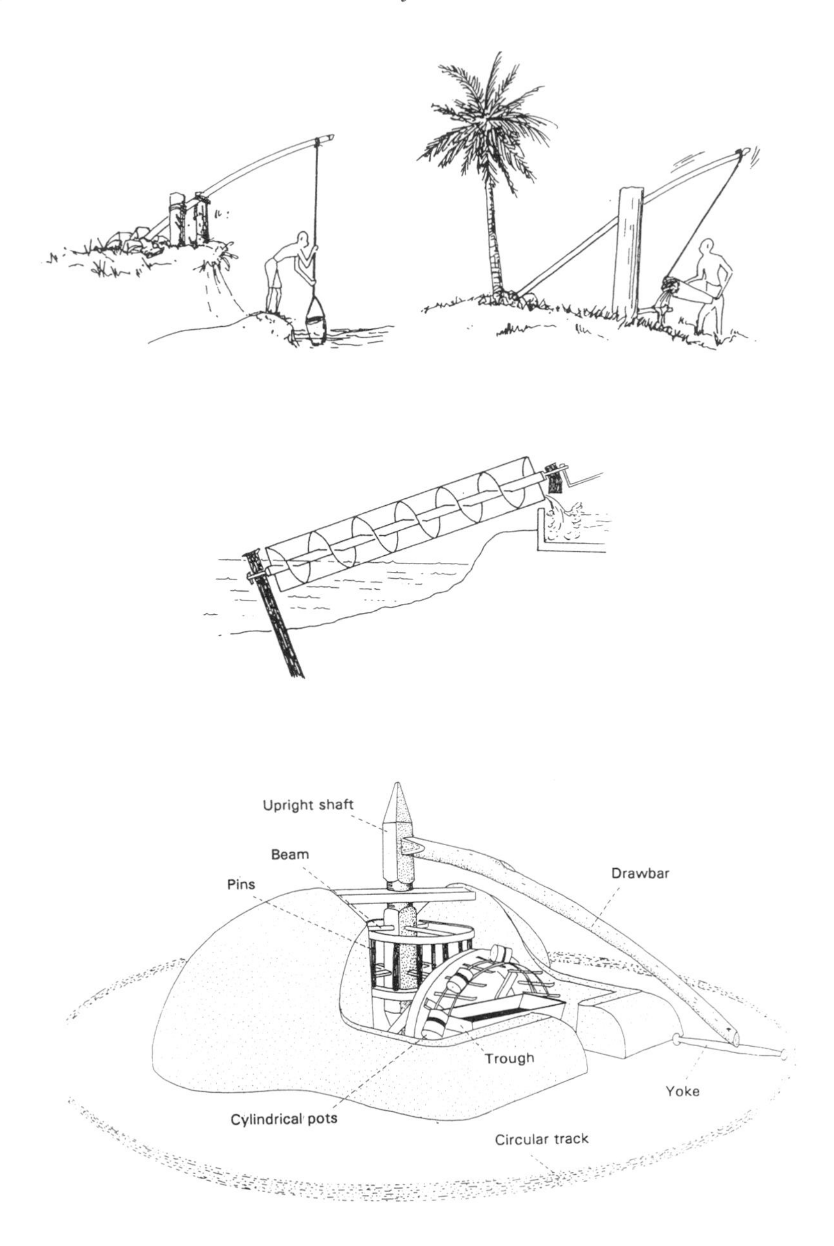

Figure 3.8 Ancient water-lifting devices. **Top:** *Shadouf;* **Center:** Archimedes' screw *(tambour)* **Bottom:** Animal powered *saqia.*

purpose of studying that mystery-shrouded ancient land. Apart from its dubious military or political results, this expedition made a great contribution to knowledge. It initiated the systematic study of Egyptian history, known as "Egyptology," and led, among other achievements, to the discovery of the Rosetta Stone and the deciphering (by Jean-François Champollion) of ancient Egyptian hieroglyphics. More relevant to our topic is the fact that French agricultural experts surveyed the Nile Valley and conducted a census there. They reported that the population of Egypt in the early 1800s was less than 3 million. Historical evidence suggests that the population had fluctuated over the millennia from perhaps 1.5 to 2.5 million. So the population at the beginning of the nineteenth century was not much different from what it had been throughout prior history.

The present population of Egypt is nearing 60 million. This means that 20 times as many people as in ancient times are now striving to eke out a livelihood on practically the same soil and water resource base. Then, Egypt was able to export food and it helped to sustain the Roman Empire. Now Egypt imports half its food needs. Worse yet, its soil—legendary for its durability and productivity—is now undergoing progressive deterioration. We shall have more to say about the predicament of modern Egypt in a subsequent chapter of this book.

Large sections of the Middle Eastern deserts are practically devoid of perennial water resources. However, although the term *desert* is derived from the Latin word for "abandoned" or "deserted," not all deserts are totally useless wastelands. Some desert subregions of the Middle East were, in fact, settled by extraordinarily diligent and ingenious societies, who proved that civilization can be established even in extremely difficult circumstances. Evidence of such civilizations can be found in North Africa, the Arabian Peninsula, Jordan, and, notably, in the Negev of southern Israel.

The name *Negev* denotes dryness in the original Hebrew. As deserts go, it is rather small, constituting only a minuscule part of the great desert belt of North Africa and Southwest Asia. Being on the fringe of this belt, much of the Negev is not extremely dry. The mean annual rainfall ranges from 250 mm in the northwest to about 25 mm in the far south, occurring in the winter months, November to April. The distribution of rainfall within the rainy season is irregular, and the total seasonal amount fluctuates widely from year to year.

The historical importance of the Negev, along with neighboring Sinai, derives from its geographical position as a land bridge connecting Asia with Africa. Hence it has always served as an avenue of traffic between the continents. The advantages of controlling the area, however, were often offset by disadvantages. The routes that made possible both trade and access to cultivable areas in times of peace were the same ones followed by invading armies in times of war. Moreover, desert nomads were ever ready to plunder the settled land and its inhabitants. Thus, to the difficulties posed by the paucity of water and the fragile environment was added the requirement of constant vigilance against encroachment by hostile forces.

The Israelites arrived in the Negev at the end of the Bronze Age (circa 1200 B.C.E.), and their presence there continued into the Iron Age. King Solomon and his Judean successors—Asa, Jehoshaphat, and Uzziah—

established villages, fortresses, and trade routes in the deserts of Judea and the Negev, and linked their kingdom with the copper mines of the Arava Valley and with the seaport of Eilat on the Red Sea. Concerning King Uzziah, the Second Book of Chronicles mentions as a major achievement that he "built towers in the wilderness and hewed out many cisterns, for he had much livestock . . . for he loved the land."

After the destruction of the Judean Kingdom by the Babylonians, a new nation entered the Negev and built a magnificent civilization there, the achievements of which excite the admiration of visitors to the region to this day. The new masters of the Negev were the Nabateans. Originally nomadic traders of Arabian extraction, they settled in the Negev and in time became superb architects and engineers, as well as expert hydrologists and diligent cultivators.

The Nabatean domain lay astride the ancient trade routes between Arabia in the south and Syria in the north, and between the Orient and the Mediterranean world. These were the routes along which camel caravans transported choice commodities—spices, silks, ivory, frankincense, myrrh, and medicinal herbs—that were as prized in antiquity as are perfumes and drugs today. The demand for spices was even greater then than now, not so much because our taste in food has become blander, but because—in the absence of refrigeration and other modern means of food preservation—food could quickly become inedible without a heavy dosage of spice.

Caravans passing through the desert needed stopping places to rest and obtain water and provisions. To secure and supply their trade, the Nabateans therefore had to establish and maintain regularly spaced bases along their main routes, with secure sources of water. These bases gradually grew into self-supporting villages and eventually into cities, and the Negev became more densely populated than ever before. Although the Nabateans' capital, the fabled red city of Petra, was built in the Edomean mountains (in southern Jordan), their population was centered in the Negev, where they built six major cities and numerous smaller villages. To maintain a population of tens of thousands, the Nabateans perforce had to develop agriculture to ensure a subsistence for their people. In this task they were undoubtedly aided by the example of their predecessors. But the Nabateans excelled all previous efforts.

The monopoly over the trade routes enjoyed by the Nabateans ended some time in the first century of the Christian era, when the Romans discovered that the seasonal monsoon winds made it possible for them to sail through the Red Sea to India and back. They were thus able to trade directly for the coveted spices and aromatics. Soon afterward, the Nabatean settlements had to face their greatest test of survival. Having lost their lucrative function as caravan stations, these settlements in the Negev had to become self-sustaining or disappear. The Nabatean King Rabel II (70–106 C.E.) is described on the coins and inscriptions of his day as "he who brought life and deliverance to his people." He very probably earned this distinction by emphasizing the improvement of desert farming practices, by which alone the Nabatean people could thrive in the desert. The same populace continued even after the Romans annexed the region and made it a frontier province of their empire.

After the division of the Roman Empire and the establishment of Byzantium, the entire eastern realm of the empire enjoyed a period of stability and prosperity. The Negev became still more densely populated, and the technical achievements of the era surpassed even those of the Nabateans when they were independent.

The eclipse of the Byzantine golden age in the Negev came very abruptly in the seventh century C.E. The population dwindled following the Muslim conquest in 636, which severed the region's links to the Mediterranean world. Desert nomads took over and ushered in a long period of retrogression and poverty. Where thousands once prospered, a few hundred now eked out a subsistence. Monuments were pried apart or crumbled gradually into haphazard heaps of stone. Cisterns were choked by dust, and strongly built dikes were loosened by time and left unrepaired. Complete farm systems were abandoned and allowed to disintegrate. Overgrazing the dry stream beds caused erosion, so the formerly wide bottomlands irrigated by water-spreading dikes became narrow gouged-out gullies. Thus, the experience of generations of diligent desert farmers was lost. The casual visitor to the Negev finds it difficult to understand how the ancients could have developed so grand a civilization in the midst of such barrenness. Only a careful study of their techniques can reveal the answer.

Since perennial rivers are totally absent in the Negev and even springs or proper locations for digging shallow wells are few, the major source of water could only be the collection of surface runoff obtainable from sloping ground during infrequent winter rains. That practice is known as "water harvesting." The first imperative of desert settlement was the provision of potable water for humans and livestock. This was done by means of cisterns, which are artificially constructed reservoirs filled by surface flows during infrequent rainstorms.

The early cisterns were undoubtedly leaky and inefficient. Building efficient cisterns became possible only with the advent of watertight plaster, made of burned and slaked lime. In addition, the ability to recognize suitable rock formations (such as soft marly chalk, which could be hewed out readily and was not as fissured as the hard limestone also prevalent in the region) and to collect and direct overland flow by means of channels was crucial[12].

Where cisterns could be located alongside natural streams, they were filled by flash floods. However, most cisterns in the Negev were built on hillsides and depended on the direct collection of runoff. Many hundreds of such cisterns were built in the Negev, and they are clearly discernible landmarks even today. A typical one resembles a giant necklace, with the glistening white pile of excavated rock hanging as a pendant from the two collection channels that ring the hill and curve down its sides from opposite directions. To parched travelers through the desert, to whom these cisterns would beckon from afar, no sight could be more gladdening.

Throughout most of the Negev (except in the northernmost area), rainfall alone is insufficient for dependable cropping. In the rare locations where perennial springs were available, they could serve for supplementary irrigation.[13] Where no such sources existed, farming depended on the collection and utilization of runoff from winter rains falling on sloping land.

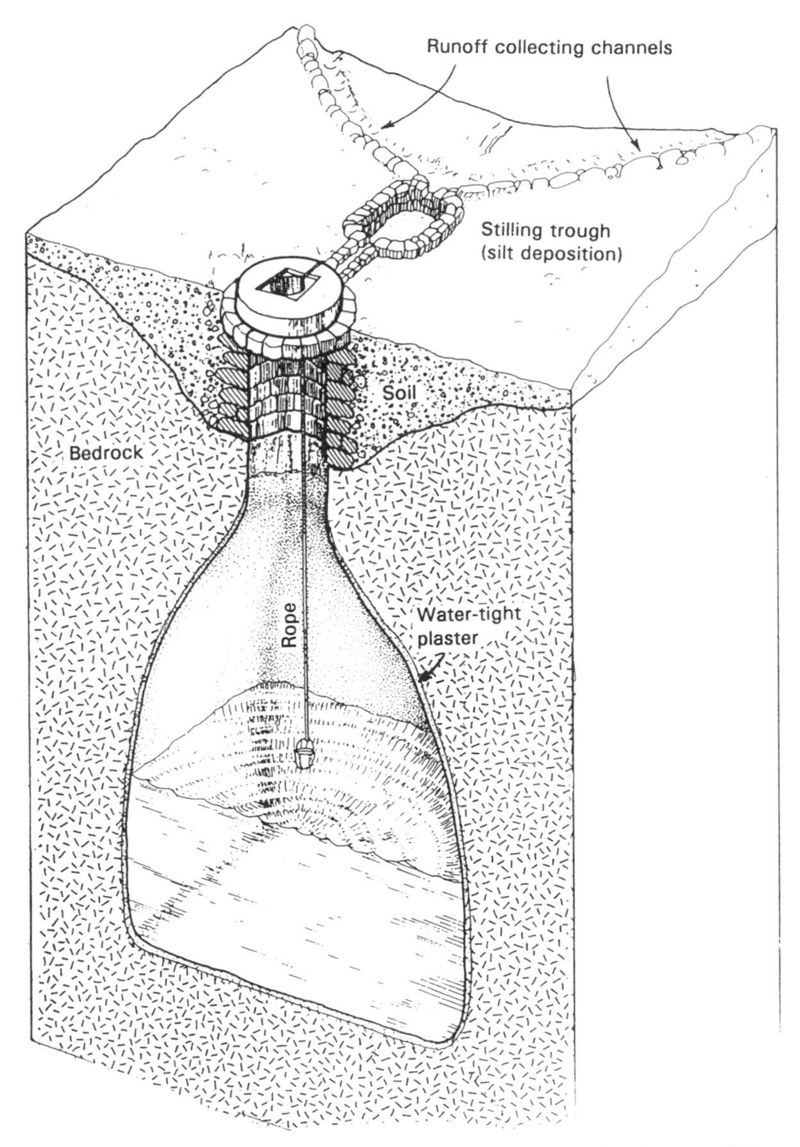

Figure 3.9 A hillside cistern excavated in the bedrock and fed by runoff from a sloping catchment.

That runoff was gathered and directed to bottomland fields for periodic soakings, so as to accumulate sufficient moisture in the soil for crop growth. Although the average winter rainfall in the Negev is only about 100 millimeters, the desert farmers were able to gather and concentrate sufficient runoff from the barren slopes to develop intensive agriculture in the depressions and bottomlands. These narrow and winding strips of runoff-watered land, however, constituted less than 5 percent of the total area in the northern Negev highlands subregion.

This ingenious type of desert agriculture has been called runoff farming. Whereas farmers in more humid regions aim to have the soil absorb all the rain where it falls and thus to prevent runoff, the desert farmers worked on the opposite principle. Their aim was to reduce the penetration of rain into the soil on the slopes so as to produce the maximum possible runoff. They then collected this runoff from a large area of slopes and directed it to a relatively small cultivated area in the bottomlands.

The cultivated area was usually divided into small field plots, which were leveled and terraced to ensure efficient spreading of water as well as both soil and water conservation. The oldest version of runoff farming probably consisted of terracing the small creek beds that collected the runoff naturally. This terracing had the effect of transforming the entire length of each creek into a continuous stairway, with stairs perhaps 10–30 meters wide and 20–50 centimeters high. The terrace walls were designed to spread the flood and to prevent erosion. The slowed-down cascade of the flood from one terrace to the next could thus irrigate the field plots sufficiently for a crop to be grown. Distinct groups or series of terraced plots having definable catchment areas and surrounded by stone walls formed integral farming units of perhaps several hectares of cultivated land. The remains of hundreds of such farm units are spread throughout the Negev highlands, most commonly in the environs of the principal ancient towns.[14]

Detailed observation of ancient runoff farm units reveals that each unit was served by a particular and well-delineated portion of the watershed. An elaborate system of conduits was constructed to collect runoff from specific sections of the adjacent slopes, not merely for each farm or set of fields, but indeed for each terraced field within the farm. The complete farm unit thus comprised both the slope catchment (the runoff-contributing area) and the bottomland fields (the runoff-receiving area). The larger the catchment (watershed), the greater the expectable water supply and the corresponding area that could be irrigated. Clearly defined catchment areas, allocated to serve particular farm units, constituted "water rights."[15] Thus, early in the history of the Middle East, control over land was equated with control over the water yielded by that land.

The fraction of rainfall shed by the land as runoff varies, of course, from rainstorm to rainstorm and from year to year. Gentle showers contribute practically no runoff, whereas intense squalls might yield 30 percent or more of the rain as runoff. The amount also depends on the nature of

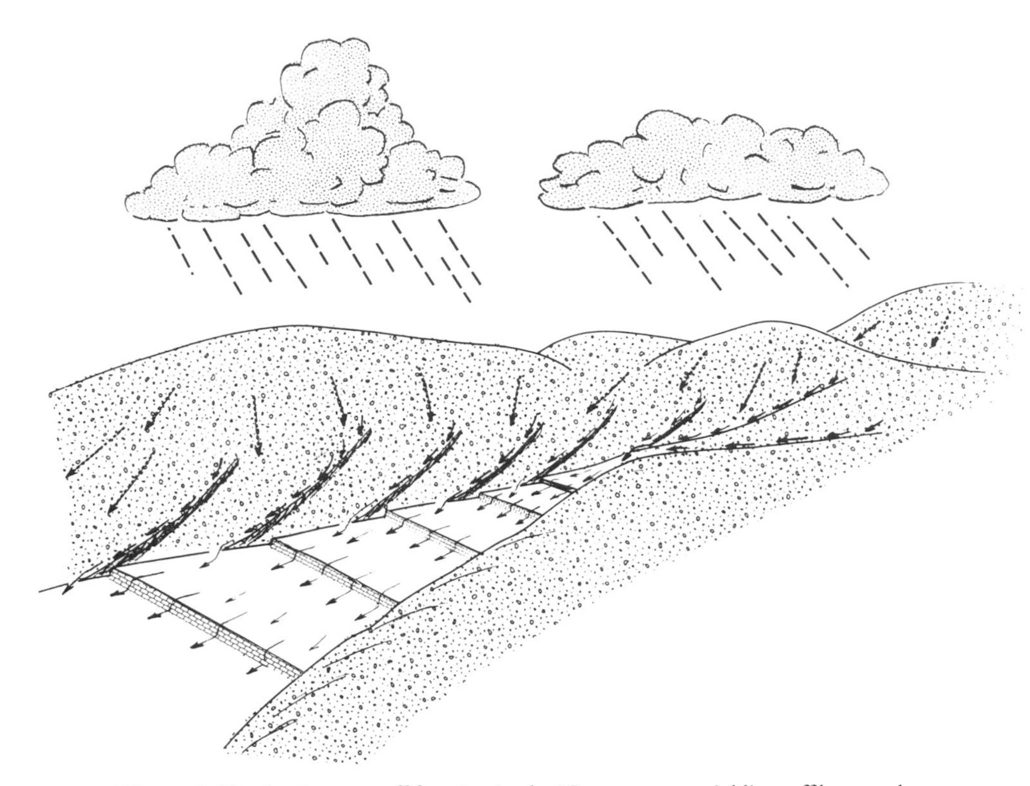

Figure 3.10 Ancient runoff farming in the Negev: water trickling off barren slopes was directed to terraced fields in the wadi bed.

the watershed, with steep bare slopes tending to shed more of the rainwater than relatively flat areas with a denser plant cover. Above all, the total runoff yield of a watershed depends on the quantity of rainfall, which is notoriously unstable in a desert environment. As a "normal" seasonal average, natural runoff yield does not generally exceed 15 percent of total rainfall, and in seasons with small rainstorms it may be less than 5 percent. (The agricultural term *yield* is appropriate, since the desert farmers actually harvested their supply of water from the slopes, depending almost entirely on the collection of runoff to fill cisterns and to irrigate terraced fields located in the bottomlands.) So, even with all the alertness, ingenuity, skill, and diligence they could muster, the runoff farmers of the Negev operated a risky business and had to face new uncertainties each season. In view of this, it is all the more remarkable that the ancient Negevites could sustain so large a population and operate on so grand a scale. During its period of maximal development in the Byzantine era, the system of runoff farming in the Negev evidently supported tens of thousands of people and encompassed practically all of the usable bottomlands (the vales and wadi beds) in the northern Negev highlands.

The Negev runoff farmers did more than merely gather natural runoff. There is clear evidence that they enhanced it artificially, by an ingenious manipulation of the land surface. The hillsides in the Negev, as in many other deserts, are naturally strewn with a "desert pavement" of stones and gravel, and this covering inhibits and detains the flow of runoff over the surface. Noticing this, the ancient Negevites deliberately cleared the stones off the slopes to enhance and facilitate runoff.

I discovered this in the course of my doctoral research on the water regime governing the soils and vegetation of the Negev, conducted in the early 1950s. In part, the research was impelled by the puzzling presence throughout the northern Negev highlands of countless heaps, mounds, and strips of gravel on barren hillsides, particularly in the vicinity of the old towns of Shivta, Ovdat, and Nitzana. Why had the dwellers of this region in antiquity devoted such enormous efforts to raking the gravel off the surface of the slopes adjacent to their fields?

Previous explorers of the region thought they knew the answer. These mounds had long been called *tuleilat el einab* (meaning "mounds of the grapes") by the Bedouin, if only because they are reminiscent of the gravel heaped around their grapevines by farmers in the humid hill country of Judea and Samaria (who do this to protect the soil under the vines from evaporation and weed growth and to allow the grapes to ripen on clean dry gravel rather than on soil). European travelers through the Negev early in the twentieth century took the Bedouin allusion to the local gravel

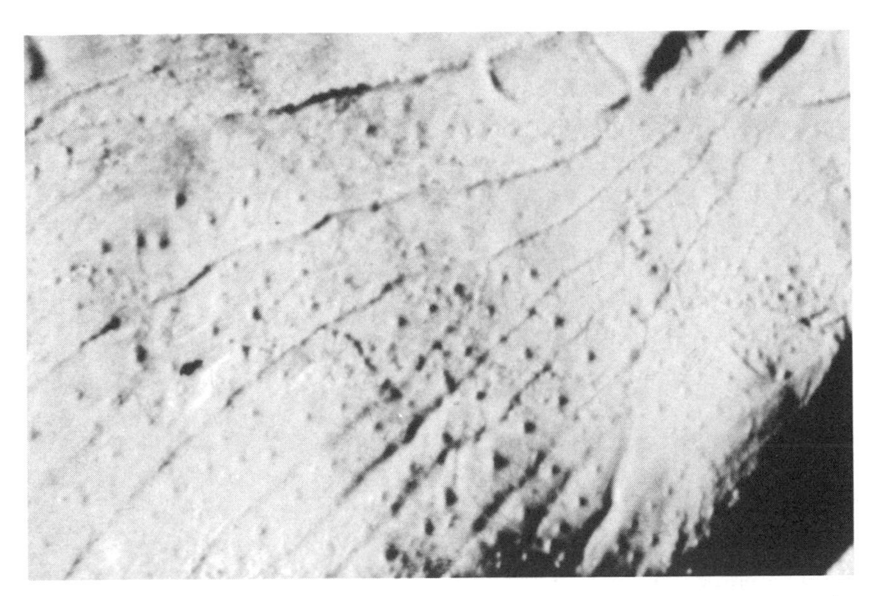

Figure 3.11 Aerial photograph of ancient runoff farming in the Negev. Note the gravel mounds on the slope at left, and the channels directing runoff to fields in the wadi bed.

mounds all too literally and accepted as fact that the ancients did indeed grow grapevines on those slopes. But, if so, how could the vines have received enough water in so dry a region? The visitors conjectured that the gravel mounds, by cooling at night, could condense enough dew from the air and drip that dew into the soil to sustain plant roots growing beneath the mounds.

Although that conjecture was totally baseless, it was taken seriously by countless writers who cited and re-cited one another as evidence, even as "proof," that dew-collecting "aerial wells" are capable of sustaining crops in the desert. Others, who came to doubt the fantasy of dew collection, surmised that the ancients might have intended to induce erosion from the hillslopes so as to fill their wadi terraces with soil. That, too, is a fanciful explanation, since it is generally lack of water, not of soil, that constrains the production of crops in the Negev. Moreover, gravel mounds are prevalent not only in catchments adjacent to terraced fields but also in those serving cisterns, where the inwash of silt would be a distinct disadvantage.

Examining the mounds, we found first of all that the cavities between the stones were choked with dust, so they could not permit the free passage of air required to condense significant amounts of dew. Then we calculated theoretically how much moisture the desert air holds and what volume of that air would be needed to provide the water requirements of grapevines. That calculation showed that the local air could not possibly provide enough water. (The only way the mechanism could possibly work would be with a steady infusion of moisture-laden air from sea breezes such as occur in the highly specific conditions of the Atacama Desert along the Pacific coast of South America.)

Next we dug under the mounds, analyzed the soil there, and found it to be highly saline—much too saline for any known crops to grow in it. This seemed to rule out the "aerial well" hypothesis in principle, but we still decided to try it in practice. So we built several types of mounds, with clean gravel as well as with dust-filled gravel, each on top of a sloping metal tray to catch the condensed dew in special containers. But there was none to catch.

So, why did the ancients undertake the herculean effort to build all those mounds, over an area of hundreds of square kilometers? Suddenly, the answer became obvious: removing the gravel that covers the ground surface exposes the finer soil that lies underneath, and that soil (called *loess*) has a tendency to form a surface crust when subject to the beating action of raindrops. The crust acts as a surface seal: it inhibits the infiltration of rainwater and increases the rate of runoff. I had proven that effect experimentally in the extensive research that culminated in my doctoral dissertation, completed in 1956. My good colleague Naftali Tadmor then corroborated my conclusion regarding the function of the mounds by examining the geometric pattern of the mounds, using a set of aerial photographs, and we wrote a joint paper on the topic.[16]

Still wishing to make absolutely sure, I set up full-scale field trials to measure runoff yields from slopes that had been cleared of gravel, with or without gravel mounds, in comparison with slopes that had a natural, undisturbed gravel cover. The results were conclusive: the practice of removing the surface gravel had the effect of increasing the runoff yield by 8 percent to 20 percent.

In addition to inducing and collecting the localized runoff from slopes and conveying it to individual farm units, the ancient Negev dwellers carried out larger-scale works to divert floodwater from regional streams onto adjacent flat lands. Regional streams drain large watersheds. Their generally gravelly stream beds may remain dry for many months, only to gush forth suddenly with an astonishing torrent of water. At times, the flash flood may occur without warning in a rainless stretch of the wadi, as a consequence of an unseen cloudburst over a distant range of mountains. In the words of the prophet Elisha (circa 865 B.C.E.): "Thus saith the Lord: Make this streambed full of hollows. . . . Ye shall not see wind, neither shall ye see rain, yet that streambed shall be filled with water, and ye may drink, both ye and your flocks" (II Kings 3:17).

After the departure of the Byzantines in the seventh century, the Negev was overrun by nomadic herders who did not have secure tenure to the land and therefore did not enjoy the stability needed to establish permanent settlements based on intensive land and water husbandry. Neglected, the old waterworks fell into disrepair. For many centuries, the

Figure 3.12 An experiment to measure the effects of stone removal on runoff water yield from a hillslope in the Negev desert.

historical techniques of desert agriculture had been largely forgotten. However, research carried out over the last few decades has shown that those techniques are still applicable, not only in the Negev but also in wide areas in the Middle East, Africa, and similar regions.

When the great Prophet Muhammad, founder of the faith of Islam, died in Medina (Arabia) in 632 C.E., his disciple and father-in-law, Abu-Bakr, was chosen to be his successor (called *Khalifah* in Arabic, or Caliph in English).[17] He, in turn, was followed by Umar (634–644). Inspired by their new faith and obeying the call to *jihad* (holy war), the Muslim Arabs then broke out of the fastness of their desert peninsula and turned with great fervor, sword in hand, toward the north. In little over a decade, they captured Byzantium's Middle Eastern possessions (Palestine, Syria, Egypt, and Cyrenaica), and in the following decade they absorbed the whole Sassanid Empire of Persia. Then, within a century, Muslim soldiers extended the domain of Islam from Spain in the west across all of North Africa and the Middle East to the borders of China in the east. Those spectacular conquests brought together the diverse cultures of many lands, and from their combination would grow a new civilization matching those of Greece and Rome in cultural and material achievements.

Because the Arabian conquerors were not numerous,[18] they allowed local customs to continue for a time with little interference. In particular, Jews and Christians were respected under Muslim rule as "People of the Book." However, the taxes exacted from non-Muslims were higher than those levied on Muslims, and the latter were given preferential treatment in many other ways as well. Gradually, over a period of decades and centuries, even without forcible conversion, a great number of the non-Arab people of the Middle East adopted the Arabic language and most of them became Muslims.[19]

As the majority of the people of the Middle East were gradually "Arabized," the term *Arab* acquired a more general meaning. Originally, it pertained only to the *Bedouin* who were descended from the tribal people of the Arabian Peninsula. In time, it was extended more loosely, to people of whatever original extraction who had adopted the Arabic language and culture. Included among the latter were descendants of numerous groups of people who lived in the region prior to the Arab invasion. Thus, the Arabic language and the Islamic religion were overlaid upon disparate ethnic communities whose prior languages included Aramaic, Hebrew, Greek, and Egyptian; and whose multifarious prior religions included paganism, Christianity, and Judaism.

For several centuries, the Caliphate of Damascus under the rule of the Umayyads, and particularly the Caliphate of Baghdad under the Abbasids (the most famous of whom was Harun al-Rashid, noted for the splendor of his court), restored the ancient irrigation systems of the region, especially of Mesopotamia. The Fatimids, who ruled in Egypt, did the same in their domain. As the Muslims extended their rule to include Spain, they established an irrigation-based economy there as well.

The glory of Arab Mesopotamia (called *Iraq*, meaning "rooted" in reference to the rootlike or arterylike twin rivers) did not last. In 1256, the Mongols attacked Iraq from the east, flooded the camp of the defending Abbasid army by breaching the dikes of the Tigris River, then pillaged Baghdad and burned its mosques, palaces, schools, and libraries. The prosperity and cultural ascendancy of Mesopotamia were destroyed again and are yet to recover in modern times. Either due to war or to environmental degradation—more likely both—other areas in the Middle East suffered as well. In the centuries that followed, practically the entire Middle East came under the control of the Ottoman Empire and sank into a long period of stagnation.

4

Modern States

Why are nations so agitated?
Why do peoples brood in vain?
Psalms 2:1.

In the long course of history since the rise and demise of the ancient civilizations described in the preceding chapter, much has happened in the Middle East. Armies invaded and were repulsed, battles were won and lost, cities were built and destroyed, nations coalesced and dissolved, competing religions and languages clashed and intermixed, technical innovations and cultural developments took place—all these and more. What has not changed is the pattern, set in motion long ago, of environmental and water resource exploitation, the results of which are etched indelibly in the face of the land.

Every one of the insidious human-induced scourges that played so crucial a role in the decline of past societies has its mirror image in the contemporary Middle East. But it seems that the mirror is warped and the problems it reflects are magnified. Salinization, erosion, denudation of watersheds, silting of valleys and estuaries, overuse and pollution of water resources—all have occurred more intensively and on an ever-larger scale. Added to the old problems are new ones, including pesticide and fertilizer residues, domestic and industrial wastes, and finally the threat of climate change. These problems are greatly aggravated by the mounting pressure of a multiplying population.

A factor that further worsens the problems and makes them extremely difficult to redress is the arbitrary political division of the region into competing states, imposed upon the Middle East by the folly of the post–World War I colonial powers. When the crucial time came to redraw the map of the region, the British and French, who considered themselves the most advanced (and hence most justified) nations on earth, acted selfishly and myopically. As they carved up what they believed to be a tabula rasa, their main purpose was to serve their own colonial interests. In so doing, they largely ignored the region's prior history, physiography, and hydrology, as well as its indigenous ethnic and religious composition. They denied the rights of some distinctive communities to autonomy while forcing disparate ones into incompatible associations. Consequently, lines contrived as if on an empty map by self-serving outsiders have become hardened and practically immutable national borders, all too often thwarting rather than promoting regional cooperation.

Especially serious was the manner in which those who drew the map of the modern Middle East disregarded the issue of water. With the exception of Lebanon, none of the states they created was provided with independent water resources, and no mechanisms were put in place for coordination in the utilization of internationally shared resources. The Tigris-Euphrates basin was sectioned irregularly and placed in the domains of three states, while the Jordan River basin was divided among four states. Long after the British and French lost their power and withdrew, the people of the region continue to suffer the consequences.

The germs of discord introduced by the colonialists continue to fester today among deprived communities vying for independence and a secure livelihood. Commentators on the current conflicts in the region often suggest that the discord arises out of age-old rivalries over resources. However, they seldom note to what extent those rivalries were exacerbated early in the twentieth century (very recently, as the long history of the region goes) by the arbitrariness with which the European powers set the political boundaries.

To understand how this tangled situation came about, and how it affects the region's environment and development, we must review the course of events that took place in the Middle East during the decisive early decades of the twentieth century. Until the end of the nineteenth century, the Middle East appeared to be unified and quiescent, seemingly stable under the long-term rule of the Ottoman Turks. Most of the Arabic-speaking Middle East was at least nominally part of that vast and amorphous empire.

The Ottomans formed a dynastic regime, not a nationality. Originally a fierce band of Central Asian Turkish warriors who first galloped onto the stage of Middle Eastern history in the thirteenth century, they established an empire that, in the name of Islam, reached across North Africa and penetrated into Europe all the way to the gates of Vienna. Although this once-great empire had begun to shrink by the early twentieth century, it still ruled over some 30 million people in the Middle East, comprising a dozen or more different nationalities. It was a shaky empire, held together by loose bonds of tradition and an inefficient administration.

The surface uniformity of the Ottoman Empire was, however, illusory. In reality, the empire was a racial and ethnic stew, simmering with discontent over the long perpetuation of an increasingly anachronistic status quo.

Though the region's disparate peoples had been largely Islamized and Arabized over the preceding millennium, their adoption of the ruling religion and of Arabic culture did not erase their heterogeneity. An unblended juxtaposition of polyglot ethnic and religious elements continued to harbor ancient, albeit largely suppressed, tribal rivalries. The region that once led the Western world spiritually, intellectually, technologically, and even militarily, had stagnated and seemed forgotten by history.

What began to jar the region out of its apparent stasis was the intrusion of the Western powers and of Western culture, which challenged and eventually overturned the existing regime. This intrusion, slow at first, was manifested in the takeover by Britain of Cyprus (1878) and of Egypt (1882); the penetration of German interests into the Ottoman Empire; and the growing cultural influence of France in the Levant. The seductive notions of the enlightenment (secularism, progress, democracy, along with the lures of the industrial revolution and of materialism) began to penetrate the urban elite of the region through commerce, literature, and education.

The explosive event that brought about the complete and abrupt restructuring of the entire region's political and economic status was the First World War. That ruthless war, begun in Europe, soon spread into the Middle East, where it constituted a historical watershed sharply sepa-

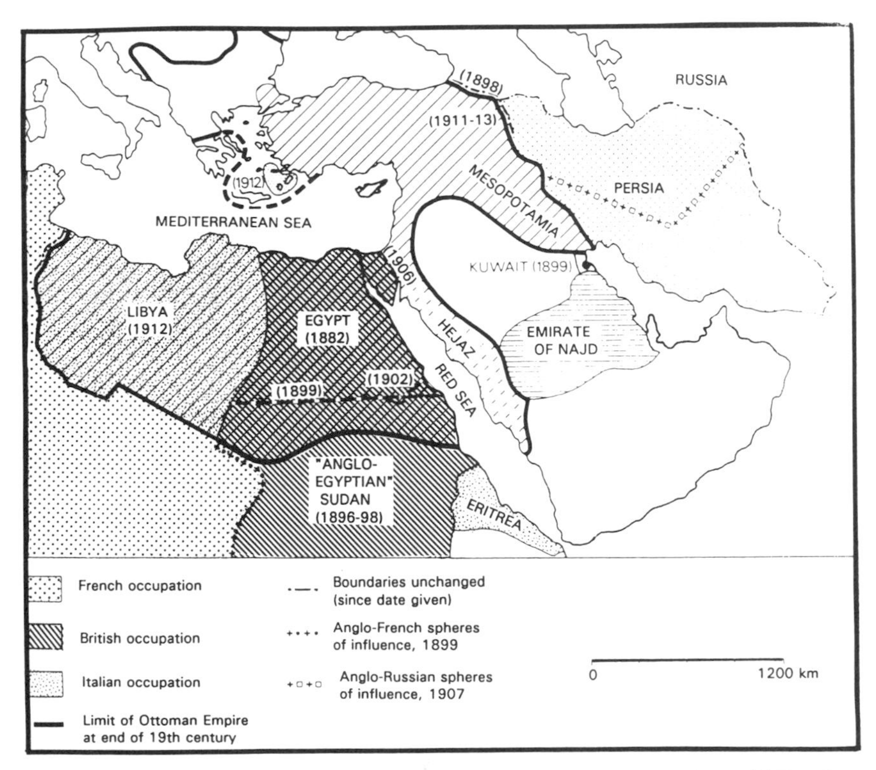

Figure 4.1 The political map of the Middle East on the eve of World War I.

rating the old order from an entirely new one. Hence it marks the true beginning of the modern history of the region. The changes wrought in the course of that war and its aftermath (the period from 1914 to 1923) consisted of the dismantling of the Ottoman Empire and the redrawing of the region's map and its boundaries. New states were formed where none had existed before while old nations were ignored or denied. The entire process was carried out swiftly and with little regard for geographical, hydrological, or environmental realities.

The collapse of the Ottoman Empire, in 1918, seemed to herald one of the rare moments in history when fresh starts beckon and dreams can become realities. The Arabs, Jews, Armenians, Kurds, Christians, Druzes, Greeks, Italians, Egyptians, as well as the French and British, all had their aspirations and claims to parts of the Middle East. For some, however, the opportunity was lost. The victorious British and French seized the moment to redesign the region. Ultimately, however, their machinations gained them nothing: within a generation they were to lose their new position of power in the region they had expended so much effort to control. Yet their legacy remains. The borders of the Middle East today are largely the same lines drawn on the post-Ottoman map in the early 1920s by colonial civil servants.

The new states carved out of the old Ottoman Empire in the wake of the First World War included the countries now known as Turkey, Lebanon, Syria, Iraq, and Egypt (the latter having long preexisted as a nation but not as an independent political unit). In addition, new entities were recognized that eventually gave rise to national states. Among these were the territories of Palestine and Transjordan, in which the Kingdom of Jordan and Israel later were formed (along with the territories of the West Bank and Gaza); the sections of Arabia known as Nejd and Hedjaz, which were unified by the Saudis to form Saudi Arabia; as well as the Persian Gulf Emirates and Yemen. The aim of the the Western powers was to perpetuate their dominance over the region: France to control Lebanon and Syria, and Britain to control Iraq, Jordan, Palestine, and Egypt (and to "influence" the other states as well).

These profound changes occurred so rapidly that the peoples of the region could hardly assert themselves effectively in determining their own fate. The subsequent turbulent history of the Middle East has been marked by the painful attempt of its nations to adjust to, or rectify, the new order (or disorder) imposed upon them by the imperialist European powers. Since those foreign interests were often contradictory, as were the internal interests of the disparate peoples of the region itself, the resulting interactions were often contrived haphazardly. Hence many of the region's fundamental problems, including the availability and development of water resources, have remained unresolved to this very day.

The great conflagration of World War I had ignited regardless of the Ottomans, but soon engulfed them as well. The Ottomans had joined the war on the side of the "Central Powers" (Germany and Austria) and against the "Allied Powers" (Britain, France, Russia, and later Italy). That fateful decision had been made by a group of brash military officers called "The Young Turks." These nationalistic officers resented the humiliating encroachments upon the once-expansive Ottoman realm by the Russians in the east and by the British and French in the south and west, and they hoped by joining the war to restore their empire to its former glory.

With the entry of the Ottoman Empire into the war, the Allied Powers began to plot its total demise and the division of its territories among themselves. In March 1915 they signed a secret agreement in which they promised Russia what was considered the greatest prize, namely the city of Istanbul (Constantinople) and the Straits of Bosporus and Dardanelles connecting the Black Sea to the Mediterranean, thus granting Russia the status of a major naval-imperial power. Only the Bolshevik Revolution of 1917 and the subsequent withdrawal of now-Communist Russia from the war prevented the Russians from realizing their promised war booty. Britain and France, in turn, agreed to divide most of the remaining Asian portion of the Ottoman Empire between them.

Britain's main imperial interest was to establish geographical and political continuity, indeed a land bridge, from its pivotal base in Egypt guarding the Suez Canal to the "Crown Colony" of India and beyond it to the British possessions in Southeast Asia. That Asian land mass would also connect, via Egypt as a fulcrum, with the eastern African territories subject to British control (Sudan, Kenya, Uganda, the formerly German-ruled Tanganyika, the Rhodesias, and South Africa)—indeed a grand design. Thus the British hoped to ensure their communication and transportation links with their far-flung eastern empire, and to exploit the resources of those territories. A particular focus in this grand ambition was Mesopotamia, where the British considered developing (as they had in Egypt and Sudan) a major agricultural center for the production of wheat and cotton.

Territorial continuity and agricultural production were not the only considerations, however. A new factor had entered the equation of power politics in the Middle East. It was the discovery in the first decade of the twentieth century of petroleum deposits in southern Persia, and the possibility of finding additional supplies in Mesopotamia. The British were particularly interested, because they had just then converted their navy from dependence on coal to the use of petroleum fuels. They also wished to block Russia's expansion southward in Persia toward the same oil fields. For this reason, the British were willing to grant France a stretch of territory along the northern tier of the Fertile Crescent, to constitute a further barrier against Russia's southward expansion.

An interesting aspect of World War I in the Middle East was the Arab Revolt, which began in western Arabia in June 1916. Though from a strictly military viewpoint its significance may have been marginal, its political implications were far reaching. The revolt was initiated by the Hashemite family, headed by Hussein Ibn Ali, the Sharif (chief magistrate) of Mecca. Two of his sons played leading roles in that revolt: Emir (Prince) Feisal and Emir Abdallah. Their main force was recruited from among the

Bedouin tribes of Hedjaz (western Arabia), and originally numbered fewer than 3,000 men. Initially, the revolt was a localized affair, but after it was joined by T. E. Lawrence, a British liaison officer later mythologized as "Lawrence of Arabia," the rebellion was coordinated with the main thrust of the British-led forces in Sinai and Palestine. The Arab Revolt was extended northward from Arabia to become a regional campaign advancing through Transjordan as far north as Damascus. The charismatic personality of Lawrence lent that campaign a romantic aura and helped to give it historical significance.

The small Jewish community then living in Palestine also became involved in the war, mainly on the side of the Allies. An underground organization acting against the Ottomans was crushed, but some of the young men expelled by the Ottomans at the onset of the war, together with Jewish volunteers from other countries (including the United States) joined the British forces, first individually, later as a recognized Jewish Brigade. Among them was David Ben-Gurion, who eventually became the first premier of Israel.

The decisive British thrust into Palestine took place under the command of General Edmund Allenby in October 1917. At the end of that month, Australian-New Zealander (ANZAC) forces breached the Turkish defenses near Beersheba, then advanced to Gaza and Jaffa. General Allenby personally led the triumphant procession into Jerusalem on the ninth of December. It was a prideful occasion for the British—the first time since the defeat of the Crusaders by Saladin, eight centuries before, that a Christian force had captured the Holy City. (Alas, this time, too, their sojourn there would be ephemeral, no more than 30 years.)

The army under Allenby continued its advance northward through Galilee, while the Arab forces with Lawrence advanced in parallel fashion in Transjordan. On October 1, 1918 the Emir Feisal galloped into Damascus and by the end of the month the British had taken Homs and Hama and reached Turkey itself. The "Young Turks" who had led the Ottoman Empire into the war were deposed, and an armistice agreement was signed.

Simultaneously with Allenby's thrust in Palestine, a separate campaign was waged by another British expeditionary force, organized from India, against the Ottomans in Mesopotamia. In the late summer of 1918, their expeditionary forces there (mainly from India) reached the northern town of Mosul. That operation assumed particular importance when the vicinity of Mosul was found to contain oil.

Now came the time for Britain and France to divide up the spoils of the Great War. The leaders of France felt that their country deserved a generous portion of the conquered territory, since France had borne the brunt of the bloodiest fighting along the Western Front in Europe. On the other hand, the British, with their victorious armies already deployed throughout the Middle East, had effective possession of the territories to be divided.

The French had begun to pressure the British as early as 1915 for an explicit commitment regarding the postwar disposition of the Ottoman Empire. Sir Marc Sykes was appointed by the British to negotiate with the French representative Charles François Picot over the future allocation of territories. In the spring of 1916, they both traveled to St. Petersburg to coordinate their plans with the Russian government, then still allied with the Western powers. Their secret agreement, known as the "Sykes-Picot Agreement," divided the main territories of the Fertile Crescent between France and Britain according to three criteria: (1) Lands to be annexed outright by each of the powers as part of its empire: France was to obtain a strip of land along the eastern coast of the Mediterranean, as well as a part of Anatolia with its center at Adana; Britain was to be given direct rule over southern Mesopotamia together with a continuous stretch of land from Mesopotamia to the Mediterranean including the Bay of Haifa. (2) Lands where the Arabs were to be granted self-rule, but under the "influence" of one of the Western powers: the central regions of Syria, including the cities of Damascus, Homs, Hama, and Halab (Aleppo), as well as a strip of territory eastward to include northern Mesopotamia, would be in the French sphere of influence. (3) An international zone: such would be in principle the status of Palestine with its holy sites (principally Jerusalem).

The importance of the Middle East to the Western powers was greatly enhanced as a consequence of the First World War. Its strategic location as a land bridge and naval corridor linking Europe with southern and southeastern Asia had long been realized. The discovery of petroleum in southern Iran in 1908, and the granting of oil extraction rights to the British by the Shah, made the region even more vital for the British Empire. Shortly afterwords, apparently large additional deposits of petroleum were discovered in the vicinity of Mosul, and these induced the British to add northern Mesopotamia to their sphere of influence in contravention of the Sykes-Picot Agreement, which had originally allocated that territory to the French.

While the Great Powers were thus machinating, the peoples of the region were demanding recognition of their national aspirations. For the Arabs, who constituted the majority in the region, the demise of the Ottomans was an opportunity to reestablish their political, military, and cultural ascendancy in what they had long regarded as their rightful domain. But the Arab nationalists were not alone in this moment of awakening. A collection of minority nationalities and sects, long ignored or denied, now sought all at once to assert their distinctive identities. Included among these were the Christians in Lebanon; the Shiite Muslims in various communities in Iraq, Syria, and Lebanon; the Druzes in Syria and Lebanon; the Kurds in northern Mesopotamia, western Persia, and southeastern Anatolia; the Turks as a distinct nation emerging from the ruins of Ottoman Empire; and the Jews returning to their biblical Promised Land after two millennia of dispersion.

A casual reader might consider the historical summary given in this chapter to be a digression and distraction from the main topic of this book. In fact, however, that history is highly relevant. The seminal events that took place in the wake of the First World War shaped the subsequent pattern of the entire region, especially as regards the inequitable distribution and utilization of natural resources in general and of water in particular.

The process by which the Allied Powers determined what states and boundaries to create where none had existed for hundreds of years was a surprisingly casual and hasty exercise. Those who decided the fate of the Middle East had little time to consider, nor sound information by which to anticipate, the potential consequences of their decisions.

The British themselves were not of a single mind on how to manipulate the situation to their own advantage. Two separate agencies of the British Empire were acting for a while at cross purposes. The one, called the "Eastern School," was based in India and was responsible for British policies in Mesopotamia and the Persian Gulf area. It was impelled by the old imperial tactics of "divide and rule," aimed at expanding and perpetuating British rule. The alternative approach was that of the "Western School," centered in the "Arab Office" in Cairo. It aspired to base future British interests not on direct rule, as in the past, but on an alliance with and hence support for Arab nationalism. Members of the latter group, most notably T. E. Lawrence, sought to cultivate the ruling Hashemite family of Hedjaz, headed by Hussein, the Sharif of Mecca. At the same time, members of the rival British group proceeded on a contradictory path to arm and encourage a tribal chieftain from Nejd named Ibn Saud, who was Hussein's implacable enemy.

In the British corridors of power, the Western School eventually prevailed over the Eastern School, but that was scant comfort for Sharif Hussein, who turned out to be the loser in his rivalry with Ibn Saud. It was the latter who, in the wake of the war, took control of most of the Arabian Peninsula and banished the Hashemites from Mecca and the entire Hedjaz. To compensate that clan for the loss of their home base and to reward them for their part in defeating the Ottomans, the British then tried to grant Hussein's sons—Feisal and Abdallah—alternative domains. They first tried to establish Feisal in Damascus as king of Syria, but after he was expelled from there by the French in 1919, the British crowned him king of newly created Iraq. His brother Abdallah, originally slated for the position now given to his younger brother, had to settle for a smaller domain called Transjordan (then a province of mandatory Palestine), first as Emir (Prince), and after the Second World War, as king. Abdallah's grandson, Hussein, named after his great-grandfather, is now the king of that country, since renamed The Hashemite Kingdom of Jordan.

In November 1917 British Foreign Secretary Arthur James Balfour issued his famous declaration, favoring the establishment of a Jewish national home in Palestine.[1] That declaration was later sanctioned by the

League of Nations as part of the terms of the mandate granted to Britain over Palestine.

The postwar Middle Eastern scene was thus a tangled web of contradictory aspirations, rights, demands, and promises made and remade only to be abrogated or reinterpreted. Conflicts smoldered and were occasionally rekindled between secular modernists and religious traditionalists, between nationalists and foreign occupiers, between French and British imperialists, between Muslims and Christians, and—increasingly—between Arabs and Jews.

While the British and French still haggled over the division of the spoils of the war, the old-fashioned version of unbridled imperialism had become anachronistic. The ideal of self-determination for all peoples was being promoted by United States President Woodrow Wilson. The Russian Revolution, too, challenged "imperialistic capitalism." The victorious powers were therefore constrained from setting themselves up as absolute rulers of the territories and nations they had conquered. Rather, they needed to find a new mode for justifying and exercising their hegemony.

The League of Nations, although limited in power and thwarted by America's failure to join it, made an effort to limit imperialism. Seeking an interim formula for the transition from colonial imperialism to independence, the League of Nations devised the concept of "mandates" (first conceived by President Wilson and by South Africa's Jan Smuts). According to this concept, nations freed from the yoke of former colonial masters but adjudged not yet "ready" to function independently would be administered by more "advanced" nations whose "sacred trust" should be to provide "tutelage" toward independence. Those who conceived this paternalistic notion undoubtedly intended it as a temporary arrangement, but the imperialistic powers that received the mandates over the territories of the former Ottoman Empire proceeded to rule over the "natives" as if they were to remain in charge indefinitely.

The Conference of San Remo, held in April 1920, approved the regional arrangement. France was granted the mandate over Syria, from a part of which they carved out a new entity to be called Lebanon. Britain received the mandate over newly created Iraq, as well as the mandate over Palestine, which initially included Transjordan. The entire agreement was ratified by the League of Nations in the summer of 1922.

The artificial division of the mostly Arabic-speaking Middle East into "mandated" states did not please the Arab nationalists. Their aspiration was to constitute a single, large, totally independent nation. But the European powers paid no heed to the dream of Pan-Arabism.

For a time after the end of the war, the new boundaries drawn on the map of the Middle East were tentative. The final borders between British-administered Palestine and French-administered Syria and Lebanon were not set de facto until 1920, and only sanctioned formally in 1923. The border between Syria and Iraq was set in December 1920, but the fate of the oil-rich province of Mosul, over which the Turks continued to lay claim, was not made official until 1926, when it was finally included in British-controlled Iraq. In October 1921 the French and Turks agreed on the northern border of Syria. The creation of Lebanon and the determination of its borders took place in 1920. A year later the British decided to separate Transjordan from mandatory Palestine in order to establish a new Emirate under their protection, with the Jordan River as the boundary between the two entities. The boundaries between Iraq and Iran, originally drawn in 1909, and the boundary between Palestine and Egypt, drawn in 1906 (then as an administrative boundary formally within the Ottoman domain) remained as originally set. The desert boundary between Transjordan and Arabia was established in 1925 (but was modified a half century later by a bilateral agreement).

In creating Lebanon, the French had the intention of rewarding and securing the Christian communities, for which they had assumed paternalistic responsibility. Wishing to make Lebanon a more viable state, and incidentally a more powerful ally, the French greatly enlarged its territory beyond the domicile of the Christian communities there. The new state included not only the old Ottoman district of Mount Lebanon, but also the Beqaa Valley in the east, the port city of Tripoli in the north, Beirut in the center, and the districts south of the Awali and Litani rivers bordering Galilee in northern Palestine. In adding these territories to the Christian enclave in Lebanon, the French did not consider the demographic trend. The rural Muslim population maintained a much higher birth rate than the largely urbanized Christians. Moreover, many of the more cosmopolitan Christians chose over the years to emigrate. As a result, the Christian community lost its majority, which was small even at the start, within just a few decades. The resulting ethnic and religious tensions plunged Lebanon into a civil war (with international repercussions) during the 1970s and 1980s.

The British encountered their own difficulties when they attempted to reorder the territories subject to their control. The state that the British contrived in the east, namely Iraq, was and is an ill-fitting composite of the former Ottoman provinces of Basra (predominantly Shiite), Baghdad (mainly Sunni Muslim), and Mosul (including the mountain-dwelling Kurds). The juxtaposition of such disparate units created tribal and religious conflicts that remain unresolved to this day.

To secure strategic control over the province of Mosul with its petroleum reserves, the British disregarded the national aspirations of the Kurds, who constituted the majority there. Another problem resulting from the contrived boundaries of Iraq and Syria was the irrational separation of the headwaters of the Tigris and Euphrates in Turkey from the downstream reaches of the rivers in those new states. A potential conflict was thus created not only between upstream Turkey and each of its two downstream neighbors but between the latter two as well. No arrangement was put in place at the time for the equitable allocation of those rivers among the riparian states, which occupy irregular portions of the watersheds and stream beds.

The country then known as Palestine[2] existed as a distinct entity in the historic consciousness of the Jews, who called it *Eretz Yisrael* and regarded it as the origin of their nation, the place where their religion and culture evolved. The same land was venerated as *Terra Sancta* by the Christians, who viewed it as the birthplace of their Savior and the site of their holy places. The Muslims similarly venerated the city of Jerusalem, which they called *Al Quds* ("The Holy"), and particularly the Temple Mount with its mosques (the Dome of the Rock and El Aqsa). They did not originally regard the country per se to be a political or national entity, but rather a part of the greater continuum of the Arab domain. Later, however, the Arabs of that country (mostly Muslims but some Christians as well) came to regard themselves as a nation, the Palestinians, with attachment to their specific homeland.[3]

In 1918, at the start of the British administration, the Arab population of Palestine numbered 573,000, of whom nearly 90 percent were Muslim and about 10 percent Christian. The number of Jews then in Palestine was less than 80,000, but it was soon to increase significantly through the return of the thousands exiled from the country by the Ottomans during the war, and through immigration, mainly from eastern Europe.[4]

In July 1920 the British Government, operating under the League of Nations' mandate,[5] appointed a civilian high commissioner to oversee the civil administration of the country. In a highly symbolic gesture, their first appointee to that position was a Jew, Sir Herbert Samuel. Ironically, it was he who was called upon, in August 1920, to declare the separation of Transjordan from Palestine, thus reducing the area available for Jewish settlement to one-fourth of the originally mandated territory. However, the creation of the desert province of Transjordan as a separate entity did not go nearly far enough to satisfy the awakening national aspirations of the Palestinian Arabs.

During the formative years 1918–20, while the fateful decisions regarding the future map of the Middle East were hanging in the balance and the attitudes of the Arabs and the Jews regarding Palestine had not yet congealed into implacable conflict, there were several attempts at compromise and cooperation. One of the most interesting of these was the agreement made between Emir Feisal (then Lawrence's ally and the principal Arab spokesman at the peace conference ending the First World War) and Dr. Chaim Weizmann (then president of the Zionist Federation).

The first meeting took place at Feisal's desert encampment between Aqaba and Maan on June 4, 1918, and it was followed by several more meetings. Their discussions culminated in an agreement, signed by Feisal and Weizmann on January 3, 1919. It affirmed the kinship and ancient bonds existing between the Arab and Jewish peoples, and called for realization of both their national aspirations through the closest possible collaboration in the development of the greater Arab State and of a Jewish entity. In effect, Feisal accepted the principle of the Balfour Declaration that Palestine become a Jewish national home, within boundaries that were to be negotiated at the conclusion of the Peace Conference in Paris, provided the rest of the Arab Middle East achieve its own independence.[6] Jewish immigration into the country and intensive settlement of the land were to be encouraged, on the condition that the rights of the Arab farmers and their economic progress were ensured. Weizmann, on his part, undertook to have the Zionist movement assist in the economic development of the Arab population of Palestine, guarantee religious freedom and access to the holy sites for all religions, and cooperate in managing those holy sites.

The Feisal-Weizmann agreement, for so long considered a mere curiosity, may yet prove to have been prophetic. As the century nears its end, the time has come for the two nations, too long locked in mortal combat and mutual vilification, to rediscover each other. The reciprocal recognition by Israel and the Palestine Liberation Organization, signed in Washington in September 1993, is a first step in this direction.

An important problem in establishing the boundaries of Palestine was where to draw the line dividing the country from Lebanon in the north and Syria in the northeast. Several tentative lines were drawn before the "final" boundary was determined by compromise, and not to any side's complete satisfaction. The secret Sykes-Picot protocol of May 1916 had set the northern boundary of Palestine roughly from the northern shore of the Sea of Galilee to a point just north of Acre on the Mediterranean Sea. When that agreement became known, it met with the objection of the Zionist Organization because it would have left several Jewish settlements in Upper Galilee, and incidentally all the sources of the Jordan River, outside the intended Jewish National Home. The British, in any case, tried to modify the Sykes-Picot agreement on the contention that Russia's departure from the war in 1917 had rendered its provisions obsolete.

In February 1919 the Zionist Organization submitted a proposal to draw the line north of the now-Lebanese village of Nabatiyeh, so as to include in mandatory Palestine the lower course of the Litani River as well as the Hasbani tributary to the Jordan. The Zionist proposal also included the Golan Heights, which in the Second Temple and the Mishnaic-Talmudic periods (roughly from the third century B.C.E. until the seventh century C.E.) had been inhabited by Jews. However, the British and French agreed in September 1919 that the British would withdraw to a line linking the northern shore of Lake Huleh with the Mediterranean Sea at Ras El-Naqura. This line, too, would exclude some Jewish settlements.

When the British withdrew, Arab nationalists tried to prevent the French from occupying the area. They treated the Christian and Jewish villagers of the area with hostility, suspecting them of collaboration with the French. Isolated from their community in the southern part of Palestine, the Jewish villagers were caught in the vise between the warring sides. In March 1920 Arab fighters attacked the Jewish settlements of Tel-Hai, Kfar Giladi, and Metulla, killing a number of the Jewish defenders and forcing the remainder to evacuate the area temporarily. The fate of those settlements then became a powerful rallying cry for the Zionist movement, so when the French entered Damascus in July 1920 and quelled the Arab resistance, the Jewish settlers determinedly returned to their villages and demanded their inclusion in mandatory Palestine. Their demand in this case happened to coincide with the British interests.

Consequently, in December 1920, the British renegotiated the boundary issue with their French allies and rivals. The two powers then redefined the border between Palestine and Lebanon to include within Palestine the area now known as the "finger of Galilee," with its Jewish villages and, more significantly, with the spring of Dan, the largest of the three upper sources of the Jordan River. Still, the boundary was not final. For two more years, the British retained control of the Golan Heights. The final border agreement between the domains claimed by the two European powers was signed in March 1923, and it provided for the evacuation of the British from the Golan Heights area and its inclusion within Syria.

The upper sources of the Jordan were thus split among four countries: Dan Spring in Palestine (today in Israel); the western tributary, Hasbani, in Lebanon; the eastern tributary, Banias, in Syria; and the Yarmouk River between Syria and Transjordan (now the Hashemite Kingdom of Jordan). Forty years later, this division would become a major point of contention.

An old legend describes the origin of the Jordan River, which rises at the foot of stately Mount Hermon, called *Jebel ash-Sheikh* ("Chieftain Mountain") in Arabic. When that mountain was young, its melting snows gave rise to three separate streams. They were prideful and boisterous streams and they made their way in different courses, each claiming to be the most important one. None, however, could survive alone in the desiccating heat of the parched valley that was their common destination. Arrogantly, each of the three streams commanded the others to be its tributaries, but none was willing to be subservient and deviate from its own course. Finally, the three appealed to a venerable sage to mediate their dispute. The sage sat upon a mound of earth and listened to their petty, competitive contentions. Then he pronounced his verdict. "Foolish little streams," he admonished them, "you must forego your vanity and unite immediately as equals. Only by joining forces can you become an important river and prevail over the valley's heat."

That sage has been remembered ever since as "the Judge," called *Dan* in Hebrew and *Qadi* in Arabic. The mound (*tel*) upon which he sat is still known as Tel Dan, or Tel el-Qadi. And the stream that flows from the bottom of that mound is quickly joined by its two sisters, which together form the River Jordan, whose name means literally "Descending from Dan."

The last political entity established in the Middle East, the Emirate of Transjordan (later to become the Kingdom of Jordan) was an unplanned improvisation by the British. Under the Ottomans it had largely been a neglected territory, its sole importance being its location astride the route of the *haj* ("pilgrimage") from Damascus toward Mecca. At the time of its separation from Palestine, the population of Transjordan numbered no more than 200,000—divided between nomadic Bedouin tribes and settled *fellahin* (cultivators) along with a few town dwellers.[7]

The purpose of the British[8] in awarding this entity to Emir Abdallah was to assuage the disappointment of their Arab allies, the Hashemites, who had aspired to rule over the entire Arab-speaking region. At the same time, the British attempted to retain effective control over the Emirate of Transjordan by overseeing its administration, and especially by commanding its army, the Arab Legion. Abdallah proved to be an astute governor, who managed to consolidate his rule, enlarge his domain, turn it into a kingdom, and perpetuate his family's control over it. However, he himself was assassinated by a Palestinian nationalist in 1950.

The situation was quite different in Egypt, where the movement toward national independence was much more advanced. The tendency on the part of Egyptian nationalists during the early part of the century was to stress the distinct and unique historical roots of Egyptian society and culture, rather than identify with the Pan-Arab movement. It was only later, following the Second World War, that Egypt developed a strong identification with the regional movement toward Arab independence, and began to play a central role—in its capacity as the most populous Arab country—in the political and military struggles of the region. Despite the rising tide of Egyptian nationalism, the British tried to perpetuate their power under the guise of a "protectorate." They actually succeeded in maintaining control through the Second World War, only to be forced to withdraw after it. Their demise was demonstrated most dramatically by their loss of the Suez Canal and their failure to retake it in 1956.

The European powers did not include the Arabian Peninsula in their designs during the First World War, if only because it seemed so desolate and barren. No one had yet imagined the enormous importance that desert region would acquire only 20 years later. Commercial quantities of oil were first discovered in the Persian Gulf district of Arabia in 1938, on the eve of the Second World War, and extraction began after the war, in 1946. Had they an inkling of the potential petroleum reserves hiding under Arabia's sands, the European powers would probably not have left Arabia to its own devices, but would have drawn a very different map for the region.

In the wake of its defeat, the once-great Ottoman Empire was reduced to the single country now known as Turkey. The victorious Europeans were bent on imposing severe punitive terms on their defeated adversary, even to the point of dividing up the Turkish homeland itself. Their intention was to separate out the ethnic Armenians and Kurds, who were concentrated in eastern Anatolia, and to recognize them as independent nations. That territory includes the headwaters of the Tigris and Euphrates Rivers. However, the Turks were able to reorganize their army after 1918 and to foil those designs. Their new leader was the charismatic Mustafa Kemal, who had earlier defeated the British at the battle of Gallipoli and who came to be known as Kemal Ataturk, "Father of the Turks." He eliminated the Sultanate, repudiated the Ottomans' pretension to leadership of the Islamic world, and concentrated instead on building an exclusively national, secular, and modernistic Turkey, relatively uninvolved in the affairs of the Middle East and looking more toward Europe. In a highly symbolic act, he outlawed the old custom of forcing women to wear veils in public, and substituted Latin script for the Arabic alphabet that had earlier been used for the Turkish language.

Like the First World War, the Second World War (1939–45) began in Europe but soon spread into the Middle East. This time, however, Britain won a Pyrrhic victory. Its triumph led to debacle. Britain emerged from the war in a much weakened state and was consequently forced to relinquish its control over Palestine (1948), Egypt (1956), Jordan (1956/57), and Iraq (1958). Similarly, France reluctantly granted independence to Syria and Lebanon in 1948. The myth of centuries, the European dream of perpetual empire and world ascendancy, evaporated. The forces of nationalism were strengthened throughout the region, and in later years religious fundamentalism reawakened and is now powerful once again. In the meanwhile, the population has increased dramatically, as has the pressure on the region's fragile resources, especially land and water.

Following the British decision to withdraw from Palestine, the United Nations considered the future of that country. In November 1947 the General Assembly approved the partition of Palestine into a Jewish state and an Arab state, which were expected to enter into an economic union. In accord with that decision, the Jewish community of Palestine declared the establishment of the State of Israel on May 14, 1948. The Palestinian Arabs, however, refrained from doing the same in the territory assigned to them by the U.N. Instead, armed groups of Palestinians, aided by the armies of the neighboring Arab states, attacked the new state. The Arab Legion of Transjordan then seized the West Bank and the Egyptian army seized Gaza.

After months of bitter fighting, a truce was declared and an armistice agreement was mediated by the United Nations. King Abdallah of Transjordan then reached an unofficial modus vivendi with the Israelis, establishing a more or less stable truce. King Abdallah annexed the mountainous territories north and south of Jerusalem (with the principal cities of Nablus in the north, Hebron in the south, and East Jerusalem in the center), calling them the West Bank of his kingdom, which he renamed the Hashemite Kingdom of Jordan.

Unfortunately, the armistice did not lead directly to peace. The smoldering hostility between the Arabs and Israel ignited again and again. Open wars were fought in 1956, 1967, 1973, and 1982. Numerous lesser clashes flared up between those wars and since. Active rivalries over the allocation of the region's meager water resources and charges of usurpation by one side and the other greatly aggravated hostilities. The war of 1967, in fact, was triggered in part over the Syrian attempts to divert the Jordan River waters. As a result of that war, Israel took control of Sinai and Gaza from Egypt, the West Bank from Jordan, and the Golan Heights from Syria.

In an effort to reverse the consequences of the 1967 war, Egypt and Syria attacked Israel on Yom Kippur in October 1973. On the Egyptian front in that war, water featured, quite literally, as a weapon. Although water is known to have been used as a means of destruction in many places throughout history,[9] it has seldom been so decisive in battle as it was along the Suez Canal on that fateful day.

In the years preceding that war (between 1967 and 1973), the Israeli Army had built strong defenses on its side of the Suez Canal, designed to prevent any Egyptian attempt to reconquer the Sinai Peninsula. Those defenses consisted of a massive, continuous earthen embankment, laboriously constructed along the entire length of the Suez Canal's east bank. Inside and behind that embankment were regularly spaced concrete and stone bunkers, altogether a Maginot-like fortification then known as the Barlev Line and purported to be practically impregnable. The embankment was so high and steep, and the sand so soft, that no combat vehicle such as a tank coming from across the canal could possibly climb it.

Early in 1973, while I was serving as professor and head of soil and water sciences at the Hebrew University of Jerusalem, officers of the Israeli army approached our scientists for advice. They had noticed a series of puzzling exercises by the Egyptians on their branch of the bypass canal (where the Suez Canal splits into two branches to allow northbound ships to bypass southbound ships). Egyptian bulldozers were piling up mounds of sand, then squirting the mounds with jets of water drawn from the canal by powerful pumps on floating barges. What could they be up to? Might they be intending to breach the Israeli defenses this way? Surely, the officers scoffed, so massive a structure could not be breached by something so flimsy as a jet of water! Nevertheless, they wanted our people to look into the matter.

The scientists were skeptical at first, but their trials and calculations indicated that water could indeed do the job. The desert sand was so loose and erodible that concentrated jets of water, if continued long enough, might saturate a section of the embankment, turn it into fluid mud, and cause it to slump. These conclusions were reported to the officers with recommendations that they take whatever precautions were needed. But when one of those officers was encountered some months later and asked whether they had acted on that advice, he dismissed the query with a wave of his hand. "Don't worry," he said complacently, "we can counter whatever they try to throw at us."

Alas, his complacency was self-defeating. Not many weeks later, on the Jewish Holy Day of Atonement, when the Israeli positions were weakly manned, the Egyptians attacked with a barrage of water from floating water cannons. Within hours, they punched through the embankment at several locations so their amphibious tanks could then ford the canal, ride through the gaps, and fan out behind the Israeli frontline fortifications.

The Israeli army had been overconfident and paid dearly for its mistake. Paradoxically, however, what began as a misfortune may have held the seeds of eventual good fortune. It seems, in retrospect, that it was precisely their initial success in that war that redeemed the pride of the Egyptians sufficiently to enable their President Sadat to come to Jerusalem in 1977 and initiate a peace agreement with Israel. That landmark agreement, signed in 1979, provided for Israel's withdrawal from the Sinai Peninsula—including the air bases, oil fields, farms, towns, and tourist facilities that Israel had developed there.

Some years later, in 1984, I was sent to Egypt by the U.S. Agency for International Development to serve as an irrigation advisor. On one of my field trips to the East Salhia development area, the ministry assigned a young hydraulic engineer to accompany me. As we chatted at lunch in Ismailia, he began to reminisce about his part in the 1973 war. I then realized that it was he who had been in charge of the water attack that broke the Israeli defenses! When I asked him to show me how he did it, he quite willingly drove me in his jeep along the western bank of the canal and proudly pointed out the gaps in the old embankment. And when I revealed to him that I had been aware of the water tactic at the time, he laughed good-naturedly. We became the best of friends after that.

Peace between Israel and Egypt was relatively easy to achieve because the two states are separated by an empty desert and do not compete for vital common resources such as water. The situation is quite different between Israel and each of its other neighbors—Syria, Jordan, and Lebanon. These states have population centers that are in much closer proximity, without the benefit of geographical separation to serve as a security buffer zone, and they also compete directly over shared water resources.

Even more complex are the issues in contention between Israel and the Palestinians, who have suffered most from the prolonged violent conflict. Hundreds of thousands of Palestinians have been displaced from their villages and towns and have lived the lives of refugees in dire circumstances for nearly half a century. Those remaining in the territories have endured the oppression and humiliation of military occupation, and—for too long—the denial of their national rights. No account of the region's history, however brief, should ignore the human plight of these people, who have been so victimized. Their condition has generated intense hatred and fueled violent revenge, which in turn has provoked counterviolence and made the feud progressively more difficult to resolve. Yet, as great as the difficulty to achieve reconciliation and true peace between the Israelis and the Palestinians might be, greater still is the imperative need to do so. Fortunately, the first real steps toward the needed reconciliation were taken by Israel and the Palestinians in 1993.

Important among the problems yet to be resolved is the competition with Israel over limited water resources, as the Palestinians throughout the West Bank, and even more so in the Gaza Strip, suffer increasingly from an acute shortage of water.

The strategic importance of the Middle East has grown enormously during the course of the twentieth century, as the industrial world shifted from reliance on other sources of energy to primary and crucial dependence on petroleum. At the century's beginning, no one could have guessed that at the close of the same century the Middle East would be found to contain more than two-thirds of the entire world's known petroleum reserves and would become the world's most important petroleum-supplying region.

The majority of the people of the Middle East, however, are yet to enjoy the benefits of those resources. Not only are these resources unequally distributed, but they have also been largely squandered in nonproductive investments and in the futile pursuit of military "security" or supremacy. Moreover, the nations of the region have thus far been denied the advantages of progress and development by the repeated disruption of war. Only now that the region has embarked on a course of peace can the human and natural resources—the latter including essential water resources—be harnessed toward realization of the region's great economic and cultural potential.

5

The Twin Rivers

He turneth rivers onto the wilderness
And watersprings to the thirsty land.
Psalms 107:33

The adjective "exotic," in common parlance, conveys the sense of something strange, enigmatic, intriguing. But when applied by geographers to rivers it simply means "from another region." The two greatest rivers of western Asia, Tigris and Euphrates, originate in one climatic and topographic zone and end up in quite a different one. They begin, scarcely 30 kilometers from each other, in a relatively cool and humid zone with a rugged landscape of high mountains and deep gorges, raked by autumn and spring rains and visited by winter snows. From there, incongruously, the two rivers run separately onto a wide, flat, hot, and poorly drained plain. In their middle courses, they diverge hundreds of kilometers apart, only to meet again near the end of their journey and discharge together into the Persian Gulf.

From source to sea, the Euphrates (Hebrew *Prat,* Arabic *Furat,* Turkish *Firat*) traverses a distance of 2,700 kilometers, of which some 40 percent are in the modern state of Turkey, 25 percent in Syria, and 35 percent in Iraq. Its twin, the Tigris (Hebrew *Hiddekel,* Arabic *Dijla,* Turkish *Dicle*), has a total length of 1,900 kilometers, of which about 20 percent lie in Turkey, 78 percent in Iraq, and only 2 percent along the pointed northeastern corner of Syria known as the "Duck's Beak".

Both rivers wend their ways circuitously over two-thirds of their courses through the gorges of east Anatolia and the valleys of the Syrian and Iraqi plateaus before descending into the arid plain of Mesopotamia, literally the "land between the rivers." Along their upper courses, at elevations ranging from 2,000 to 3,000 meters above sea level, the rivers cascade tumultuously through spectacular gorges. They continue more tranquilly through the plateaus of north Syria and Iraq, where they cut deep beds in the Tertiary rock so that their courses have remained stable over the millennia. Here the twins are separated by a triangle of limestone desert known as the Jazirah (the "island," or peninsula). In this section they descend from elevations of about 400 meters to little over 50 meters above sea level where they enter the alluvial plain. In that vast plain, the once-swift mountain rivers snake their way sluggishly as lazy, braided streams.

A feature common to both rivers is the heavy concentration of suspended sediment (silt) that they carry at flood-time: as much as 3 million tons of eroded soil from the highlands in a single day. Little of this sediment reaches the sea, however. Most of the material settles en route and is responsible for the great deposit of alluvium that fills the Mesopotamian Plain.

This great alluvium-filled depression (physiographically, an extension of the Persian Gulf), having a length of 800 kilometers and an average width of 200 kilometers, constitutes the combined delta of the twin rivers. Here the surface of the land is so flat—its average slope toward the sea being only 5 centimeters per kilometer—that it greatly impedes the natural drainage of the area.

From the head of the plain to the gulf, the rivers meander slowly. The loss of velocity and turbulence causes the rivers to deposit their silt and thereby to build up their beds and banks above the level of the plain, a characteristic that makes these rivers prone to overflow their banks seasonally and to change course capriciously. Throughout history, the twin rivers have bifurcated and altered their beds on numerous occasions. Here and there, their earlier courses can be traced by the lines of *tels* (mounds), representing ancient settlements established alongside their banks. The tels are especially numerous in the southern portion of the plain, and they suggest a pattern of prior occupation practically unrelated to the present system of waterways.

Each of the rivers is joined in its middle course by substantial left-bank[1] tributaries: the Balikh and the western Khabur, which flow into the Euphrates; and the eastern Khabur, Great Zab, and Little Zab (fed by snowmelt from the high ranges of Iraqi and Iranian Kurdistan), as well as the Uzaym and Diyala (from the Zagros Mountains of Iran), which augment the Tigris. Because the tributary inflows to the Tigris are more numerous and copious than those to the Euphrates, the Tigris is a swifter stream. On its way it passes through the homeland of the ancient Assyrians, and the prominent remains of their three great capitals—Nineveh, Calah, and Ashur—still overlook the middle course of the river. The Tigris also runs through the original Muslim sites of Samarra and Baghdad.

The raised levels of both river beds above the alluvial plain facilitate the diversion of water for the purpose of irrigation, to which Mesopotamia owes its prodigious agricultural productivity. That advantage is countered periodically by the tendency of the rivers to overflow spontaneously and to inundate the countryside; hence the local inhabitants have made repeated attempts to build artificial embankments—dikes, or levees—to contain the rivers. Still, the rivers crest occasionally and break through the confining dikes to cause devastating floods.[2]

The excess flows dissipate themselves naturally in extensive marshes, which have formed along the lower segment of the plain. The largest of these marshes is the Hawr al-Hammar, through which the Euphrates flows before uniting with the Tigris about 160 kilometers above the head of the Persian Gulf.

Those marshes, dense with reeds and cattails, are the abode of the *Madan,* a distinct people called the "Marsh Arabs," who subsist on the rich biota of the wetlands. For countless centuries, these marshland dwell- ers have utilized the local reeds (*qasab*) for fodder as well as for weaving

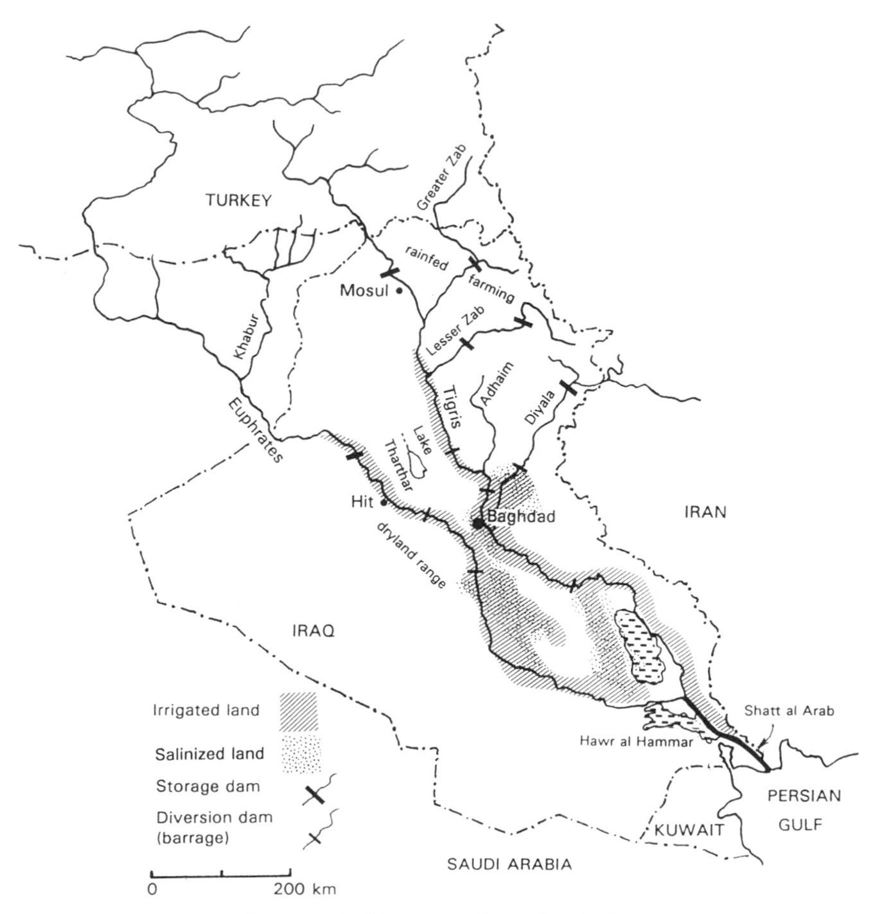

Figure 5.1 Dams and irrigation in Iraq.

baskets and mats. They use reeds to build their unique shelters and large assembly halls on floating reed-mat islands. They catch fish in the surrounding shallow waters, in which they also raise water buffalo for milk, yogurt, meat, and hides. Their mode of transportation is similarly characteristic: they move between houses and villages by means of high-prowed *mushhuf* (reed canoes), which they manipulate through the practically stagnant waters by using of long wooden poles. Along the banks of the rivers themselves grow poplar and willow trees, together with thickets of tamarisk and mesquite.

Until recently, the notion prevailed that the head of the Persian Gulf once extended much farther inland, and must have receded southward over time as more and more sediment was deposited. This notion has apparently been refuted by geologists who detected a gradual subsidence of the Mesopotamian basin at a rate sufficient to compensate for the accretion of sediment and to maintain the valley at a nearly constant level over a long period. Accordingly, they concluded that the coastline could have remained almost unchanged in historic times. However, the question of just how far the original Sumerian cities were from the sea remains in contention.

An extremely diverse assemblage of peoples lives in the catchment of the Tigris and Euphrates. In the lowlands live peasant cultivators (*fellahin*) who irrigate lands along the banks of the rivers, nomadic desert Bedouin who graze their flocks in the arid plains, seminomadic *Madan* who reside in the swamps, and increasing numbers of urbanites who inhabit the rapidly growing cities. In the highlands of eastern Anatolia live Kurdish mountaineers. Those highlands, now sparsely inhabited, were once home to a sizable population of Armenians and are still named the Armenian Mountains in many atlases, though the Armenian people were expelled by the Turks early in the twentieth century.

The headwaters of the Euphrates comprise two confluent streams, the Murat and the Kara Su, which drain the high lava plateau to the northwest of Lake Van. Where those streams meet, at Keban, the Turks have built the rock-filled Keban Dam, which spans a deep gorge. From there, after cutting through the range of Guneydogu Toroslar, the river swerves southward through the foothills of the Commagene district to approach within 160 kilometers of the Mediterranean. In Syria, however, it turns leftward (to the southeast), across a barren and thinly populated section of Syria. The mean natural flow of the Euphrates is about 30 billion cubic meters per year (BCM/Y) at its entrance into Syria from Turkey, and 32 BCM/Y on leaving Syria (after having taken in two tributaries, the Balikh and the Khabur). Henceforth, over the remaining 1,000 kilometers of its course in Iraq, it obtains no further increment of water.

In Iraq, along the narrow floodplain between the Syrian border and the town of Hit, a riparian system of cultivation is practiced that still uses waterwheels to lift water for irrigation. Large-scale irrigation begins below Hit, where the plain widens and diversion canals are used to convey water onto the adjacent lands. Along the way, barrages and sluice gates have been built in an ongoing effort to control and regulate the water supply. Toward the south, however, as the river rises above the plain, its course becomes unstable and difficult to control. Below an-Nasiriyah, the main stream breaks into numerous channels that feed their waters into the marsh known as Hawr al-Hammar. From there, the waters of the Euphrates run into the Shatt al-Arab, where they combine with the waters of the lower Tigris and discharge into the Persian Gulf as a united river.

The upper Tigris originates from a small mountain lake, Hazar Golu, south of Elazig in eastern Turkey—scarcely 30 kilometers from the headwaters of the Euphrates—and flows through the basaltic district of Diyarbakir. The Tigris is a less circuitous river than the Euphrates, and it flows more directly toward Mesopotamia. After skirting the northeastern corner of Syria, at its junction with the eastern Khabur, the Tigris enters Iraq. At this point, its mean annual flow is between 20 and 23 BCM/Y. In Iraq it collects an additional 25 to 29 BCM/Y from its left-bank tributaries, for an average total of roughly 50 BCM/Y.

The two main tributaries of the Tigris are the Great Zab and Little Zab, which join the river downstream from Mosul, past the ruins of ancient Nineveh. Because the Tigris flows alongside and parallel to the Zagros mountain range, the streams draining the western slopes of those mountains descend steeply and swell the river with silt-laden water during the annual snowmelt season. Continuing its course southward, the Tigris flows through the al-Fathah gorge. Some 100 kilometers beyond that it enters the head of the alluvial plain near Samarra. A barrage at this point diverts surplus water into the Tharthar Depression to the west for the purpose of downstream flood protection. South of Baghdad, the Tigris is confined between artificial levees, the overspill from which flows into a marsh. At al-Kut, some 320 kilometers south of Baghdad, there is a barrage that directs part of the flow through the Shatt al-Gharaff (an ancient bed of the river) to irrigate a wide area between the two main rivers, with the surplus water merging with the Euphrates at its entrance into the Hawr al-Hammar.

The main stream of the Tigris beyond al-Kut is split into several channels that distribute water for rice-growing areas and also feed extensive marshes on either bank. At al-Qurnah, the Tigris finally joins with the Euphrates (after the latter's emergence from al-Hammar) to form the Shatt al-Arab. Along the Shatt, a strip of land with datepalms is irrigated from the river. Just before discharging into the sea, the Shatt is joined on its left bank by the large Karun River, which flows southward from the Iranian province of Khuzistan.

At its entrance into the delta, the bed of the Tigris is at a slightly higher level than that of the Euphrates. Before reaching Baghdad, however, their relative elevations are reversed, with the latter river rising some 10 meters above its parallel sister river. Still another reversal occurs farther downstream above al-Kut. In ancient times, the hydraulic engineers of Mesopotamia utilized these differences in elevation to draw water from one river while discharging drainage into the other.

The climate of Mesopotamia is characterized by hot summers and cool winters. In the mountainous headwater districts, freezing temperatures prevail in the winter, and much of the precipitation falls in the form of snow. As the snow melts in spring, the rivers are in spate, augmented by seasonal rainfall, which reaches its maximum between March and May.

Southeastern Turkey, as well as northern Syria and Iraq, enjoy a milder climate and considerably more rainfall than the Mesopotamian Plain. In those districts, winter grains (wheat and barley) are grown by rainfed farming (without irrigation), though supplementary irrigation—where applied—raises yields and allows multiple cropping. Since the river beds generally lie below the level of the land in those areas, mechanical lifting devices or modern high dams are needed to draw water out of the rivers. Typical crops grown under supplementary irrigation are vines, olives, and temperate-region fruits.

In the Mesopotamian Plain itself, however, the annual rainfall is very scanty, rarely more than 200 millimeters. The summer months are exceedingly hot and dry, with midday temperatures approaching 50 degrees Celsius and with daytime relative humidity as low as 15 percent. Consequently, the evaporative demand is very high, and crops require intensive irrigation. The main crops in the southern plain are wheat, barley, millet, rice, and dates, all of which are totally dependent on irrigation.

The rivers have two distinct surge periods: a minor rise from November to the end of March, due mainly to rainfall; and the main rise in April and May, due mainly to snowmelt. During the latter phase, the rivers may carry a torrent 10 times as great as during the low-flow period of late summer. From the standpoint of agriculture, the rivers are high at an unfavorable time of year for most crops: too late for winter crops, too early to sustain summer crops. Thus, irrigation by direct inundation is not generally practical (except for rice). Rather, water must be diverted from the rivers during the period of high crop demand, an operation that requires considerable engineering and skilled water management.

The great volume of water carried by the ferocious floods endangers the levees within which the rivers are normally confined. The primary requirement of river control is therefore to maintain an effective system of diversion and storage, both to prevent destructive inundation of the land and to retain the floodwaters for subsequent irrigation during the main growing season.

Evidence of some of the most ancient waterworks in the world has been found in lower Mesopotamia. Viewed from the air, the Mesopotamian Plain appears to be a tangle of shifting stream beds and ancient irrigation channels. Clearly, all these channels were not in simultaneous use but rep-

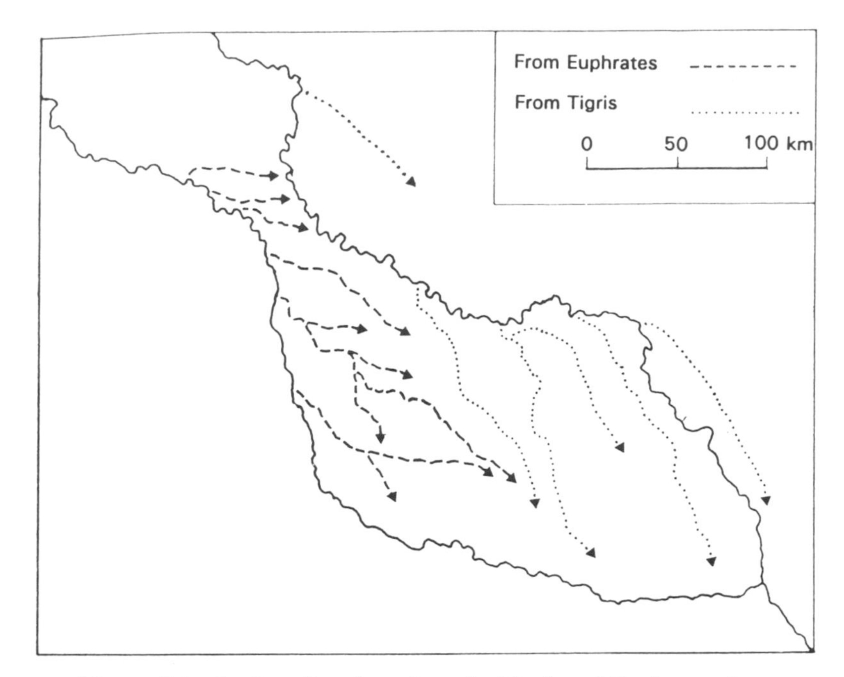

Figure 5.2 Ancient diversions from the Tigris and Euphrates rivers.

resent the successive works of diligent societies over a period of millennia. Some of the old canals are now reused and extended, while others are neglected and clogged with silt. The five great canals that diverted water from the Euphrates and on which the fertility of central Mesopotamia depended during the early Muslim (Abbasid) period all have their modern counterparts, but their overflow can no longer be drained into the Tigris, as in former times, because of the rise in the level of that river.

Perhaps the most impressive of the ancient works is the great Nahrawan Canal. Apparently built in the sixth century C.E., this canal was nearly 300 kilometers long and about 30 meters wide, and it conveyed water from the Tigris eastward to extend the area of irrigated land almost to the Persian frontier. Its construction may have been impelled by the gradual salinization and waterlogging of the lands between the rivers. Beginning at a point just north of Samarra and following the path of an old bed of the main river, the Nahrawan Canal collected the waters of the Uzaym and Diyala tributary rivers and carried them southeastward.

This impressive system, along with other extensive waterworks, was evidently abandoned following the Mongol invasion during the twelfth century C.E. That invasion has been characterized by some historians as one of the most destructive episodes in the entire history of civilization, for it laid waste to the entire Mesopotamian domain and left it desolate for centuries. There is evidence, however, that the enormous fertility of Mesopotamia during the early Muslim period (made famous by the fables of the great Caliph Harun al-Rashid of the late eighth century C.E.), had already begun to decline spontaneously during the tenth century, two hundred years prior to the Mongol invasion, owing to the old scourge of salinization—the same curse that had earlier caused the demise of the Sumerians.

In modern times, numerous barrages and diversion canals have been built for the purpose of controlling the flows of both rivers and drawing water from them. The picturesque water-lifting devices—powered by humans, animals, or river flow—are now being replaced by petroleum-powered pumps.[3]

The efforts to rehabilitate the ancient land of Mesopotamia began in earnest only in the twentieth century, with the construction of the Hindiyah Barrage on the Euphrates just prior to the First World War. Later, following the discovery of petroleum and the introduction of modern motorized pumps, it became possible to draw water from the rivers in much larger quantities and thereby to enlarge the irrigated areas. Since the level of water in the rivers is so variable, however, an effective water supply system required permanent storage. Therefore, numerous large and small dams have been built in modern Iraq. Among the larger reservoirs are Lake Habbaniyah, Lake Abu-Dibbis (having a total storage capacity of some 45 billion cubic meters), and Lake Tharthar, which occupies a formerly saline depression (with a storage capacity of about 30 BCM). The latter lake, which also serves for flood control, lies between the Tigris and Euphrates in the upper part of the Mesopotamian Plain. This reservoir also facilitates water transfers from one river to the other.

In the great irrigated lands of lower Mesopotamia, the age-old plagues of waterlogging and salinization are more prevalent than ever. Even where the irrigation water contains a low concentration of salt, its evaporation in hot weather necessarily leaves a saline residue, which tends to accumulate in the topsoil wherever it is not leached away. The necessary leaching is made especially difficult by the rise of the water-table, caused by flood-irrigation as well as by seepage from unlined canals. In Iraq, where an estimated 60 percent of the irrigated land already suffers from some degree of salinity, more and more land is being rendered sterile and abandoned. According to some reports, the total area under irrigation shrank from about 7.5 million hectares in the early 1970s to about 6 million in the 1980s. (The actual extent of productive land under irrigation may be smaller yet, as the figures cited apparently include marginal or even submarginal agricultural lands already strongly salinized.)

Despite its great agricultural potential, Iraq has in the last decade become a net importer of grain to feed its population of 20 million (expected to reach 26 million by the year 2000). Therefore an imperative need has arisen to replace the old piecemeal salinity-prone irrigation system with a coordinated scheme of rehabilitation and sustainable water management, including the provision of regional drainage.

Figure 5.3 Sampling the soil in an area of extreme salinization.

In December 1992, notwithstanding the residual tensions following the Gulf War of 1991, Iraq finally completed its 40-year effort to provide comprehensive drainage. Now lower Mesopotamia no longer lies between two rivers. A "Third River" has been created: a vast drainage canal that flows midway between the Tigris and the Euphrates. The canal, with its feeder channels and drainage pipes, was planned long ago by American engineers seeking ways to reclaim vast tracts of Iraq's agricultural land that had become barren through salinization. The only way to desalinize such land is to flush away the salts with sweet water. But the water thus flushed through the soil becomes saline, and must be drained away from the land. The solution was to convey all that water to the sea via a single regional canal.

Work on the regional canal began in 1953, and has continued sporadically over the decades since, though punctuated by wars, rebellions, and international sanctions. At various times, Dutch, German, South Korean, Chinese, and Russian contractors were all involved, each in turn, but did not complete the task prior to the Gulf War. Following the war, the Iraqis themselves took over the project, using nearly $2 billion worth of commandeered construction equipment and materials left behind by the foreign companies after international sanctions forced them out of the country.

The main canal, now ready, is 565 kilometers (350 miles) long and wide enough to accommodate 5,000-ton barges carrying cargo from the Persian Gulf to Baghdad. It was designed to drain 1.5 million hectares of

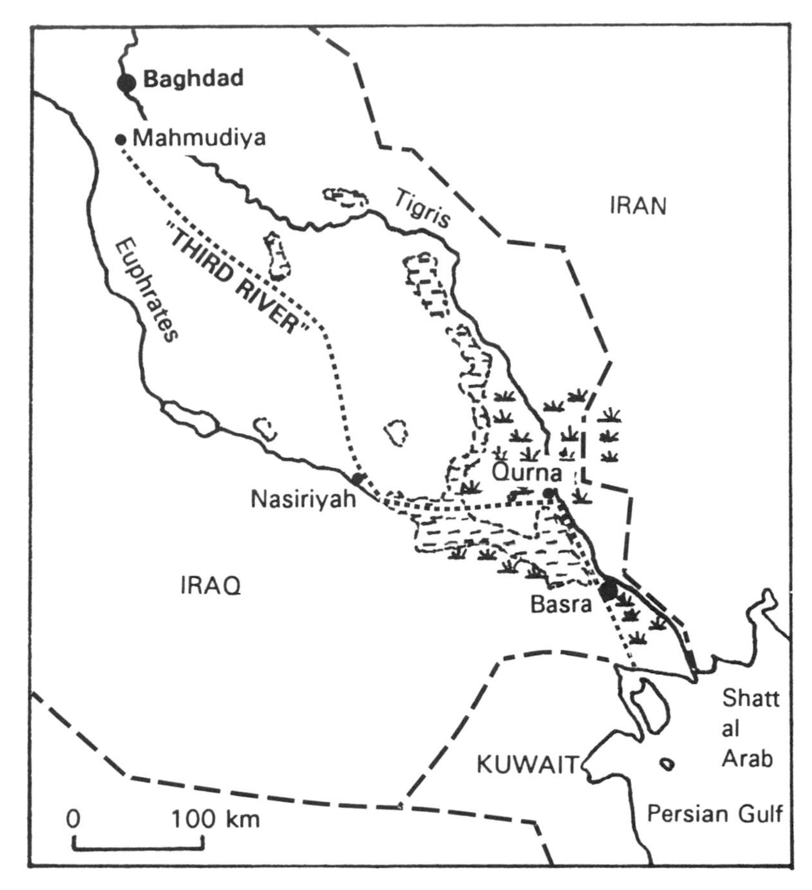

Figure 5.4 The "Third River"—recently completed drainage canal in southern Iraq.

land, allowing Iraq to increase its domestic food output significantly and thus help achieve its aim of economic independence. The waterway runs between the twin rivers from Mahmudiya, just south of Baghdad, to Qurna, crossing the Euphrates by tunnel at Nasiriyah. At its southern end, its saline waters are discharged into the Basra "river"—another man-made channel running parallel to (and west of) the Shatt al-Arab, which was blocked as a result of the Iraq-Iran war of the 1980s.

The regional drainage project, though planned by Western engineers and initially supported by Western aid, became controversial after the Gulf War,[4] when the suspicion arose that it had a hidden purpose. Some observers suspect that the present Iraqi regime's additional motive in draining the reedy swamps is to ferret out the Shiite dissidents and other opponents of the central government who have been hiding in those thickets (as had fugitives from authority since time immemorial). Draining so large an area threatens to destroy an entire ecosystem as well as the way of life of the

Marsh Arabs. United Nations officials have criticized the canal as "an environmental crime." An even more compelling ulterior motive for draining the area may be to allow access to large deposits of petroleum, possibly as great as those of Kuwait, that apparently underlie the marshes. If the Iraqis carry through with this design, the ancient land of the Zigurrats may yet become a modern land of oil derricks and refineries.

The Iraqis deny any intention to drain the area's natural marshes. Although the Third River cuts across some 40 kilometers of Hawr al-Hammar, the largest marsh, Iraqi government spokesmen have claimed that dikes were built to prevent both its drainage and the intrusion of the saline water from the canal into it. They ascribe any possible lowering of the water level in the marshes to the Turkish and Syrian damming of the upper Euphrates, and to the Iranian diversion of streams that feed the Hawr from the east.

Similar to Egypt's position regarding the Nile basin, Iraq claims longstanding historical rights to utilize the waters of the twin rivers. Moreover, owing to its extremely arid climate it depends on those rivers more crucially than do the upper riparians—Syria and particularly Turkey. However, Iraq's position as a downstream state places it at a strategic disadvantage relative to the position of the upstream states, which can exercise more direct control over the headwaters.

The natural annual discharge of the Euphrates at Hit is about 30 billion cubic meters on average, whereas that of the Tigris at Baghdad is just under 40 billion. The discharge varies greatly from year to year. Moreover, the discharge varies along the length of each river, depending on the amount of water augmented by tributaries versus the amount abstracted for irrigation and evaporated from reservoirs above each site. The lowest annual discharge ever measured in the Euphrates was 16.8 billion cubic meters, and the highest was 47.5. Available records indicate that the Tigris also exhibits considerable variations: from a minimum of 16.9 to a maximum of 58.7 billion cubic meters. The average flow of the two rivers combined has been estimated at about 70–74 BCM/Y. These are rough estimates, however. As yet, there is little uniformity or consistency, and even less international cooperation, in measuring the quantities of flow and utilization along the courses of the two rivers.

The claims and counterclaims of the three riparian states are complex. Most of the water feeding both rivers is generated in the territory of Turkey.[5] Syria contributes to the flow of the Euphrates via the two tributaries that run within its boundaries (Balikh and western Khabur). Iraqi territory contributes practically no water to the Euphrates, but does contribute substantially to the tributaries of the Tigris. Altogether, Turkey provides about two-thirds of the combined flow, Iraq about 20 percent, and Syria less than 10 percent. The issue in contention is how to weigh historical rights against proportionate contributions to flow, taking into consideration such associated factors as the real needs of each country. Among those are the availability of energy (e.g., petroleum); the need for hydroelectricity; the feasibility of developing economic alternatives to irrigation-based farming; the efficiency of water use; and the size of each country's population.

Assigning relative weights to those disparate factors in order to establish criteria for the equitable allocation of the rivers' waters among the riparians would be an exceedingly difficult task even if the contenders were willing to submit their claims to impartial adjudication. In the absence of such willingness, the issues remain in contention and may lead to armed conflict. Thus far, no serious negotiation (let alone agreement) has yet taken place among the riparians directly. On the contrary, each of the riparian states has been active in "developing" and utilizing its section of the rivers independently, disregarding the rights, concerns, or works of its neighbors.

Preliminary efforts at coordination have, all too often, been honored in the breach. In 1987 Turkey and Syria signed a Protocol of Economic Cooperation, under the terms of which Turkey undertook to release a yearly average of more than 500 cubic meters per second at the Turkish-Syrian border. In the event that the monthly flow should fall below that level, Turkey agreed to make up the difference during the following month. Note that the above flow rate totals 15.75 BCM/Y. Subsequently, in 1990, Syria signed an accord with Iraq by which it would retain 42 percent and release to Iraq 58 percent of the annual flow it receives from the Euphrates. In 1991 and 1992, however, both countries protested repeatedly against Turkey's actions limiting the flow of the Euphrates.

The upper riparian states—Turkey and Syria—have been developing their separate multipurpose projects for flood control, perennial storage, irrigation, urban supply, and hydropower generation. Altogether, the storage capacity of reservoirs built or planned by Turkey (estimated at 90 billion cubic meters), Syria (about 15 BCM), and Iraq (about 100 BCM) greatly exceed the annual flows of the two rivers (estimated at 74 BCM). Part of the reason for this excessive and uncoordinated investment in storage facilities is the lack of mutual trust among the three countries. Consequently, each is attempting to appropriate as much as possible of the common water resources. In the process, the three countries are actually diminishing the potential size of the resources by increasing the amount of evaporation from the large reservoirs they have built or are about to build. The annual evaporational loss of water from dams is already estimated to exceed 2 billion cubic meters in Turkey, 1 billion cubic meters in Syria, and 5 billion cubic meters in Iraq (where the climate is most arid), for a total of some 8 BCM/Y.

By the time the three countries expect to complete their competitive plans for the Euphrates (perhaps by the year 2015), they will need as much as 36 BCM/Y, considerably more than the mean annual flow of that river.

The intense nationalistic feelings that pervade the Middle East and thwart an accommodation among its peoples manifest themselves in many ways. A visitor to any country in the region must be careful not to offend the sensitivities of the local people. Just how delicate that task can be became all too apparent to me in connection with a visit to eastern Anatolia a few years ago. I was invited there by a Turkish engineering firm to render advice on developing modern irrigation in the area to be served by the Ataturk Dam. It is an area of contrasting land forms, with broad valleys and plateaus, windswept hills long denuded of their original forest cover, and snowclad mountains.

When, in conversation with my professional colleagues, I casually—and quite innocently—referred to those mountains by their common geographical name, the "Armenian Mountains," the faces of those present froze. Their entire attitude had changed. Inadvertently, I had committed a practically unforgivable faux pas. I hadn't realized this until later, when my otherwise amiable host, who felt personally responsible for me, apprised me of his feelings in no uncertain terms. These are Turkish mountains, not Armenian, he declared. But Armenian is their internationally accepted geographical designation, I had the temerity to comment. No foreign geographers have the right to name *our* mountains, came the stern reply.

Turkey's ambitious Southeast Anatolia Project is known by the Turkish acronym GAP (for *Guneydogu Anadolu Projesi*). It aims to develop six of Turkey's poorest provinces, altogether constituting nearly 10 percent of the country's area. The provinces under development border Syria and Iraq and encompass the headwaters of both the Tigris and the Euphrates. For their size and potential, they are sparsely inhabited, with a population smaller than 6 million. It is, however, an ethnically sensitive population. Fully half the people are Kurds, and a militant group of separatists among them considers the district to be part of its nation's rightful domain, which it calls Kurdistan. The rest of the people are ethnic Turks (some 40 percent) and Arabs. Of the area's current working population, about two-thirds are engaged in farming.

In embarking upon their great undertaking, the Turks hope to transform a semiarid plateau into their country's breadbasket by means of irrigation.[6] They also hope to promote the development of industry there and elsewhere in Turkey, utilizing the abundant hydroelectric power to be generated by the projected dams. Turkey's state planners aim thereby to raise the living standards of the impoverished inhabitants and integrate them into modern Turkey's economy and society. Another obvious aim is to weaken the troublesome Kurdish separatist movement, not only by alleviating the local people's economic plight but also by attracting more ethnic Turks into the area, thus diluting the Kurdish preponderance there.[7]

The latter aim, however, has incurred the vehement resistance of the Kurdish nationalists, who strongly resent what they consider an invasion and colonization of their homeland, and an attempt to deny them independence. Moreover, although GAP promises to provide great eventual benefits, in the short run it forces the displacement of perhaps 250,000

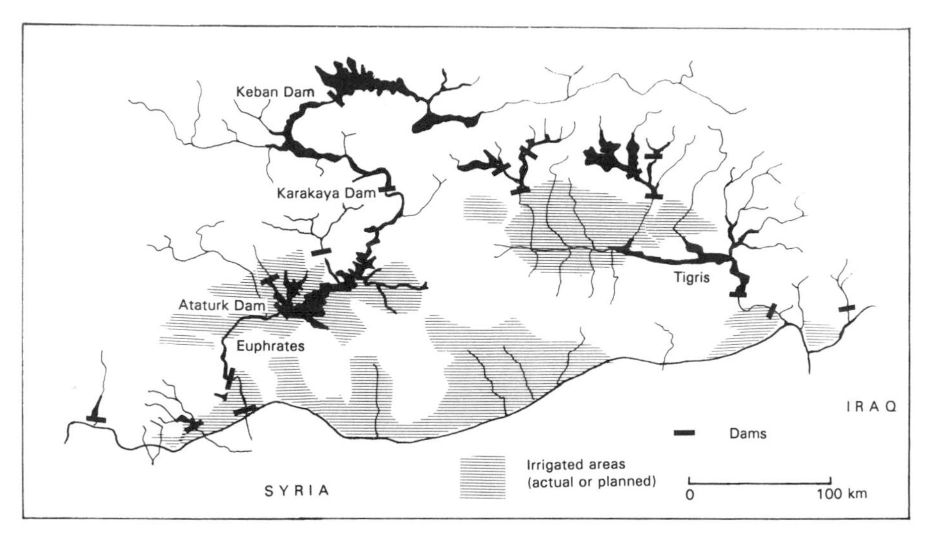

Figure 5.5 Turkey's Southeast Anatolia Project (GAP).

people from lands to be inundated by the dams. Consequently, a bitter civil war has broken out in the district, resulting in costly delays of the government's entire development program.[8]

The total cost of the GAP program is estimated to be well in excess of $20 billion. Most of the cost must be borne by Turkey itself, since the World Bank has made its international financing conditional upon a comprehensive agreement among all the riparian states of the Tigris-Euphrates basin. The financial (and technical) difficulties constitute another reason for the delay in implementation of this huge undertaking. In principle, the program is meant to include both rivers, but thus far the Turks have worked only along the Euphrates. They intend to develop the Tigris headwaters at a later stage.

The GAP plan calls for the construction of 80 dams, 66 hydroelectric power stations with a total capacity of 7700 megawatts, and 68 irrigation projects covering up to 2 million hectares. Among the principal dams are the Keban, the Karakaya, and—most important—the massive Ataturk. Keban Dam was begun in 1965 and completed in 1974. Its construction was delayed by the porous nature of its substratum, a fissured limestone that required the injection of liquid cement and a cover of clay to render it impervious. This dam has created a lake covering 680 square kilometers, with a storage capacity of about 30 billion cubic meters. It is intended primarily for the generation of electricity rather than for irrigation. Its present reported production of 1240 megawatts serves the cities of Ankara and Istanbul. Karakaya Dam was built downstream of the Keban Dam. Completed in 1988, it is designed to generate 1800 megawatts. Its lake covers nearly 300 square kilometers.

Ataturk Dam is the linchpin of the entire GAP complex. One of the largest dams in the world, its dual purpose is to generate 2400 megawatts of electricity and to store up to 82 billion cubic meters of water. It is designed to provide 10 billion cubic meters of water per year—via tunnels, pipes and canals—for the irrigation of 500,000 hectares of land in the vicinity of Haran. Begun in 1983, this dam was originally scheduled to be completed in 1992. Its storage capacity is to be 48.4 billion cubic meters of water and to inundate an area of 877 square kilometers. The total cost of the Ataturk Dam alone is likely to exceed $3 billion.

The filling of Ataturk Dam has already caused international problems. In early 1990, Turkey blocked the flow of the entire Euphrates for one month to begin impoundment. Although Turkey had tried to compensate its downstream neighbors—Syria and Iraq—by releasing more water into the river during the preceding months, those countries suffered serious deprivation. During the period of curtailed flow, Syria not only experienced crop losses but also was forced to reduce its production of electricity, and—because many of the pumps drawing water from wells along the coastal area of Syria are electrical—there were even shortages of drinking water. Iraq also suffered crop losses. Consequently, both countries demanded that Turkey cease impeding the natural flow of the river. A threatened confrontation was averted when the Turks renewed the river's flow. However, that near-crisis was a harbinger of what is likely to happen in the future, especially in dry years, if the countries concerned do not enter into a basin-wide water-sharing agreement soon. As of early 1994, there were no signs that such an agreement is near.

Several smaller dams, hydroelectric installations, and irrigation projects are also included the GAP complex. Among them are the Birjik and Karkemish dams on the Euphrates, and a series of land developments. Altogether, the complex is expected to provide for the irrigation of more than 1 million hectares, requiring some 12–15 billion cubic meters of water per year (including losses in conveyance and evaporation). A fraction of this amount will return to the river in drainage, but the bulk of it—together with the amount of evaporation from the open reservoirs—will be lost to the downstream states. Turkey has promised to release no less than 15.5 billion cubic meters per year of Euphrates water to Syria, but that offer has not been accepted by Syria as sufficient, nor has it been anchored in any binding commitment.

In addition to the Euphrates development now being implemented, the Southeast Anatolia Project also envisages a series of dams and irrigation developments on the Tigris River, to be implemented after the year 2000. When built, those dams will abstract an estimated 5–7 billion cubic meters of water, constituting about one-third of the upper Tigris' natural flow into Iraq. The dams are designed to generate over 2200 megawatts and to provide water for the irrigation of about 550,000 hectares.

As Turkey extracts more water, less will remain for the two countries downstream, and their already fierce rivalry with Turkey and with each other is certain to intensify.

Even more than Turkey (which has abundant alternative water resources), and more than Iraq (which has independent control of tributaries to the Tigris, as well as alternative economic resources—e.g., petroleum), Syria depends crucially on the Euphrates to develop its agricultural economy. The Euphrates is Syria's largest river and most important water resource. Its natural flow into Syria is about 30 billion cubic meters per year on average, compared to only 3–4 BCM/Y supplied by all other sources in Syria. Apart from having the Euphrates and its tributaries (the Balikh and the Khabur) that flow within its territory, Syria has access to a short segment of the Tigris. However, Syria cannot withdraw substantial amounts of water from the two rivers without affecting its downstream neighbor, Iraq, and thereby risking violent conflict.

Syria's other water resources are much smaller. The Orontes (Asi) River flows into Syria from Lebanon, and the Yarmouk River flows along Syria's boundary with the Kingdom of Jordan. In addition, two small streams serve the valley of Damascus, and several smaller ones flow seaward. Apart from its surface water resources, Syria has several aquifers, the safe yield of which has been estimated at between 1.5 and 2 BCM/Y. Water is drawn from these aquifers by means of thousands of wells, located in the vicinities of Damascus, Aleppo, and along the Syrian portion of the Orontes River basin. Many of the wells have been overdrawn and have begun to exhibit progressive salinization.

Syria's economy has lagged in the last few years. Although Syria had long been an exporter of agricultural products and still has great agricultural potentialities, it has become a net importer of food. The problem derives in part from the mismanagement of the country's water resources. Moreover, the Syrian government has encouraged population growth, believing that it adds to the country's strength and enhances its claim to leadership in the Arab world. However, with the continuing deterioration of the rural environment, Syria's ability to feed its growing population has declined. The regime's huge expenditure on arms has diverted capital from productive investments. Finally, the water and power development schemes have to date failed to fulfill expectations.

The damming of the Euphrates, Syria's greatest engineering project that was begun with great fanfare in 1974, has thus far been a disappointment. The Tabqa Dam, which formed the giant reservoir named Lake Assad[9] (in honor of Syria's president), was built with the aid of Soviet engineering and financing. It was supposed to irrigate some 400,000 hectares, generate electricity, and make the entire region prosperous. After nearly 70,000 indigenous Bedouin were displaced, it turned out that some of the imported generators were faulty and that power generation was greatly affected by the seasonal variation in the river's flow.

An even worse debacle occurred when water was actually brought to the land, with the hope of making a barren area productive. Unfortunately, the soil to be irrigated contained large amounts of gypsum, a soluble mineral composed of calcium sulfate. When irrigated, the gypsum in the soil

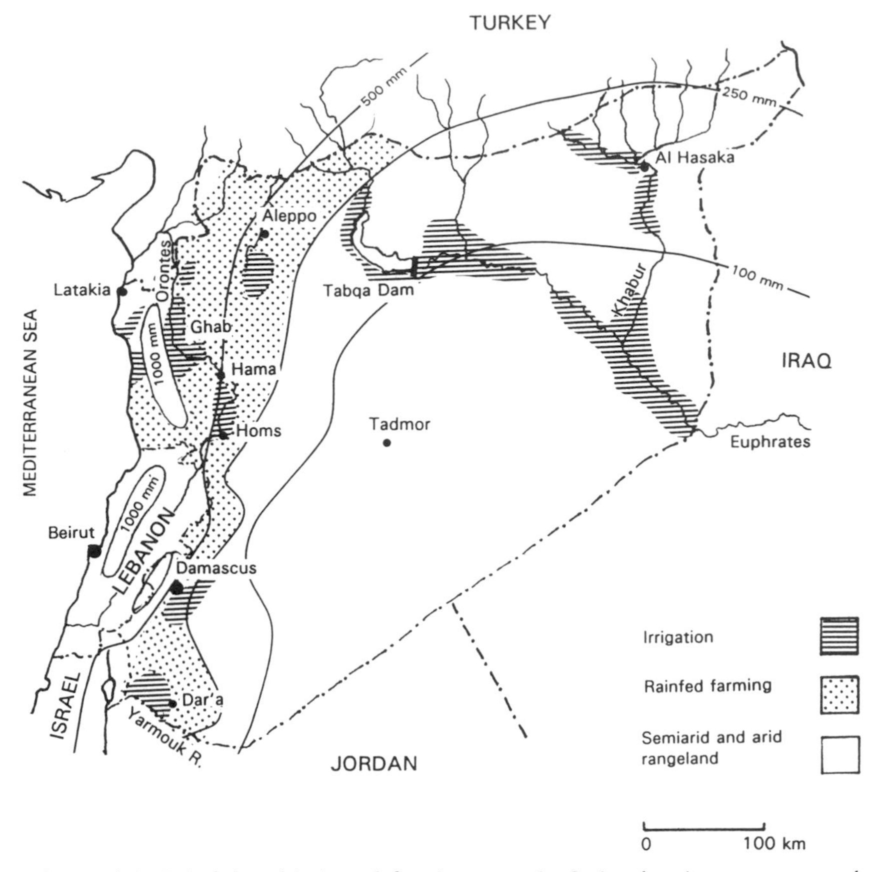

Figure 5.6 Rainfed and irrigated farming areas in Syria, showing mean annual rainfall (mm).

tended to dissolve. Consequently, the land surface—so carefully and expensively leveled for irrigation—subsided irregularly, turning a smooth area into a patchwork of hummocks and depressions that thwarted the effective distribution of water over the surface. Several of the canals collapsed due to seepage. Moreover, the concentration of dissolved gypsum in the root zone affected some crops adversely.

The construction of the Tabqa dam in Syria in 1973–1974 provoked an aggressive response from downstream Iraq. In 1975, during the filling of the dam's reservoir, the Iraqis complained that the Euphrates delivered less than one-third of its normal inflow to their country, and that they had suffered the failure of 70 percent of their winter crop as a consequence. The Syrians denied that they had deprived Iraq of its normal water usage, and claimed that in any event Iraq could compensate for any shortfall by withdrawing more water from the Tigris. Iraq insisted that its rightful share of Euphrates water amounted to 16.1 billion cubic meters. The two countries then closed each others' airline offices and airspace, and sent troops to their common border. Following Soviet and Saudi mediation, Syria agreed to release more water from the Tabqa dam and the crisis was averted. During the drought-prone 1980s, however, Iraq repeatedly accused Syria of withholding Iraq's rightful share of water. The problem between the two is far from being resolved, and is in fact complicated by Turkish activities, which further diminish the discharge of the Euphrates River.

Most worrisome to Syria is the upstream damming of the Euphrates by the Turks, giving them control over the flow of water to Syria's prime source of electric power and irrigation water. When the Anatolia plan is fully implemented, Syria stands to lose about 40 percent of its potential supply from the Euphrates. Syria's population, currently about 14 million, is expected to exceed 22 million within 20 years. The projected dwindling of the Euphrates' inflow, along with the already worrisome depletion of Syria's groundwater resources, may cause severe food shortages within the next decade. Turkish development threatens not only to reduce the volume of flow into Lake Assad but also to pollute the water with salts and agro-chemical (pesticides and fertilizer) residues.

Tabqa Dam has a storage capacity of some 12 billion cubic meters. Plans have been announced to raise the level of that dam so as to increase its storage capacity by about 14 BCM. Such an increase will also entail greater seepage and evaporation losses, to the detriment of downstream Iraq. Other projects being implemented or planned include an additional dam on the Euphrates River upstream of Lake Assad, and a series of dams on the Khabur, on the headwaters of the Yarmouk in the south,[10] on the Orontes,[11] and on the minor streams flowing toward the Mediterranean in the west. These projects are designed to answer Syria's growing needs for electricity and water for urban, industrial, and agricultural development.

Of Syria's total area of 185,000 square kilometers, about one-third is potentially arable. Only a fraction of that is actually cultivated, and less than 5 percent is under irrigation. The total irrigated area in Syria is officially reported to be 863,300 hectares. This includes about 309,000 hectares on the Euphrates, nearly 208,000 on the Khabur and the Tigris, 230,000 on the Asi, 27,000 on the Yarmouk, and the remainder in the Damascus basin and elsewhere. The government still intends to expand irrigated areas on the Khabur, Sajur, Balikh, and Tigris rivers as well as on the Euphrates, to attain a total irrigated area of 1.4 million hectares. In view of Turkey's expansion of its own irrigation, Syria's plans appear to be unrealistic.[12]

Syria's effort to expand irrigation rapidly has foundered because of several other unforeseen obstacles. The fact that the land originally designated to be irrigated was found to be unsuitable is but one manifestation of the lack of professional and administrative expertise. It now seems doubtful that the Syrians will be able to achieve their goal of doubling the area under irrigation within a decade or two, unless they receive considerable foreign assistance.

If Turkey fulfills its declared promise to release at least 15 BCM/Y of Euphrates water, and if Syria uses, say, 6 BCM of that, then the amount remaining to Iraq may be but a fraction of its historical share. Owing to both Syria's and Turkey's withdrawals, Iraq may lose as much as 80 percent of its Euphrates inflow. Particularly vulnerable is the grain-producing area of northwest Iraq. Conflict over water will then be more than likely between Syria and Iraq, or (what with the rapidly shifting alliances and hostilities of the Middle East) possibly between both and Turkey.

Of course, there is no certainty that Turkey will indeed implement its plans to the full extent, but even partial implementation will reduce the flow of the rivers owing to evaporation and seepage from the new reservoirs as well as from irrigated lands. Under the circumstances, the absence of any comprehensive water-sharing agreement among the riparians, indeed the absence to date of any progress toward such an agreement, creates a dangerous situation for all concerned.

The situation of the three riparians along the twin rivers exemplifies how tangled and how fateful the rivalry over shared waters among neighboring nations can be. It is a classic dispute, pitting the prerogatives of an upstream country against those of a midstream and a downstream country; the territorial sovereignty of one country against the historical rights of another; the aspiration of ethnic minorities against centralized authoritarian states; the need for economic development against the requirement to protect the environment; short-term sectoral interests against the longer-term national needs; and the concerns of particular nations against those of the regional community of nations. The challenge facing the nations directly involved and the international community at large is to find ways to resolve such rivalries equitably and peaceably.

The instability and volatility of life in the region was recognized by the ancients. In the Mesopotamian epic of Gilgamesh, the sage Ut-napishtim muses: "Do brothers divide shares forever? Does hatred persist forever? Does the river forever rise up to bring on floods? . . . Since the days of yore there has been no permanence."

6

The Mighty Nile

And there was famine in all lands but in the land of Egypt there was bread.
Genesis 41:54

For thousands of years the people of Egypt have owed their very existence to a river that flowed mysteriously and inexplicably out of the greatest and most forbidding desert in the world. Had the river failed to flow, even for one season, all Egypt would have perished. Hence the ancient Egyptians regularly sent emissaries upstream to observe and report the coming of the annual floods. And yet, clever though they were, they remained totally ignorant of the source and cause of the Nile's lifegiving floods upon which they depended absolutely.

Not knowing whence the stream came meant having no assurance that it would continue. Living in a state of uncertainty, the Egyptians could only be sustained by fatalism or faith. Indeed, they deified the river, praised its powers, and prayed for its benevolence. (They associated the river with the god Apis—Hapi—represented by the figure of a chubby man with large breasts and a clump of papyrus reeds adorning his head.) And, though from time to time it turned whimsical, alternately miserly or superabundant, the river never withheld its waters completely. Each year, as regularly as the cycle of heavenly bodies, it surged and overflowed its banks and poured forth its dark brown waters over the land of Egypt at the very end of summer, the driest and hottest time of year, in perfect timing for the autumn planting season. Thus, wondrously, the river created the most fertile of all lands.

So, for millennia, the Egyptian people lived by and of the Nile: a good flood could provide a good harvest, whereas an insufficient or untimely flood could spell famine. Although they could neither control nor predict the river, they kept very careful records of its variations with the aid of

"Nilometers"—gauges formed by graduated scales cut in vertical stone walls on the river's banks. Today we know the origin of the floods but are not much better at understanding the cause of their variation over a period of years, or of forecasting their magnitude in the years to come.

About 457 B.C.E., Herodotus, the Greek historian and geographer who characterized Egypt as "the gift of the Nile," puzzled over the anomalous surging of this river, which—unlike any other major river then known—flowed from the south northward. So he set out to follow the course of the Nile in an effort to discover the secret of its beginning. He apparently reached as far as the first cataract at Aswan before turning back, more puzzled than ever regarding a river that seemed to gush out of an utterly dry desert. In the third century B.C.E., another Greek explorer, Eratosthenes, provided a nearly correct description of the river's course up to the confluence of its two main tributaries (the Blue Nile and the White Nile) and speculated that the rivers originate in lakes. Other Greek geographers, including Bion, Dalion, Simonides, and Strabo[1] all investigated the Nile but none reached much farther upstream than today's Khartoum.

In the year 66 C.E., Roman Emperor Nero sent two centurions to explore the land of Nubia (today's Sudan) and seek the ultimate source of the Nile. The farther they advanced upstream, the more difficult became their journey. Finally, having traversed the terrible desert, they encountered a most incongruous obstacle—an impenetrable morass (the Sudd swamps). Defeated, they, too, were forced to turn back. Later in the same century, the geographer Marinus of Tyre wrote of a legendary Greek traveler named Diogenes[2] who journeyed inland from the East African coast for 25 days and arrived at two great lakes and a snowy range of mountains whence he claimed the Nile draws its twin sources. Relying on that legend, Ptolemy, the Greek geographer and astronomer who lived in Alexandria during the second century C.E., drew a map showing the upper Nile, sight unseen, as rising from two round lakes, both located in Central Africa and fed by the waters shed from the *Lunae Montes*—"Mountains of the Moon" (most probably the mountains known today as the Ruwenzori Range).

Through the seventeen centuries that followed, even while Asia and the Americas and Australia were explored and charted, the origins of the Nile, hidden in the recesses of Central and Eastern Africa, remained almost as much an enigma as they were to the ancient Egyptians. Finally and painstakingly, the actual sources were discovered in the late nineteenth century, not much more than a single century ago. Yet it was not until the 1960s that a detailed survey of the upper gorges of the Blue Nile was completed.

Today we still do not know all that needs to be known about the two principal rivers and the many smaller tributaries that contribute to the Nile. Unlike the nearly identical twins of Mesopotamia, the two principal headwaters of the Nile are very different from each other: they arise in contrasting climatic and physiographical areas and are characterized by disparate hydrological regimes. The one source, called the White Nile, flows out of Lakes Victoria and Albert in the tropical rainbelt of Central Africa, with relatively little interseasonal variation.[3] The other source, called the Blue Nile, flows from the highlands of Ethiopia and is strongly seasonal, subject to the annual monsoons streaming in from the Indian Ocean.

The name "Nile" (Greek *Neilos,* Latin *Nilus,* Arabic *Nil,* ancient Hebrew *Yeor*) may have been derived from the Semitic root *nahal,* meaning valley or stream.[4] The ancient Egyptians named the river *Ar* or *Aur* (preserved in the Coptic *Iaro*), meaning "black" in reference to the color of the dark silt carried by the river during its flood stage. That dark silt, deposited over the river's floodplain, has given the land itself its name, *Kham* or *Khami.* In modern Arabic, the river is variously termed *an-Nil, al-Bahr* (the "sea"), and *Bahr an-Nil* (the "sea" of the Nile).

With a total length of about 6,650 kilometers, the Nile is the longest river in the world. It drains an area estimated at 3,350,000 square kilometers—fully one-tenth of the African continent. Its catchment includes parts of nine states: Tanzania, Burundi, Rwanda, Zaire, Uganda, Kenya, Sudan, Ethiopia, and Egypt.

The geological history of the Nile is rather interesting. In the mid-Tertiary period, some 30 million years ago, it was evidently a much shorter river, probably originating at the stream now known as the Atbara (which drains the northwestern highlands of Ethiopia). South of it rested a separate basin with a vast landlocked lake, known as Lake Sudd. Still farther south lay another drainage basin containing Lake Victoria. Geologists have conjectured that at some later stage, perhaps 25 million years ago, the southern lake system formed a new northward outlet, which sent its excess waters flowing into the former Lake Sudd. As the latter lake filled with sediment, its level rose, causing its waters to spill over into an upper tributary to the Nile. Thus was the Lake Victoria system linked to the river draining the western Ethiopian highlands toward the Mediterranean. Evidence of the previous extent of Lake Sudd can be found in the great alluvial plain of Sudan's Gezira,[5] with its thick blanket of dark clay.

Although the White Nile has several headstreams, it emerges as a major river only at its departure from Lake Victoria, which lies between Tanzania, Uganda, and Kenya in Central Africa's tropical rainbelt. Victoria is the third largest lake in the world, covering an area of 68,800 square kilometers. Being at a high elevation of nearly 1,150 meters above sea level, it is in effect a plateau reservoir, with most of its water supply due to direct rainfall (about 1,600 millimeters, distributed evenly throughout the year) and only about 14 percent coming from streams—the main one being the Kagera. Most of the lake's water is lost by evaporation. Only some 18 percent of the lake's water supply discharges into the so-called Victoria Nile, at a rate of about 2 BCM per month—winter and summer alike. Near its outlet, the river originally ran over a series of cascades called Ripon Falls. However, that stretch of the river is now submerged by the Owens Falls Dam, built in Uganda in 1954. The dam has transformed Lake Victoria in effect into a controlled reservoir, permitting the storage of surplus

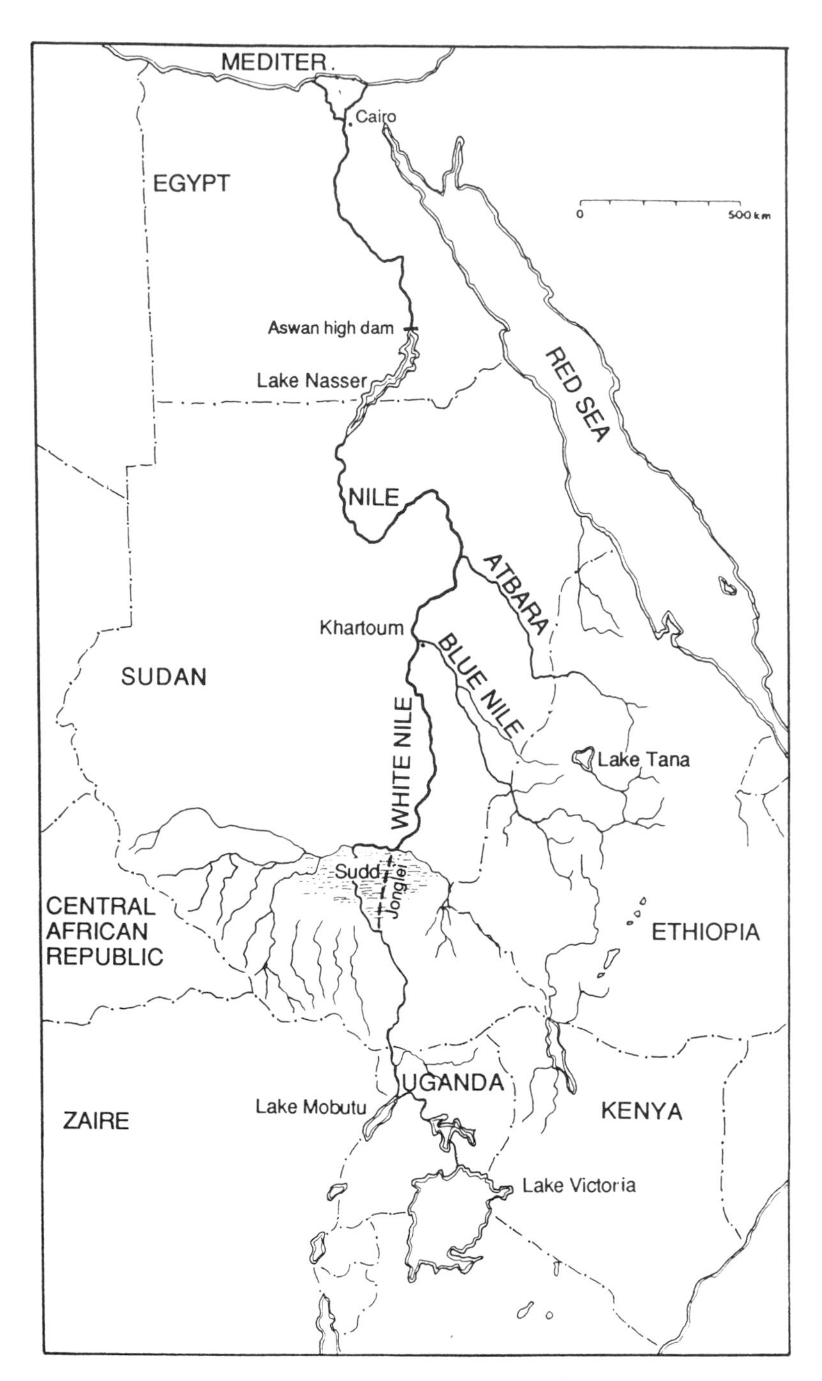

Figure 6.1 The Nile and its major tributaries.

water in rainy years so as to meet water deficits downstream in relatively dry years.

The Victoria Nile next passes through shallow Lake Kyoga, then turns west, descends over Murchison Falls into the East African Rift Valley, and enters Lake Albert (elevation 619 meters above sea level), which is bounded by Uganda and Zaire. The Victoria Nile contributes about two-thirds of the total inflow into Lake Albert, the remainder being received from rainfall and from tributaries (the main one being the Semliki, which flows from Lake Edward, elevation 913 m). Unlike Lakes Victoria and Kyoga, Lake Albert (renamed Lake Mobutu by Zaire) is a deep, narrow lake with steep mountainous banks, hence its loss of water by evaporation is relatively low. The river emerging from that lake and known as the Albert Nile carries some 22 billion cubic meters per annum. It is wider and slower than its parent stream, and is fringed with swamps.

At its entrance into Sudan, the river becomes rapid again, cascading over steep slopes and gorges, so here it is given the name Bahr el-Jebel ("Mountain River," or "Mountain Sea" in Arabic). Along this segment of its course the White Nile is augmented by torrential tributaries that add some 17 percent to the water received from the great lakes. Less than 200 kilometers downstream, at Rejaf, the river spreads over a wide and extremely flat clay-lined plain (about 135,000 square kilometers in extent), giving rise to a patchwork of swamps and ponds, overgrown with tall grasses and papyrus reeds and known as the Sudd (literally meaning "barrier"). The area within the Sudd that is actually inundated varies seasonally and has expanded and contracted over a period of decades. The tangled mass of vegetation here clogs the course of the river and makes it non-navigable.[6]

Numerous other streams converge into the Sudd: from the west comes Bahr el-Ghazal ("Sea of the Gazelle"), draining a subtropical savanna zone of summer rainfall; and from the east comes the Sobat (fed by its tributaries the Baro, Gilo, and Pibor), draining the southwestern Ethiopian mountains. Together, these and other streams feed the swamps and compensate, more or less, for the enormous amount of water lost in evaporation from the Sudd, estimated to total at least 20 billion cubic meters per year.

The Egyptians and Sudanese plan to prevent a substantial fraction of that loss by digging a canal (called the Jonglei Canal) through the Sudd for the more rapid through-flow of the river. Such a canal can indeed produce great benefits to water users downstream, but only at great environmental cost. It would inevitably drain a sizable portion of the marshes, thus reducing the habitat of the area's unique flora and fauna, as well as the livelihood of its indigenous inhabitants. Emerging from the Sudd near Malakal, the river carries an annual discharge of about 25 billion cubic meters. That discharge varies little during the year because of the buffering effect of the large swamps and lagoons of the Sudd. In Sudan the river carries the title White Nile (*Nil al Abyad* in Arabic), apparently because of the milky suspension of organic matter that it has acquired from the decomposing vegetation of those swamps. The White Nile continues northward for about 800 kilometers before joining the Blue Nile to form the United Nile. Along that long stretch the White Nile is a placid stream with a very gentle slope (grading down only about 5 centimeters per kilometer of run) and marshy banks that lose water by evaporation and seepage. Consequently, the White Nile's somewhat diminished yet steady discharge amounts to approximately one-quarter of the united river's average annual flow below Khartoum. That fraction decreases to just one-fifth after the united river is joined by the Atbara.

The Blue Nile (*Nil al-Azraq* in Arabic, so-called because of its content of dark sediment) differs markedly from the White Nile. It drains the lofty volcanic mountains of western Ethiopia, descending from an elevation of 2,000 meters above sea level. The river flows from the southern end of Lake Tana (elevation 1,700 m) through a spectacular curving gorge marked by a series of rapids. The stream discharging from that lake is practically silt-free, but the steep tributaries that join it—the Rahad and the Dindar, which also originate in Ethiopia—swell the river seasonally, following the summer monsoonal rains, with silt-laden runoff. That silt begins to settle over the plain of Sudan (elevation 400 m), where the river loses much of its initial velocity.

The Blue Nile begins to rise in June, following the onset of the monsoonal rains over Ethiopia. Those rains continue for about five months and total 500 millimeters on average. Because of the steep, bare, and relatively

Figure 6.2 The Blue Nile as it descends from Lake Tana in Ethiopia.

impervious nature of the landscape, much of the rain is shed as runoff, flowing into the river. The maximal flood stage of the Blue Nile arrives at Khartoum (where it meets the White Nile) in late August or early September. During the flood season the Blue Nile delivers an average of 10 billion cubic meters per month. When the Blue Nile crests it pushes back the waters of the White Nile, in effect turning the latter river into a quiescent lake, which often spreads and backs up to the Jebel Aulia Dam, some 70 kilometers upstream. The interface between the darkish waters of the Blue Nile and the whitish waters of its sister river can be seen shifting seasonally up and down the course of the latter, depending on the relative flows of the two. During the summer months the Blue Nile predominates, and accounts for some 90 percent of the united river's flow. During the winter and spring months (from December to June), however, the Blue Nile delivers no more than 0.5 billion cubic meters per month, so it is the White Nile that sustains the flow with its delivery of about 2 BCM per month. When a drought takes place in Ethiopia, as it did during the 1970s and again in the 1980s, the Blue Nile's proportionate contribution to the united river's total yearly flow diminishes markedly.

The last tributary of the Nile is the Atbara, which enters the main river some 320 kilometers north of Khartoum. It drains the high peaks of the northwestern Ethiopian mountains, descending from an elevation of 2,000 to 3,000 meters to the Sudanese plain (elevation approximately 400 m). Like the Blue Nile, the Atbara is a seasonally variable stream, swelling in the wake of the monsoon to become a torrential and muddy river, with a mean flow exceeding 2 billion cubic meters per month. However, unlike

Figure 6.3 Erosion in Ethiopia: The source of the Nile's fertile silt.

the Blue Nile, which is a perennial stream, the Atbara shrinks during the dry season to little more than a series of stagnant pools.[7]

After its junction with Atbara, the Nile wends its way northward for more than 2,500 kilometers while receiving practically no further addition of water. Through most of that length, the river traverses the bleached desert, with only the narrow strips of its banks tinged with green. En route, the river slices through several cataracts, thereby becoming unnavigable, as it swings through a wide S-shaped curve before resuming its northerly direction. At Wadi Halfa, a valley long inhabited but now submerged under Lake Nasser, the river crosses the modern border between Sudan and Egypt.

Inside Egypt, the Nile flows sinuously in a relatively narrow, flat-bottomed bed, generally incised into the underlying bedrock (which consists of sandstone and limestone formations—except at Aswan, where granitic rocks outcrop). The maximum flood stage normally arrives at this point in mid September, when the flow rate may exceed 700 million cubic meters per day. Of that flow, the Blue Nile accounts for 68 percent, the Atbara 22 percent, and the White Nile 10 percent, on average. The minimum flow rate normally occurs in early May, and may dwindle to no more than 45 million cubic meters per day, contributed mainly by the steady flow of

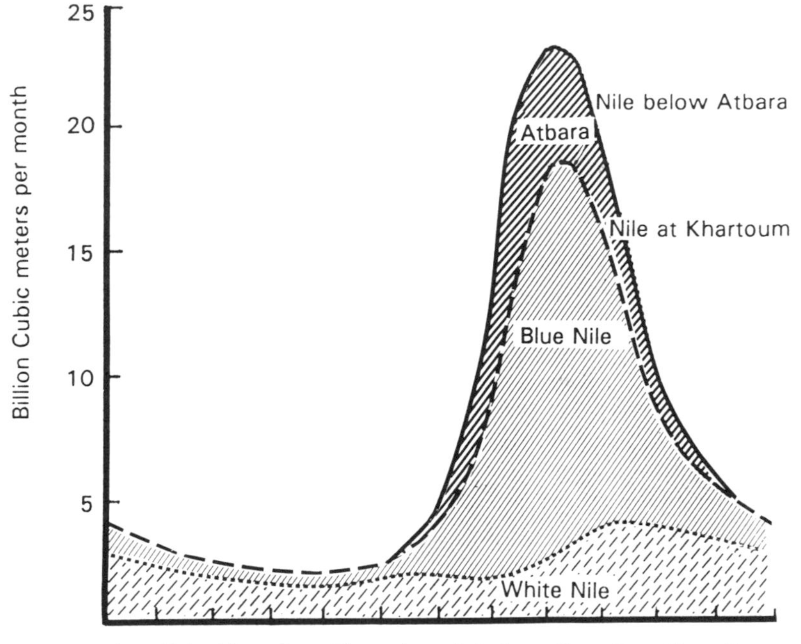

Figure 6.4 Relative contributions of the White Nile and Blue Nile and of the Atbara to the united Nile.

the White Nile. The united Nile's annual delivery to Aswan generally ranges from 80 billion to 90 billion cubic meters. However, depending on the vagaries of the monsoonal weather over Ethiopia, the variation can be much greater. Since modern measurements were begun, the yearly discharge at Aswan has varied from a high of about 137 BCM in 1879 and 126 BCM in 1885 to a low of 45.5 BCM in 1913 and only 37.5 BCM in 1987 (following the severe and prolonged drought that prevailed in eastern Africa and the Sahel during the early and mid 1980s). Prior to the construction of the High Dam and the provision of perennial water storage, years of such low flow would have spelled immediate disaster in Egypt.[8]

The annual flow of the Nile at Aswan (averaged over the period 1900–1959) is about 84 billion cubic meters. Approximately 86 percent of that flow originates in Ethiopia: the Blue Nile contributes 59 percent, the Atbara 13 percent, and the Sobat (a tributary of the White Nile) 14 percent. The remainder (14 percent on average) originates in the East African equatorial plateau.

From Aswan north the river is entirely navigable, and it descends gently at a gradient of about 1:14,000. Its floodplain is seldom more than a few hundred meters wide, in a valley that seldom exceeds 20 kilometers in width and is often much narrower. As the Nile tends to hug the eastern edge of the valley, the greater part of cultivable land is found on the western bank.

North of Cairo, the clifflike walls of the rocky plateau that hem in the upper Nile's floodplain disappear, and the river fans out in several distributaries within the triangular lowland that constitutes the Delta. In the first

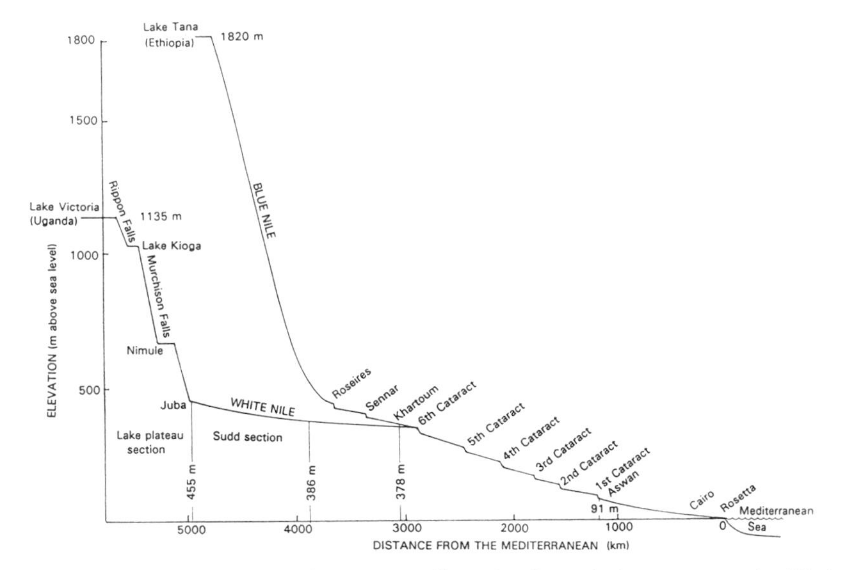

Figure 6.5 Slopes of the Nile's major tributaries from their sources to the Mediterranean Sea.

century C.E., Greek geographer Strabo counted seven distributaries. In modern times, the flows through the Delta have been regulated, so only two main distributaries, the Rosetta (Rashid) and the Damietta (Dumyat), convey most of the water seaward.

The Nile Delta—so named because its shape resembles the Greek letter delta[9]—comprises, like Mesopotamia, an ancient gulf of the sea that has been filled in by river sediment. The silt, mostly originating from the Ethiopian highlands, varies in thickness from 15 to 25 meters. The Delta, also known as Lower Egypt, is a giant triangle of land some 200 kilometers wide along the coast, with its apex at Cairo about 160 kilometers inland, altogether covering an area twice that of the Nile Valley in Upper Egypt. The surface grades gently toward the sea, falling some 15 meters from Cairo to the Mediterranean (a gradient of about 1:10,000).

The strip of Delta nearest the coast is very poorly drained, and actually consists of a series of shallow, brackish lagoons and saline marshes. Even the inland sections of the Delta are not very well drained, and—now that the level of water in the Nile distributaries is kept high—gradual waterlogging and salinization are threatening the land's legendary fertility.

Over most of its length the Nile basin is practically rainless, lying as it does in the great desert belt of the Sahara. Only the roots of the river—the Ethiopian highlands feeding the Blue Nile and the equatorial region feeding the White Nile (including southern Sudan)—receive abundant rainfall, amounting to over 1,500 millimeters per year. Central Sudan, including the plains of Darfur and Kordofan, normally receives sufficient rainfall to maintain pastures and even permit rainfed farming. But these areas, lying along the edge of the desert, are vulnerable to droughts and to frequent wind storms (called *huboobs,* meaning "gusts"), which carry enormous quantities of sand and dust.

North of where the two main tributaries meet to form the United Nile, the climate is almost entirely desertic, characterized by a dry atmosphere and exceedingly hot summers, with daytime temperatures frequently rising above 40 degree Celsius. The winter season, from November to April, is cooler (the average temperature in Cairo, for example, being about 11 degrees Celsius) but still warm enough to permit year-round crop growth. During the spring season, Egypt is frequently visited by dry and hazy southerly winds coming from the deserts and called *khamsins* (meaning "fifty"—as they are reputed to be most prevalent during the spring season, locally believed to last 50 days).

The ancient method of irrigation in Egypt consists of dividing the fields on the flat floodplain into a series of basins, separated by hand-built earthen embankments. At the time of the flood, the Nile waters, rich in silt, would naturally impound the basins and would be kept there for some weeks to soak the soil thoroughly. After the river would begin to ebb, the excess water could be drained away from the basins by breaching their embankments and returning the water to the river via ditches. The basins were then ready for the planting or sowing of autumn and winter crops.

The disadvantage of the traditional method of flood-irrigation is that it permits but one crop per year—a crop that is at the mercy of uncontrollable fluctuations in the size and exact timing of the flood. Direct flood-irrigation thus leaves the land fallow for more than half the time. Since the climate with its abundant sunshine and warmth permits the growing of two or three crops in succession, year-round, the problem is how to ensure a supply of water for irrigation during the long period between floods. In ancient times, farmers did this by carrying buckets, and later by means of human-powered or animal-powered water-lifting devices.

Perennial irrigation can be facilitated by maintaining the water in the river at a high enough level to allow convenient pumping at all times. Toward that end, engineers began to construct barrages and waterworks toward the middle of the nineteenth century, first in the Delta area and later further upstream. In 1843, during the reign of Muhammad Ali,[10] a series of diversion dams (barrages or weirs) were begun across the Nile at the head of the Delta about 20 kilometers north of Cairo. The level of water was raised upstream to supply the irrigation canals and to regulate navigation. This early project, which marks the beginning of modern irrigation in the Nile Valley, was completed in 1861 and was later extended and improved. The Zifta barrage, about half-way along the Damietta branch of the Nile, was added in 1901. It was followed in 1902 by the Asyut Dam, built in Upper Egypt about 320 kilometers upstream from Cairo, then in 1909 by the Isna barrage about 250 kilometers farther upstream, and by the Naj Hammadi Dam about 240 kilometers still farther south. Finally, the engineers were emboldened enough to attempt to control the entire flow into Egypt by building a dam as Aswan.

The first Aswan Dam was built by the British between 1899 and 1902. It was enlarged between 1908 and 1911, and again between 1929 and 1934. The latter development raised the water level by some 36 meters and increased the dam's storage capacity from 1 billion cubic meters to over 5 billion cubic meters. That amount of water seemed large at the time but could not provide Egypt with sufficient storage to tide the country over a season of low Nile flow. For this reason the Egyptians undertook to build the Aswan High Dam (called *as-Sadd al-Ali* in Arabic) about seven kilometers upstream of the old Aswan Dam. Following the completion of the Aswan High Dam, in 1970, irrigation in Egypt became entirely perennial.

Early in this century, perennial irrigation was also begun by the British in Sudan. Their primary aim was to provide a plentiful supply of cotton for the textile factories of Lancashire. They built the first dam on the Blue Nile at Sennar in 1925. That dam has a storage capacity of nearly 1 billion cubic meters and can provide water at the time of year when the natural discharge of the Blue Nile is seasonally low. Thus, it made possible the development of irrigation in the Gezira plain[11] between the two Niles south of Khartoum. The dam also produces hydroelectric power.

The success of the Gezira scheme led to the construction of additional dams and barrages in Sudan for large-scale irrigation. In 1937 a dam was completed on the White Nile, at Jebel Aulia, with a storage capacity of 2.5 billion cubic meters, for the purpose of regulating the flow of that river and to augment the flow into Egypt when the Blue Nile is low. (At the time, the administration of both Sudan and Egypt was in British hands, a fact that facilitated coordination between the two countries.) In 1964, following Sudan's independence, a dam was built on the Atbara at Khashm al-Qirbah with a storage capacity of about 1.3 billion cubic meters. In 1966 yet another dam was completed on the Blue Nile at Roseires, with a storage capacity of nearly 3 billion cubic meters.

In the 1970s, Sudan was widely perceived to be a potential breadbasket capable of alleviating the growing food deficit in the Middle East. Unfortunately, these high expectations have not yet been realized, apparently owing to poor management. The potential is still there, however.

The pattern of water demand and use in the Nile basin contrasts sharply with the pattern of supply. The paradox is that the countries contributing the most water are using the least, and vice versa. The humid equatorial zones use little of the river's water because their humid climate makes irrigation less necessary, but that is most certainly not true for Ethiopia, in which a short season of torrential rains alternates with a long dry season. Yet this country, despite its pervasive poverty and the fact that it contributes a disproportionate four-fifths of the Nile's total annual flow, uses practically none of its water. On the opposite side of the ledger, Sudan (whose territory contributes but little to the river's net flow), and Egypt (which contributes no water at all) are the principal users of the river's water and claim incontrovertible historical rights to go on using the same amount. The existing pattern of utilization seems to leave little water for the upstream countries.

Prior to the construction of the High Dam, Egypt's total water storage capacity amounted to about 9 billion cubic meters. This was hardly enough to allow the country's agricultural production to keep pace with the rapid growth of population (which began early in the twentieth century as a consequence of the improvement in sanitary standards and the increase in life expectancy). Nor could Egypt depend on the additional timely supplies of water from the reservoirs of Sennar and Jebel Aulia in Sudan, as those reservoirs were insufficient to satisfy their needs as well as the growing needs of independent Sudan. Each year the reservoirs of Egypt would empty by mid summer, long before the arrival of the flood, thus leaving the fields of Egypt high and dry and preventing the farmers from bringing their summer crops to successful harvest. The increasing discrepancy between the demand for food and the supply of it impelled the Egyptians to seek a way to augment and stabilize their water supply.

That quest was actually begun by the British, who held sway over the greater part of the Nile basin until the middle of the twentieth century. Their idea was to develop the entire basin in an integrated fashion, through a series of dams controlling the outflow from the upper lakes feeding both tributaries, and a canal for the White Nile to bypass the Sudd.[12] Several of the proposed projects have been implemented, but on a piecemeal basis rather than within the framework of an integrated basin-wide scheme. Among the implemented projects are the dams at Sennar and Roseires on the Blue Nile; Jebel Aulia on the White Nile; Owens Falls at the outlet of Lake Victoria; Hashem al-Qirbah on the Atbara; and—most important—Aswan. However, while facilitating control over the flow of the river, these dams cause considerable losses of water due to evaporation and seepage. By some estimates, these losses may amount to between 15 and 17 BCM/Y.

Prospects for the comprehensive integrated development of the Nile basin appeared to dim when British colonialism ended and the separate nations in the region began to assert their distinct national interests. In particular, after the revolution of 1952 in Egypt and the subsequent rise to power of Gamal Abdel Nasser, the new leadership there wished above all to ensure Egypt's independent water supply rather than depend on water control schemes beyond its borders. Therefore, the Egyptians proceeded with their plan to build the Aswan High Dam. Although some voices were raised in opposition to the project, President Nasser himself embraced the plan decisively and paid no heed to its critics. The grand scope and scale of the plan captured his imagination and ambition.

The main purpose of the Aswan High Dam was to free Egypt from dependence on the whims of upstream states or even of climate, by providing the country with an assured reserve of water subject to her own control and sufficient to tide her over periods of drought. In Nasser's own words: "After completion of the High Dam Egypt will no longer be the historic hostage of the upper partners to the Nile basin." Specifically, the aims of the project were to ensure a steady supply of water for intensification of crop production and extension of the growing season so as to facilitate the year-round production of two, three, or even four crops in succession. Additional aims were to insure Egypt against seasons of drought or low river flow; to irrigate new lands (to be reclaimed from the desert) and thereby to settle landless peasants; to prevent uncontrolled and potentially destructive flooding; to generate electricity for Egypt's industrialization; and—finally—to develop fishing, recreation, and tourism on the new lake formed behind the dam, to be called Lake Nasser.

The succession of events preceding and leading to the construction of the High Dam provides an instructive tale of intertwined power politics, personal ambition, national pride, and rivalry over resources, in the face of which all considerations of ultimate environmental consequences were largely ignored.

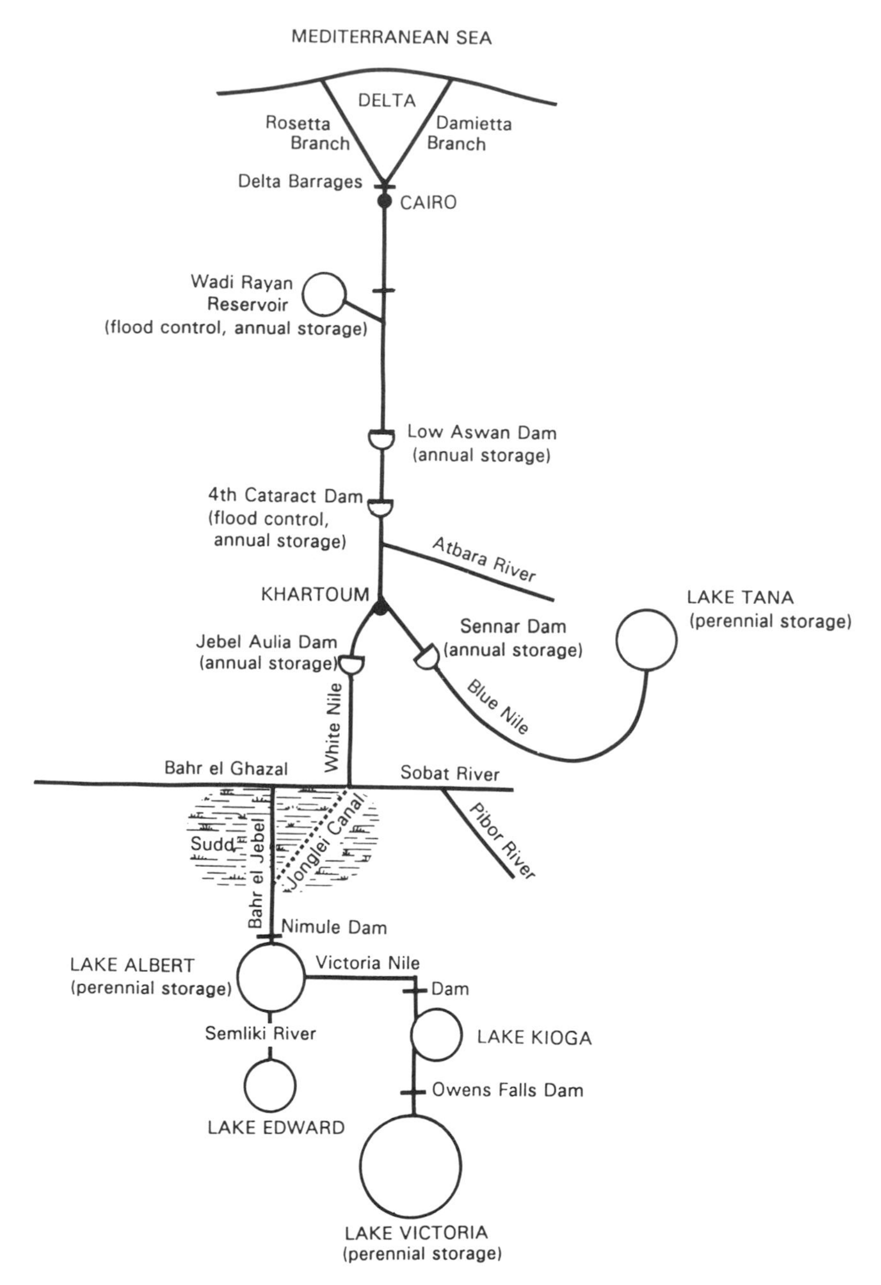

Figure 6.6 Schematic representation of the Nile basin and the possibilities for controlling its flows as perceived by British planners early in the 20th century.

The early planning of the dam was carried out in 1952 by German engineers. The West German government then had a political interest in supporting economic development in a leading Arab state, so as to counterbalance its agreement—announced that same year—to grant reparations to Israel for part of the Jewish property confiscated by the Nazis during the Second World War. After the project was deemed feasible, the governments of the United States, Britain, and West Germany were willing jointly to finance the first stage of the construction. President Nasser, however, wishing to avoid the possibility of political pressure at a later stage, demanded that the Western powers commit themselves at the outset to financing the entire project. At the same time, to demonstrate his independence, he entered into an arms purchasing agreement with Communist Czechoslovakia.

To pressure Egypt back into the Western fold, John Foster Dulles (then secretary of state under President Eisenhower) withdrew American support for the High Dam project. In defiance, Nasser then nationalized the Suez Canal, which had long been under British control. In 1956 Britain undertook, along with its partner France and with Israel (which had its own altercation with Egypt along the Gaza Strip), to attack Egypt. But because those states did so without informing the United States or obtaining its prior approval, President Eisenhower interceded and forced their withdrawal from Egypt. Rather than return to America's fold, however, Nasser then turned to its Cold War rival. The Soviet Union was willing to help Egypt in the financing and the actual construction of the Aswan High Dam, and an agreement to that effect was signed in 1958.

Work could not begin immediately, however, because the Sudanese objected to certain provisions of the original plan. Specifically, they objected to the submergence of the Wadi Halfa valley within their territory, requiring the displacement of tens of thousands of the indigenous Nubian population living there. In addition, they demanded to change the anachronistic Sudano-Egyptian treaty of 1929, according to which Egypt received 48 BCM/Y while Sudan was granted only 4 billion cubic meters. Now Sudan claimed that, in view of its population and needs, it should be entitled to one-third of the Nile's water.

The negotiations lasted one year and culminated in a new agreement, signed in October 1959. It was based on the assumption of a mean annual inflow at Aswan of 84 billion cubic meters, of which Sudan could abstract 18.5 and Egypt's share would be 55.5 BCM/Y. The remaining 10 billion cubic meters were projected to be lost to evaporation and seepage from the Aswan reservoir. Egypt also agreed to grant Sudan financial compensation for the resettlement of the displaced population from the Sudanese lands that were to be inundated by the dam's reservoir. The resettlement would take place near the confluence of the Nile with Atbara, on which an additional dam would be built to provide water for the new villages envisaged there. The agreement also allowed Sudan to build a new dam on the Blue Nile, at Roseires (upstream of the preexisting dam of Sennar),

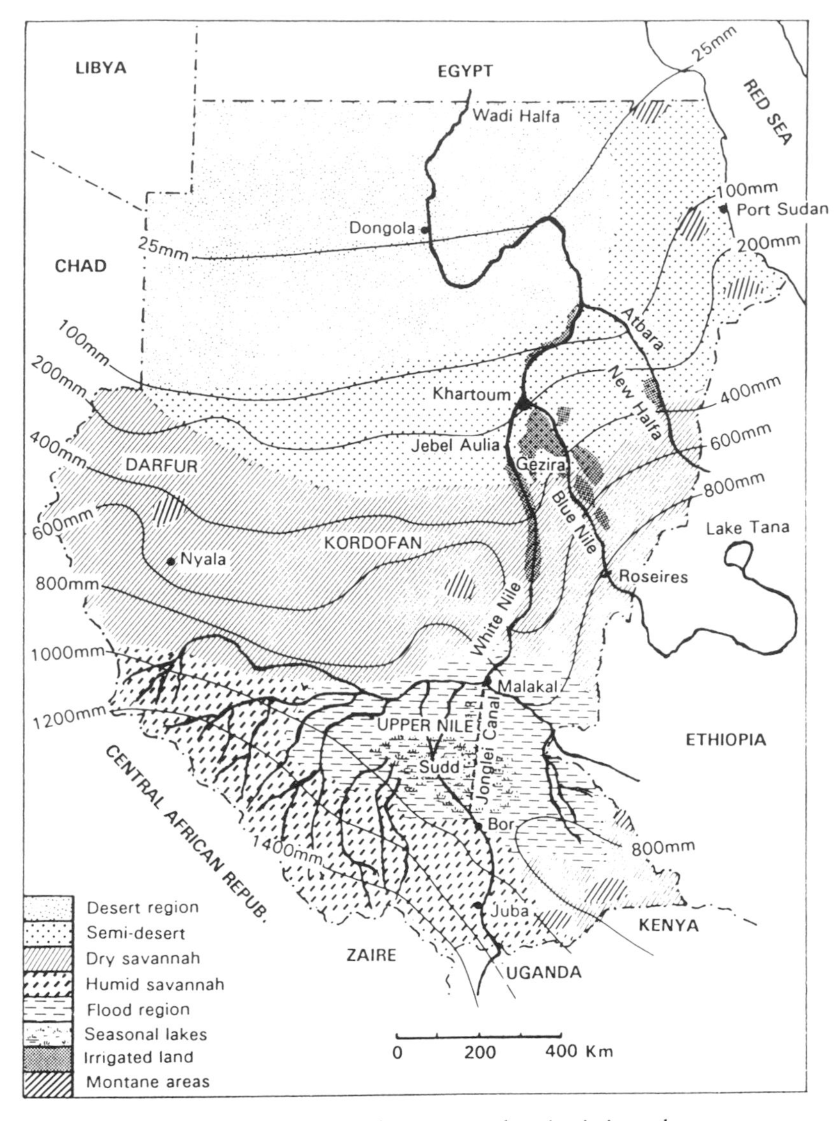

Figure 6.7 Sudan: natural regions and major irrigated areas.

to expand its irrigated lands. Altogether, the new agreement represented a considerable enhancement of Sudan's rights regarding the waters of the Nile.

The 1959 agreement between Sudan and Egypt made no provision for the rights of any other riparians. The agreement merely stated that "once other upstream riparians claim a share of Nile waters, both countries will study together these claims and adopt a unified view thereon. If such studies result in the allocation of an amount of Nile water to one or another of these territories, then the value of this amount shall be deducted in equal shares from the share of each of the two Republics." To date, this aspect of the agreement has never been tested. No other riparian countries have made formal claims to Egypt and Sudan for an allocation of Nile water. Specifically, Ethiopia—the source of most of the water that arrives at Aswan—has not yet attempted to exercise its rights under international law to an equitable portion of the Nile waters. Surely, that is not because Ethiopia has no need for the water. Hence it seems only a matter of time before Ethiopia stakes its claim, and before a situation of active rivalry results.

Actual construction of the Aswan High Dam began in January 1960. The site of the new dam, 7 kilometers south of the old Aswan dam, was chosen because the solid bedrock there could bear the massive weight of the huge structure without the danger of subsidence. (The strength of the foundation and the structural integrity of the dam were of supreme importance, since any failure of the dam in the future would be catastrophic for all of Egypt.) The first stage of the construction, completed in 1964, consisted of building temporary embankments and a channel to divert the river from the site of the dam. The second stage, the building of the dam itself with its electric generating installations, was completed in 1970.

The High Dam was built at a site where the river is about 550 meters wide and has steep banks of granite. The dam is about 1 kilometer long at the base, about 3.8 kilometers long at the crest, and nearly 1 kilometer wide at the bottom of the river bed. Its height above the bed is over 110 meters. The tunnels embedded in the dam (designed to deliver water to the generator turbines and thence to the river below) are placed about 37 meters above the foundations, in order to prevent clogging by sediment. The volume of water below that level, called the "dead storage," is about 6.8 billion cubic meters. The dam impounds up to 160 billion cubic meters, three or four times the amount of water then utilized annually in Egypt. If filled to capacity, the dam's water level would reach an elevation of 182 meters above sea level. In practice, however, the reservoir is never filled completely, and that is to leave space for possible additional floodwaters, and to minimize seepage and evaporation.[13]

Lake Nasser, as the reservoir behind the dam was named, is the largest man-made lake in the world, extending some 480 kilometers upstream from the dam with an average width of 10 kilometers. Of that length, about 160 kilometers lie in Sudan, where it inundated the Wadi Halfa valley and required the displacement of over 50,000 Sudanese villagers. The lake also submerged numerous archaeological remains, only a few of which (notably the temple of Rameses II at Abu Simbel) could be saved.

During the first few years of the High Dam, Hapi (the ancient god of the Nile) was apparently benevolent, and the flow of the Nile was plentiful. By 1978, the volume of water accumulated in the reservoir reached 110 billion cubic meters. Water appeared to be in such abundant supply that Egyptian President Anwar Sadat publicly offered to provide Nile water to

Israel. In subsequent years, however, Hapi turned miserly. During the drought of the 1980s, the inflow to Lake Nasser fell to 38 BCM/Y, only two-thirds of Egypt's alloted share. Nevertheless, Egypt ignored the drought and continued for several years to release its full measure of water from the dam, assuming that the drought would soon end and the reservoir would be refilled naturally. But the drought persisted, and the reserves were gradually depleted, so that by 1987 the quantity of water remaining in storage had fallen to a meager 24 billion cubic meters—less than one-fifth the reservoir's effective capacity and less than one-half of Egypt's normal annual water requirement. Belatedly, the release of water from the High Dam was reduced.

By early summer 1988 the level of water in Egypt's Nile dropped to the lowest point in more than a century. Tour boats that normally cruise the river's archaeological sites were left stranded on sandbanks. The drying of the Nile threatened to break the slender fertile ribbon of irrigated land that provides all of Egypt's home-grown food. If the drought had lasted one more season, Egypt would have had to drastically curtail its generation of hydroelectric power as well as its irrigation, and it—no less than Sudan, Ethiopia, and later Somalia—would require massive relief assistance to prevent starvation.

At the last moment, in August 1988, Egypt was saved by the sudden occurrence of heavy rainfall over the watershed far to the south.[14] The results mixed redemption and horror in a capricious proportion. In Khartoum, the capital of Sudan, floods coming from the highlands of Ethiopia wrought devastation as the Nile exploded from its banks and inundated the city and its vicinity. And, as the river swelled and rushed northward, it refilled Lake Nasser to the level it had in 1983, more than 16 meters above its level at the depth of the drought in July 1988. Egypt's water supply was restored, its electricity generation could be maintained, and its crops could continue to grow with the blessing of irrigation. But that was a fortuitous deus ex machina that is not guaranteed to recur when another drought—perhaps even more prolonged and severe—befalls the region.

The Aswan High Dam has brought considerable benefits to Egypt's economy, and it has saved Egypt—thus far—from the consequences of drought. As long as the lake is sufficiently full, water can be released as needed, to maximize its utility in generating electricity, promoting navigation, expanding irrigated lands, and intensifying year-round cropping. The lake itself affords possibilities for tourism and recreation, as well as for developing a fishing industry.

The negative effects are more subtle. The first of these is the blocking of the fertile silt, which, rather than be conveyed downstream by the river, now settles where it is least needed—in the farthest reaches of Lake Nasser (where the Nile's flowing waters become motionless). The silt had long fed the plankton growing in the offshore waters along the Mediterranean coast beyond the river's outlets. Deprived of this rich plankton, the once-abundant sardine fisheries have diminished greatly. Moreover, the Nile itself, running clear of silt, has increased its erosivity and has been scouring its own banks. And along the estuaries of the Delta, there is no more silt deposition, so the coast has been subject to progressive erosion by offshore currents. Hence the shoreline, which had advanced seaward throughout history, is now gradually receding landward. There is even danger that seawater will invade the inland lagoons (separated from the sea only by thin sandbars) and salinize them, thereby exacerbating the salinity hazard in the Delta. A related effect of the damming of the Nile is the deprivation of Egypt's soils, whose legendary fertility had been renewed annually by the deposition of the river's nutrient-rich silt. Egypt's farmers must therefore rely increasingly on expensive chemical fertilizers, the residues of which contribute to groundwater and surface-water pollution.

Even more serious is the effect of maintaining a nearly constant water level in the river, necessary to allow easy pumping of irrigation water throughout the year. This raises the water-table and impedes drainage, so Egypt is now experiencing the maladies of waterlogging and salinization to which it had for so long seemed immune. Egypt must now invest in developing an expensive artificial drainage system, lest it lose more and more of its best land to soil degradation.[15]

An additional indirect and unexpected result of the dam has been the spread of water-borne diseases, principally a disease known as bilharzia (Schistosomiasis). Water-loving weeds growing profusely along the banks of ditches filled with stagnant water often shelter snails that act as vectors for the spread of this debilitating disease. People, especially children, who wade in these shallow ditches are easily infected. The disease has become endemic and all too prevalent along the entire course of the Nile, especially in Egypt and Sudan, and it is extremely difficult to eradicate.

Intensified irrigation and the maintenance of a high water level in the Nile have afflicted Egypt with quite another salinity problem outside its irrigated lands. Instead of being flushed out, as they were in the past, by receding floodwaters, the salts that now remain in the groundwater are infused into the porous soil and rocks by capillary action. Water piped in to supply the needs of the expanding population of towns and villages along the Nile, and cesspools placed underground to dispose of their wastes, have further raised the water-table. As a result, there is now a constant upward seepage of salt-bearing moisture into Egypt's temples and monuments, and these salts impregnate the porous stone walls.

As they say in Egypt, "Salt is like a sleeping devil—only when it gets moisture it wakes up." When the moisture evaporates at the exposed surfaces of these ancient structures, the salts recrystallize, forcing apart the grains of the stone. The result is a flaking and crumbling of the ornately carved reliefs and inscriptions of Egypt's magnificent monuments. Salt bubbling up under the ancient wall paintings pushes the plaster off the walls, so that the exquisitely drawn and brightly colored portraits of the ancient kings, queens, and gods, as well as the vivid depictions of landscapes and scenes of daily life (notably including farming activities), are now deteriorating rapidly. If this continues for a few more decades, many and perhaps most of the reliefs and paintings adorning the ancient temples and graves will be erased from within, and only blank, pulverized surfaces of walls and columns will be left.

Some engineers have proposed installing pumps, to lower the underground water level, and digging trenches, to be filled with gravel, around the bases of temple walls to prevent water from seeping into the stone. But such a scheme is certain to be expensive and may not work as intended. Attempts have been made to remove salt stains from the artwork on walls by covering them with wet paper and absorptive clay, but the salt deep in the walls could not be removed this way, and one month after the surface was cleaned the stains reappeared. Faced with this accelerating decay, the protectors of Egypt's priceless archaeological heritage are just as stymied by this insidious and formidable enemy—salt—as are the neighboring farmers.

A major problem in Egypt is the loss of prime agricultural land to urban development. Since the building of the Aswan High Dam such encroachment has claimed at least 200,000 hectares—over half the area of new lands reclaimed at the same time and at great expense. Moreover, the country's rapid urbanization requires bricks, and these are made of the clayey soil of the Nile's floodplain. The mining of this precious soil is now being carried out on a large scale. Many farmers are unable to resist the lucrative sums offered them for their small plots of land, and are enticed to sell their birthright to building contractors. They end up bereft of the real resource that had sustained their ancestors for countless generations. The soil is dug up and carted away to brick factories, leaving pits and hollows that can never again be farmed. Though this practice has been declared illegal, it continues nevertheless. Egypt's most precious resource is thus subjected to instant erosion, which, in the absence of the annual replenishment of silt by the Nile, is truly irreversible.

The High Dam obviously has had a profound effect on the land and water regime of Egypt. In some respects, it has been a boon to the country's development, but in other respects it has been a bane. Overall, it is difficult to weigh the positive versus the negative impacts of the dam, as the various effects pertain to different aspects of the Egyptian economy and environment and to different time periods. Hence they cannot be readily quantified in common terms and compared. Suffice it to say that the benefits were immediate, whereas the drawbacks—not very noticeable at first—are likely to become increasingly important in the long run.

Quite apart from the High Dam, Egypt is facing a dilemma that may soon become a crisis. Its population, now growing at the rate of 1 million every eight months, increases its demand for water and food from year to year. At the same time, some of its soils are undergoing progressive deterioration. Yet doomsday scenarios are not inevitable. Much can be done to promote the efficient use of water and to augment water supplies. Attempts to do so generally run up against an inertial resistance to changing time-honored traditional ways. The patient and faithful farmers have always believed that thanks to God, the Nile has brought Egypt its precious water, and God willing it will somehow continue to do so, for God is a greater expert than all the planners, engineers, and economists. So they go on doing what they and their forebears have always done, working alongside their numerous children to divert water from the main canals to lesser channels running like small veins between plots of vegetables and cotton, mindless of the consequences. Inevitably, however, they will have to change.

The near-disaster of 1988 has already forced Egyptian officials to change their water management strategy. One conclusion was that the country should no longer depend so heavily on the hydroelectric generating power of the High Dam. So the government has installed additional thermal power plants. Another conclusion was that the amount of water released from the dam annually should not greatly exceed the inflow. So Egypt must conserve water more than it had in the past and scale down its ambitious plans to expand irrigation by reclaiming tracts of desert land. Finally, the most far-reaching and disillusioning conclusion was that, in view of Egypt's population growth and the probability of drought, Lake Nasser by itself cannot provide Egypt with the total hydrological security that was the original rationale for the High Dam. Nasser's dream of making downstream Egypt completely independent of the Nile's upstream riparians proved to be a delusion. In the future, sooner or later, Egypt will be required to enter into a comprehensive arrangement with its upstream neighbor for the coordinated storage, regulation, and allocation of the Nile basin's water resources.

Egypt's most serious prior attempt to reach upstream beyond its borders in order to increase its water supplies was done in cooperation with Sudan. The two countries planned to divert water from the swamps of the White Nile by digging a canal through the Sudd. Construction of the Jonglei diversion project actually began in 1978, after years of haggling between Egypt and Sudan.

The idea of augmenting the flow of the Nile by cutting a canal to bypass the Sudd swamps, to be coupled with control works to increase storage capacity, had been discussed for almost a century. The Jonglei Canal was to be part of an elaborate program for augmenting the flow of the White Nile. It includes the following projects: (1) Jonglei I is to dig a 360 kilometer canal from Bor to Malakal to convey 20–25 BCM/Y and thus to recover 4.8 BCM/Y that would otherwise be evaporated from the marshes of Sudd (where the total evaporation is about 14 BCM/Y). (2) Jonglei II envisages the construction of a parallel bypass canal and a reservoir at Murchison Falls in Uganda on Lake Mobutu (Albert), to yield an additional 4.25 BCM/Y. (3) Building canals through the Bahr al-Ghazal marshes (where the total evaporation is about 12 BCM/Y) could yield as much as 5 BCM/Y. (4) Draining the Machar marshes and the Sobat basin near the Ethiopian border (where the amount of evaporation is on the order of 19 BCM/Y) could release a further 4.4 BCM/Y. Thus, if all these projects were indeed carried out as envisaged, the augmentation of the White Nile might total 18.5 BCM/Y. These are rough estimates, however, since the hydrological data available at present is very sketchy. In any case, so grandiose a scheme is unlikely to be implemented any time soon, given the opposition of the local Nilotic people, whose distinctive pastoral way of life would be threatened.

The Egyptians were always in favor of the project, but the Sudanese were concerned over the fate of the Nilotic tribes subsisting in the Sudd, and were wary of Egyptian ambitions to control the entire Nile. A compromise was finally reached in 1974 whereby the projected diversion would drain less of the water than originally planned and would thereby maintain the integrity of a substantial portion of the Sudd wetlands. The agreement envisaged the digging of a canal having a length of 360 kilometers and a width of 38 meters, capable of increasing the outflow of the White Nile from the Sudd by about 5 billion cubic meters per year. This amount is less than one-third of the original estimate of the total potential diversion, but it is still a considerable increment of water (amounting to over three times the total annual supply of fresh water in Israel, for example). The augmented supply would be divided equally between Egypt and Sudan. The secondary benefits anticipated from the project included opening up the river's upper reaches for navigation, reducing the flooding hazard along the fringes of the wetlands, and irrigating lands alongside the canal.

In practice, the construction of the Jonglei Canal proved to be much more arduous and problematic than the parties had realized at the outset. By 1984 only about two-thirds of the task had been completed. But then further work was disrupted by the violent civil war that had broken out in southern Sudan, led by the "Sudanese People's Liberation Army." The conflict was precipitated by the strong resentment of the south Sudanese, most of whom are of Christian or animist faith and are of Central African stock, against the Arab-dominated Muslim north Sudanese who control the central government. The south Sudanese oppose the imposition of Islamic law in a region that is non-Muslim. Moreover, they consider the Jonglei project a provocative intrusion of the north into their domain and an attempt to change their environment and to deprive them of its vital resources. That bitter conflict has continued inconclusively ever since. By now, even if the rebellion were to end, it is doubtful that the neglected works can be rehabilitated without enormous additional costs.

Proponents of environmental conservation[16] are concerned lest the Jonglei Canal project, if it is ever resumed, result in the destruction of the unique habitat of the Sudd. With its abundance of water and vegetation, the Sudd has flourished as a prime habitat for birds and large herbivores, and it is a particularly important link in the chain of sites along the Nile

Valley for species migrating between the tropics and Eurasia. Included in its fauna are the endangered shoebill stork and possibly the largest number of water birds anywhere in Africa. There are nearly half a million tiang antelope, constituting one of the world's largest remaining populations of wild large mammals. Most of the wild populations of the Nile lechwe (an antelope of the genus *Kobus*) are also found in the Jonglei, as are buffalo, elephant, gazelle, hippopotamus, white eared kob, reedbuck, waterbuck, and zebra. The people of the Sudd and its environs—the Dinkas, Nuer, and Shilluk tribes (numbering 200,000 to 400,000)—have developed life patterns involving pastoral migration, fishing, and agriculture, all closely geared to the cycle of seasonal flooding and the ecological diversity of this unique habitat.

The possibility has been raised of the industrialized countries paying the Nile basin countries for the preservation of that natural habitat, thus in effect protecting the southern range of Europe's own bird life. In principle, such a trade might be similar to "debt-for-nature" swaps proposed for various other debt-burdened developing countries for the purpose of protecting their natural habitats and preserving biodiversity. In this case, water would be the principal component of the natural resource that industrialized countries would want preserved. The underlying notion is that growing awareness on the part of the global community of the importance of environmental resources will lead to the conclusion that preserving the water supply for a unique natural ecosystem may be of greater value than diverting the water for irrigation.

In considering prospects for the future, one must remember that the Nile basin is a geographical and hydrological unit. All the people living in this basin, regardless of their great ethnic, cultural, religious, and political differences, depend on the same river, and hence on one another's use of it. Much has been accomplished in the control of the river for the benefit of some of the countries in the basin. Most of the efforts to date, however, have been national and sectarian rather than regional and comprehensive. Henceforth, the efficient management of the basin as a whole will require international coordination and integration, based on the equitable sharing of the precious resources of water and energy contained in this unique river. Ultimately, all countries in the Nile basin stand to gain from cooperation in the development and sustainable utilization of the river's water supply.

The countries of the White Nile basin south of Sudan receive considerably more rainfall than either Sudan or Egypt. The economic development of the upper White Nile countries is not, for the time being, limited by water availability, so the downstream states are not threatened by a diminution of the water supply from the White Nile. However, that river normally accounts for only some 15–20 percent of the total annual flow of the United Nile into Egypt. In contrast, the Blue Nile coming out of Ethiopia delivers the great bulk of the water flowing through Sudan into

Egypt. Therefore the implications for Egypt and Sudan of any potential extractions by Ethiopia may be significant indeed. For this reason we focus our attention first on the challenges and opportunities for cooperation among the three Blue Nile riparians: Ethiopia, Sudan, and Egypt.

Although there has been some coordination between Egypt and Sudan, there has not been, so far, much joint tripartite development, or even joint research and planning, regarding the equitable allocation of the Blue Nile. Despite the threat of conflict, and despite the great potential for projects to improve the utilization of Nile waters, no serious negotiations among the riparian countries have yet begun on the legal, technical, and institutional arrangements needed to implement such projects for common benefit. In view of the growing importance of the river to all three countries, this situation seems odd indeed. It is due in part to the political upheavals and famines in Ethiopia and Sudan, and the reluctance of international agencies to enter into the fray as long as political and economic conditions seem so unstable.

Meanwhile, decisions made by each country regarding investments in water resources development, new irrigation schemes, and industrial projects will have consequences far into the future when water resources will be in much greater demand. Furthermore, the unanticipated environmental and climatic changes that began to become apparent in the 1970s have accelerated the need to make economic, political, and legal adjustments in the existing pattern of water management.

Unless international cooperation is instituted, competition over Nile water is bound to increase among the Blue Nile states as population and development pressures intensify. At the very same time that the demand for water is growing, available water supplies appear to be diminishing. Following a succession of droughts, there is much apprehension that climatic change may reduce water supplies in the future. In the meanwhile, environmental degradation (including denudation and erosion) of the upper Blue Nile catchment is already severe. More and more people in parts of the upper basin, unable to maintain their traditional mode of subsistence, are facing starvation unless they can be provided an alternative livelihood. Irrigation development is one such alternative. But uncoordinated irrigation development in the upper basin may diminish the supply of water to the lower basin.

The *Undugu* (meaning Fraternity in Swahili) group, was formed in 1983 under the aegis of the Organization for African Unity. It includes all the states of the Nile River except Ethiopia and has been supported by D. Boutros Boutros-Ghali, Secretary General of the United Nations. Yet *Undugu*'s effective powers are strictly limited.

Current use patterns in Egypt and Sudan suggest a sufficient water supply for some time to come. Of the 57 billion cubic meters of water released from Aswan annually, on average, about 33 billion are consumed for agriculture, 2 billion for municipal and industrial activities, 2 billion are evaporated downstream of Aswan, and some 20 billion are released for power generation, drainage, and navigation. Egypt has significant quantities of irrigation drainage water that could be reused, and increased efficiency of water utilization could free up even more water for alternative uses, in Egypt and elsewhere. Sudan is utilizing only about three-quarters of its 18.5 BCM/Y allocation under the 1959 Nile Waters Agreement,[17] and it, too, could do much to improve its water use efficiency. Both countries could supplement their supplies with groundwater extraction. Nonetheless, both countries are concerned that developments elsewhere in the basin could result in water scarcity in the future.

The populations of Egypt, Sudan, and Ethiopia have grown steadily since 1960. The famines and civil wars in Ethiopia and Sudan have had only a slight retarding effect on the upward trend. Moreover, this population growth is expected to continue. In 1987 there were approximately 117 million people in the three countries: 50 million in Egypt, 23 million in Sudan, and 44 million in Ethiopia. By the year 2000, unless some unforeseen disaster or an equally unforeseen social or policy change takes place, there will probably be about 67, 33, and 66 million in the same three countries, respectively, for a total of 166 million. By the year 2025, those populations might be as high as 99, 56, and 122 million, for an incredible total of 277 million (well over twice the present number!). Egypt's population today is about 10 percent larger than Ethiopia's, but if present trends continue, by 2025 Ethiopia's population may well exceed Egypt's by 20 percent or more.

Population growth will affect the demand for water in several ways. First, the larger population will need more water for human consumption, for livestock, and for industrial and commercial activities. Second, population growth will increase the demand for food, and hence for irrigation development. Unless these countries can expand food production to meet the population growth, they will need more and more food imports. Already, Egypt (once the granary of the Mediterranean world) is heavily dependent on foreign cereals. In 1974 Egypt imported 3.9 million tons of cereal grains; by 1987 this had risen to 9.3 million tons (an increase of almost 240 percent in 13 years). In 1987 cereal imports totaled about 180 kilograms for every man, woman, and child in Egypt. At this rate of growth in demand, the world's total production of cereal grains may not suffice, nor are the countries concerned likely to be able to pay for the increased imports. The burden of food imports also endangers food security and makes the needy countries extremely vulnerable to drought and famine, as well as to political pressures.

These considerations suggest strongly that water for expansion of irrigation should become increasingly vital for all three Blue Nile riparians. Of the three, Sudan has the greatest reserve of fertile irrigable soils and could well boost agricultural production greatly. Egypt, with its favorable geographical location, technological advancement, and international outlook, probably has the greatest potential for alternative industrial development. That leaves Ethiopia as the most vulnerable of the three, and it will need to press its one great advantage: being the source of the Blue Nile and hence having the most direct and unencumbered access to its waters. The problems for Ethiopia are its difficult, irregular terrain, the degraded condition of much of its land, the internal strife, and the potential threat from its more powerful downstream neighbors.

Egypt is the dominant economic and military power in the Nile basin. The per capita gross national product in Egypt is twice that of Sudan and over five times that of Ethiopia (the latter being one of the poorest nations in the world). In the late 1970s there was much talk in international development agencies about the tremendous development potential of Sudan, but progress there since that time has been extremely disappointing (in fact, the country's economy actually shrank during the decade of the 1980s). Although Egypt's performance has been better, all three economies are burdened with debt and are finding it increasingly difficult to enlist the finances necessary for investments in the water and irrigation sector. In 1988 Egypt's external debt had reached about $50 billion and required some 17 percent of exports to service.[18]

Early efforts in the 1970s and 1980s to reclaim desert lands on the fringes of the Delta and to introduce irrigation there (the so-called New Lands Scheme) produced disappointing results. The yields obtained in most cases were too low to justify the costs of initial reclamation and installations and the continuing costs of maintenance and of pumping water from the Nile and conveying it to higher elevations in the adjacent desert. So Egypt remained with a sizable surplus of water that was drained to the sea.

More recently, the prospects for expanding irrigation in Egypt have appeared more promising. Greater involvement of the private sector (rather than of government-sponsored and controlled programs) has resulted in greater productivity. Moreover, the adoption of technological advances such as plastic coverings and pressurized irrigation systems (including drip and microsprayer techniques), as well as improved crop varieties, have added to the efficiency of farming in the New Lands.[19] Finally, the reduction in the real price of fossil fuels over the last decade and the availability of natural gas in Egypt have made it less expensive to pump water to reclaimed lands. Consequently, there is greater impetus to proceed with desert land reclamation, with the assistance of the United States Agency for International Development. The Egyptian Land Master Plan has proposed that 580,000 hectares be considered for reclamation. However, Egypt's currently available water supplies can permit the reclamation of no more than about 400,000 hectares.

The estimated 6 billion to 8 billion cubic meters required annually to irrigate 400,000 hectares could come from drainage water reuse and not require that Egypt exceed its current allocation under the 1959 Nile Waters Agreement. Others estimate that the annual agricultural water use in Egypt will increase by over 10 billion cubic meters by the year 2000.[20] At the same time, Egypt's municipal and industrial requirements are also

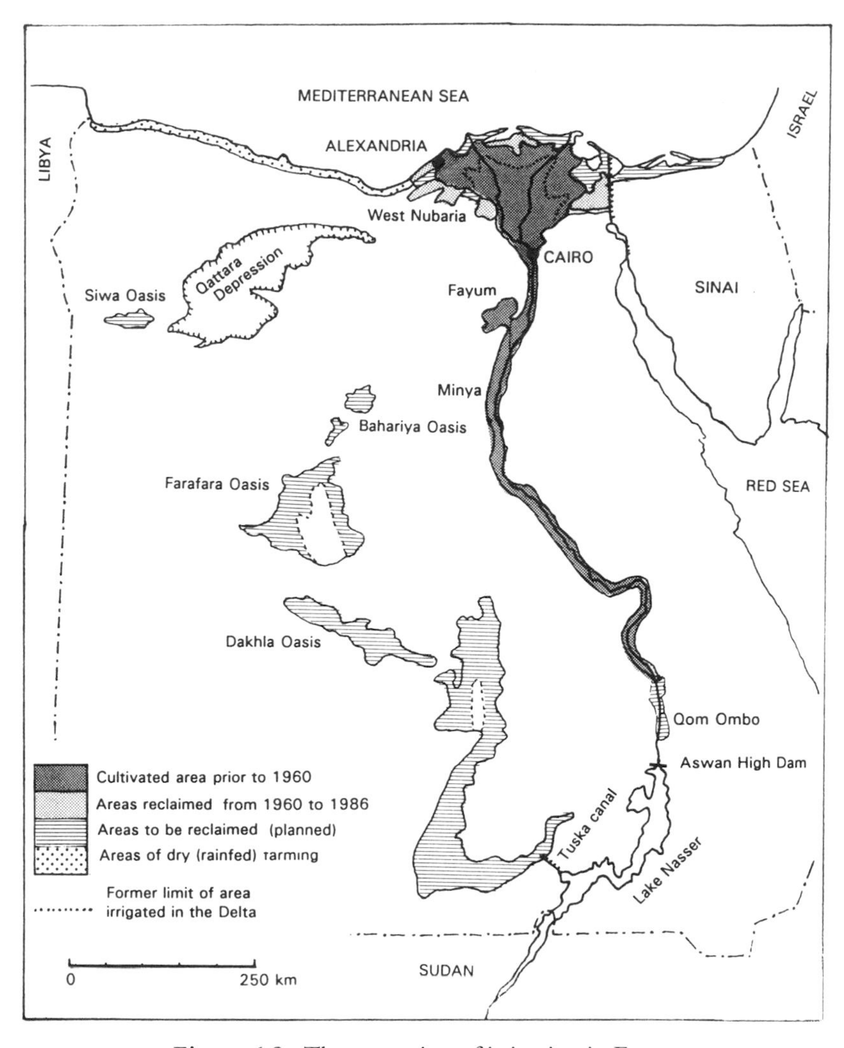

Figure 6.8 The expansion of irrigation in Egypt.

projected to increase from 4.6 billion cubic meters in 1990 to over 6 billion cubic meters by the year 2000. It therefore appears that the desert reclamation effort currently envisaged by Egypt will push the country to the limit of its current water allocation under the 1959 Agreement, leaving little if any for increasing water allocations to any upstream riparians (especially Ethiopia). The Egyptian government is convinced that the continuing rapid growth of the country's population leaves it no choice but to proceed with desert reclamation.[21]

Economists critical of this policy contend that land reclamation is feasible only under the assumption that the cost of the Nile water used in Egypt is zero; that is, that there is no opportunity cost associated with using the water for reclamation projects. From Egypt's sectarian point of view this may be true if the projects rely on drainage water from the end of the delivery system. However, upstream rivals for water can claim that the water used for desert reclamation could be used elsewhere in the system—in Sudan or Ethiopia. A further consideration is that the cost of fuel may well tend to rise in the coming years, as local sources of natural gas and petroleum are depleted. This, in turn, could raise the cost of pumping and hence the cost of irrigating reclaimed desert land.

The growth of population and economic activity in Egypt will not only raise the demand for water but also tend to reduce the quality of water supplies by increasing water pollution (from domestic sewage, industrial effluent, and agro-chemical residues) and by causing groundwater depletion and salinization. Population growth in Ethiopia, on the other hand, has led to extensive deforestation of the highlands, with consequent soil erosion. The ability of catchments to retain rainfall has thus been seriously impaired. Erosion has increased river sediment loads and reduced the useful life of reservoirs downstream. This environmental degradation is likely to continue as more people are forced to cultivate ever more marginal land.

A development of far-reaching potential importance to the future of Nile water supplies would be the construction of a series of dams in Ethiopia.[22] A serious proposal for such an undertaking was made by the United States Bureau of Reclamation in 1964, following its five-year study carried out in collaboration with the Ethiopian Ministry of Public Works. The study considered the development potential of the Ethiopian portion of the Blue Nile for irrigation and hydroelectric power generation. The proposal envisioned 33 irrigation and hydroelectric power projects, and the irrigation of about 434,000 hectares (roughly equivalent to 17 percent of the area under irrigation in Egypt), requiring about 6 billion cubic meters of water. Four major hydroelectric dams were proposed on the Blue Nile downstream from Lake Tana, with an active storage capacity of about 51 billion cubic meters (nearly equal to one year's total flow of the river) and an annual electricity production of over 25 billion kilowatt-hours (three times the actual production of the Aswan High Dam).

The crucial issue is how these dams might affect downstream flows into Sudan and Egypt. Regulation of flow in Ethiopia could effectively eliminate the annual Nile flood. The flow of water reaching Sudan would be seasonally constant and much more uniform than heretofore from year to year. The Sudanese and Egyptians are naturally concerned lest the annual flows reaching them would be reduced substantially as a result of the Ethiopian withdrawal of irrigation water and the evaporation loss from the projected reservoirs. The total reductions have been estimated at about 4 billion cubic meters (about 8.5 percent of the annual mean).

Notwithstanding the understandable anxiety on the part of Egypt and Sudan regarding possible decreases in their water supplies, there appear to be considerable potential advantages to the proposed scheme. To realize these potential advantages, however, the riparian nations would need to cooperate in a basin-wide scheme for the joint management of the river. The present lack of coordination among the three nations imposes great, albeit unacknowledged, costs on them all.

The extreme variability of Blue Nile flows has been the principal water management problem of downstream users for millennia. The proposed upstream reservoirs would regulate the flow to the possible benefit of both upstream and downstream users. Shifting the major storage of water from Lake Nasser to reservoirs located in Ethiopia would reduce the losses due to evaporation, not only because the climate in Ethiopia is less evaporative[23] than the desert climate of Lake Nasser but also because the topography of the Blue Nile gorge in Ethiopia allows for a more favorable volume-to-surface ratio (i.e., a deeper reservoir exposes less surface area for the same volume of storage than does a shallow reservoir). The reduction in evaporation losses would mean that more water could be made available to all riparian countries. By some estimates, the savings thus achievable may equal the total amount of water required by Ethiopia for all the irrigation projects envisaged in the Bureau of Reclamation program. The Egyptians are naturally reluctant to forgo their present system of managing the Aswan High Dam independently, but might be willing to modify the operation of that dam if convinced that a more comprehensive river management system would work to their advantage.

Control of the Blue Nile floods could be particularly beneficial to Sudan. The added storage upstream would facilitate expansion of Sudan's gravity-fed irrigated area. Today, Sudan has practically no perennial storage capacity and is therefore unable to store water for use during droughts. The reservoirs in Ethiopia, if managed jointly, could provide Sudan with this much-needed water security. Sudan's Roseires Reservoir, now serving a seasonal storage function, could then serve primarily to raise the water level to facilitate hydropower generation. The siltation of this reservoir would diminish greatly, as would the maintenance costs for the hydroelectric generators, which are adversely affected by the concentration of abrasive silt carried by the water. The reduced rate of silting would also help in the operation of Sudan's irrigation systems, as it would minimize clogging of distribution channels and ditches and thus necessitate less dredging.

These advantages, however, would be offset in part by the loss of the fertilizing value of the rich silt currently deposited in the irrigated fields of Sudan by the muddy waters. The retention of silt in the projected Ethiopian reservoirs will eventually cause them to clog up, but because of the considerable depth of the sites to be flooded, this clogging may not be a significant factor for many decades to come. Ideally, of course, the water development program should include improvement of soil conservation measures in the Ethiopian highlands (best achieved by revegetation of the deforested and overgrazed slopes) so as to reduce the amount of soil erosion and sediment transport into the river.

From the Egyptian perspective, potential reductions in its annual allotment of water to provide for Ethiopian needs could be compensated in large part by the water savings that would come from more efficient management of the Blue Nile. Improved cooperation in basin-wide management of the river could provide Egypt with much needed data on rainfall and climate patterns in Ethiopia, which would allow better prediction of annual river yields and improved planning for optimal utilization of the water.

Ethiopia has not yet begun to implement large-scale water works and irrigation, but it has announced its intentions to undertake such works on various occasions, notably at the UN international conference on water at Mar del Plata (Argentina) in 1977, and at the UN conference on Less Developed Countries in 1981. On each occasion, Egypt reacted negatively to such declarations (former president Sadat even threatened to bomb any diversion projects if they were built), although the amount of water that Ethiopia could realistically extract from the Blue Nile any time soon is unlikely to exceed 4 BCM/Y. Egypt has resented American assistance to Ethiopia (including the 1958–1963 hydrological study by the U.S. Bureau of Reclamation) in the past, and its press has repeatedly published reports of assistance by Israel (which Israel denies) in planning the expansion of irrigation in Ethiopia by diverting water from the Blue Nile. More recently, however, Ethiopia's relations with both Egypt and Sudan seem to have improved.

Another opportunity for cooperation exists in central Sudan. The Jebel Aulia Dam on the White Nile, about 40 kilometers south of Khartoum, was completed in 1937. It was originally designed to provide Egypt with additional water supplies for summer irrigation, but the benefit of this seasonal storage was substantially offset by evaporation losses. The total capacity of the reservoir is about 5.5 billion cubic meters, and the annual evaporation losses from it are approximately 2.8 billion cubic meters. One reason for this high relative loss is the flat topography of the site, causing the lake formed by the dam to be quite shallow and to expose a large surface-to-volume ratio. Another reason is the desertlike climate, subject to the hot and dry winds of the Sahara.

The Aswan High Dam has made the seasonal storage capacity of the Jebel Aulia Reservoir unnecessary for downstream irrigation. However, subsequent to the construction of the Jebel Aulia Dam, localized pump irrigation schemes have been developed along the White Nile. There are at present some 174 such pump schemes located on both banks of the river, stretching from 100 kilometers south of the Jebel Aulia Reservoir for a distance of 300 kilometers. These pumping schemes depend on the higher water level that results from storage of water in the Jebel Aulia Reservoir. The only purpose this reservoir now serves is to raise the water levels along the White Nile so as to reduce the lift of water necessary for the pumping schemes.

The White Nile pump schemes are currently operating at very low efficiencies. From April to September the level of water in the White Nile is low. The pump intakes are above the river level, and thus the schemes cannot operate. Only some 40,000 hectares out of the more than 180,000 hectares in the original scheme areas are currently cultivated. Moreover, the pumping equipment there is now antiquated and needs to be replaced. Plans have been made to replace those pumps with a series of electric ones capable of drawing water from the White Nile throughout the year, regardless of the river level. This rehabilitation project would thus eliminate the need to maintain the high water level behind the Jebel Aulia Dam with its associated high evaporation losses. The improvement of water supply to be made possible by such a project would greatly increase cropping intensities. The total water use in the White Nile pump schemes would rise correspondingly from 350 million to 1,600 million cubic meters per year. This increased water use could be more than offset by the reduced evaporation losses, as there will no longer be a need to keep the reservoir full.

The total capital cost of the White Nile rehabilitation project was estimated to be $279 million in 1989 dollars,[24] and was projected to yield an economic rate of return of about 20 percent, even without placing a value on the water saved from reduced evaporation losses. If the water thus saved would amount to, say, 1 billion cubic meters per year, and if it were valued at $25 per 1,000 cubic meters to downstream agricultural users, then the value of the net water savings would be approximately equal to $25 million per year. It would thus cover the total capital cost of the White Nile rehabilitation project within about 11 years. The rehabilitation of the White Nile pump schemes currently appears to be an attractive project in the Nile basin for augmenting water supplies, allowing the elimination of the large evaporation losses at very little real cost. Yet the government of Sudan has had difficulty financing and organizing this project, in large part because of the absence of international cooperation based on a basin-wide perspective.[25]

Egypt's concern is still focused on various schemes for augmenting the flow of the White Nile, notwithstanding the environmental consequences, while disregarding the possibilities of developing Blue Nile reservoirs. So far, preserving the status quo on the Blue Nile while ignoring Ethiopia's interests has seemed to serve Egypt's interests. Both Sudan and Ethiopia have been too weak and internally unstable to initiate irrigation schemes that would significantly reduce the quantity of Nile water arriving in Egypt. Moreover, those countries do not yet have the technical expertise or access to the financial resources necessary to undertake independent development projects.

Clearly, however, it will not be possible to continue disregarding the needs and interests of the upper riparians indefinitely. In the long run, Egypt is extremely vulnerable to unilateral water withdrawals, by both Sudan and Ethiopia, that might occur in the absence of a comprehensive allocation and planning framework. The worst situation for Egypt would be for those two countries to strike a separate deal on Nile waters that would exclude Egypt. Therefore, Egypt stands to gain more from increased cooperation that would ensure its interests as well as those of its neighbors on an equitable basis.

Egypt's present position is that it is willing to discuss future water development plans with upstream riparians but that its water allocation of 55.5 billion cubic meters specified in the 1959 Nile Waters Agreement is not negotiable. However, it would seem that Egypt has little to lose and much to gain from entering into discussions that impose no preconditions on any of the parties. Any concessions it may be called upon to make should be reciprocal. Egypt can expect to receive something tangible, such as agreements to undertake joint investment projects, technical cooperation, increased and more efficient water storage and regulation, as well as international financing and technical assistance.

From a river basin planning perspective, Sudan has a strong interest in developing a cooperative relationship with Ethiopia. The construction of dams on the Blue Nile in Ethiopia should benefit Sudan, provided agreements can be reached on operation of the dams during periods of drought that guarantee sufficient quantities of water to Sudan (as well as to Egypt). The completion of any of the Blue Nile dams will take many years, and it is in the interest of Sudan to begin as soon as possible.

Ethiopia suffers most from the absence of regional cooperation on the development of Nile waters. Being a poor country, Ethiopia cannot by itself finance the investments necessary to develop the headwaters of the Blue Nile. To arrest the downward spiral of ecological and economic deterioration, Ethiopia must sooner rather than later begin a process of water development to sustain its ever-increasing population. The financial and technical assistance needed from the international community can only be enlisted through regional cooperation, which must include Sudan and Egypt. From Ethiopia's point of view, waiting much longer to initiate such cooperation allows Egypt's desert reclamation program to proceed and to establish even greater rights of prior use by Egypt.

In the final analysis, reallocation of Nile water need not be a "zero-sum game" for the riparian countries. Cooperative water development can lead to increased usable water supplies for all. A powerful inducement toward international cooperation in the Nile basin (as in other shared river basins in the Middle East and elsewhere) is the refusal of international funding agencies to help plan or finance any schemes that are not agreed upon willingly by all the riparian states involved. Another compelling reason to cooperate is the common threat of a change in climate that might increase the frequency and severity of droughts. That threat can only be offset by coordinated action to augment water supplies while curbing excessive demand by modifying the pattern of water use.

7

The River Jordan

All things be healed and may live whithersoever the river cometh . . . And by the river, upon its bank, on both sides, shall grow every tree for food, whose leaf shall not wither. Ezekiel 47:9, 12.

Some 32 centuries ago, 12 tribes of desperate desert nomads crossed the Jordan River to enter the land then known as Canaan, inspired by what they believed to be a divine promise: "The Lord thy God bringeth thee into a good land, a land of brooks of water, of fountains and depths, springing forth in valleys and hills; a land of wheat and barley, and vines and fig-trees and pomegranates; a land of oil-olives and honey; a land wherein thou shalt eat bread without scarceness, thou shalt not lack anything in it; a land whose stones are iron, and out of whose hills thou mayest dig copper."[1]

The crossing of the Jordan is described dramatically in the Book of Joshua (3:14–17): "When . . . the feet of the priests that bore the Ark of the Covenant were dipped in the brink of the water—for the Jordan overfloweth all its banks during the period of the harvest—the waters which came down from above stood, and rose up in one heap . . . and the people passed over right against Jericho . . . on dry ground in the midst of the Jordan." In recent years, a mundane explanation has been offered to that miracle: as the river swerves in its meandering course, it often scours the soft marl and undermines huge blocks of earth that may collapse spontaneously and temporarily dam the entire river.

An earlier miracle, reported to have occurred at the exodus of the Israelites from Egypt, also has a natural explanation related to the region's ecology. In popular imagination, exemplified and reinforced by Cecil B. De Mille's spectacular depiction in the film *The Ten Commandments,* the parting of the Red Sea looms as a momentous supernatural event. In fact, the Hebrew Bible does not claim that the Israelites crossed the Red Sea at all. The exaggerated myth results from a mistranslation. The "sea" the Israelites are said to have crossed was called in Hebrew *Yam Soof*, meaning literally "the sea of reeds." As such, it must have been nothing more than a shallow freshwater (or brackish-water) marsh, likely formed by the overflow of one or another of the Nile's many ancient distributaries. One possibility is that an easterly arm of the Nile once flowed into the basin known as the Great Bitter Lake (at present filled with seawater by the Suez Canal). Crossing a reedy marsh would have been relatively easy for the Israelites escaping on foot, since they had nothing much to weigh them down. On the other hand, one can readily perceive how the heavily laden Egyptians, riding their narrow-hoofed horses and narrow-wheeled chariots, would bog down in the oozy mud. Thus it was that the Israelites were saved by water on their way to the dry desert.

At the same time that the Israelites entered Canaan from the east, a very different group invaded the country from the west. They were the Sea Peoples, who evidently originated from the islands of the Mediterranean. Being sailors, similar in behavior to the Vikings of northwestern Europe two millennia later, the Sea Peoples of the Mediterranean raided and plundered coastal communities. They even dared to attack powerful Egypt.[2] Following their defeat there, one of their tribes—called the Philistines—apparently became mercenaries in the service of Egypt, and settled along the southern coast of Canaan.

It was not long before the two nations—the one coming from the east and the other from the west, each seeking to expand its domain—met and clashed for supremacy. The Philistines possessed the initial advantage: they had already entered the Iron Age while the Israelites were still in the Bronze Age. With their technical superiority—based on the use of iron tools, weapons, and chariots (the ancient equivalents of tanks)—the Philistines terrorized the tribes that dwelt in the lowlands and foothills. The Israelites therefore retreated to their mountain strongholds, where horse-drawn chariots were ineffective and where skill in archery and cunning use of the steep terrain gave the mountaineers an advantage over the plainsmen. Nevertheless, the Philistines kept the upper hand as long as they were able to maintain their metallurgical advantage.[3]

The predicament of the Israelites ended when the Philistines' monopoly was broken in the days of King David. Being of the tribe of Judah but wishing to symbolize his reign over all of Israel, David captured the centrally located city of Jerusalem (circa 1000 B.C.E.), made it his capital, and from there reigned over a sizable kingdom that included Transjordan.

What made Jerusalem a suitable site for a capital city was the presence there of a perennial source of water, the pulsating karstic spring of Siloam (*Shiloah*, or *Gihon* in Hebrew, meaning "gusher"). So important was the spring to the life of the city that it was the chosen site for the coronation of Solomon, the first Hebrew king crowned in Jerusalem. David's son Solomon built the Temple in Jerusalem in order to concentrate both secular and ecclesiastical authority there. However, the centralized power of the Judeans was resented by the northern tribes, who seceded after Solomon's death to form a separate kingdom based in Samaria.

The Israelites held sway over the country for little over a thousand years. To be sure, their political control over the country that served as a corridor of travel and trade was tenuous and often challenged by outside forces, the great powers of antiquity—the Egyptians, Assyrians, and Babylonians. The Assyrians destroyed the northern kingdom of Samaria in 722 B.C.E. They also attempted to destroy the southern kingdom of Judea.

Judean King Hezekiah (circa 700 B.C.E.), anticipating the siege of the capital by the superior forces of Assyrian King Sennacherib, then undertook one of the most remarkable hydraulic works of antiquity. He decided to divert the spring via a tunnel from outside the walls into the fortified city. In haste to complete the work before the enemy arrived, the workers excavated the 530-meter tunnel from both ends. Where they met, deep inside the bowels of the mountain, they left an inscription expressing joy

Figure 7.1 Site of biblical Jerusalem, with the Temple Mount on the upper left, the structure guarding the spring of Siloam at center left, and the Mount of Olives on the right. (J. M. W. Turner, R. A., from a sketch by the Rev. R. Master.)

at the successful completion of their difficult and dangerous task. Indeed, this diversion saved Jerusalem from succumbing to Sennacherib, who eventually lifted the siege and returned to Assyria. Adventurous visitors can now wade through the tunnel and observe the work of the ancient excavators.

Little over a century later, however, another Mesopotamian invader—the Babylonian King Nebuchadnezzar destroyed Jerusalem, capital of Judea, with its sacred Temple (586 B.C.E.). Many of the people (especially of the urban elite) were exiled to Mesopotamia.[4] Yet enough returned after 60 years (under the authority of the Persians, whose King Cyrus issued a special declaration granting the Jews the right to return to their homeland) to reestablish the state of Judea. That state, small at first and for a time subjugated by the Syrian Greeks, fought for its independence during the second century B.C.E. under the leadership of the Hasmonean priests and kings, who regained control over the entire country (including Galilee in the north and Transjordan in the east).

At the end of the first century B.C.E., however, the Romans established hegemony over the region. A Judean rebellion, begun in 66 C.E., was quelled in the year 70 C.E., and the magnificent Temple in Jerusalem (rebuilt by King Herod, who reigned from 40 to 4 B.C.E.) was destroyed. The Jews recovered sufficiently to rebel once again under their leader Bar Kochba in 131 C.E., only to be finally defeated in 135 by the overwhelming might of Rome. Many Jews were exiled and scattered throughout the Mediterranean world, and the Jewish presence in the southern part of the country, particularly in the urban centers, was drastically reduced.[5]

In the wake of those bloody rebellions, the Romans wished to erase the memory of the Jews in the country theretofore called Judaea (often spelled Judea). To that end, the Romans resurrected the memory of the long-extinct Philistines, who had disappeared from the stage of history fully a millennium earlier, and renamed the country "Palestine." Still, a strong enough Jewish presence remained in the rural areas of the country, including Galilee and the Golan, to construct numerous synagogues and to develop a material and religious culture (expressed in the Mishnah and the "Jerusalemite" Talmud). The Jews persevered there in sizable numbers under the Roman Empire and later under its successor, the Byzantine Empire, and for a time even after the Arab Muslim conquest of the seventh century C.E., but their numbers dwindled especially after the Crusades in the eleventh, twelfth, and thirteenth centuries C.E.

The majority of the people of the Middle East, including the inhabitants of Palestine, were repeatedly assimilated, linguistically and culturally, by whichever power predominated over time. After the Arabic-speaking Muslims supplanted the Greek-speaking Christians of the Byzantine Empire, most of the region's inhabitants were absorbed into the Arabic-Muslim culture. (Some ethnologists have even speculated that the people who today regard themselves as Palestinians may be ethnically related to the ancient Hebrews, perhaps no less than the modern Jews who returned from Europe after two thousand years of exile to reestablish the state of Israel.)

Over the centuries, in any case, the land known variously as Israel or Palestine suffered a profound environmental change. As the forests were cleared for fuel or timber, and as the slopes were cultivated and overgrazed, more and more of the exposed soil was scoured off the hillsides and was laid down in the valleys. That sediment, together with the sand deposited along the shores,[6] clogged the outlets of many streams and caused the formation of swamps, which in turn bred malaria. The result was a degraded landscape of bare rocky hillslopes and of poorly drained valleys. By the middle of the nineteenth century, the population of the country had dwindled to a small fraction of what it had been in the past.

We have numerous eyewitness accounts of the condition of that country in the nineteenth century. One among many is that of the American writer Herman Melville. His visit to the Holy Land in 1856–57 inspired him to compose the epic poem *Clarel.* That poem records Melville's encounter with the desert, which held for him the same fascination, in its immensity and pitilessness, as the sea. In both realms, human scale is lost and the mind grapples for a larger perception of life and nature.

Melville was profoundly disturbed by the condition of the Holy Land: "No country will more quickly dissipate romantic expectations than Palestine. . . . The disappointment is heart sickening. Is the desolation of the land the result of the fatal embrace of the Deity? Hapless are the favorites of Heaven." The unrelenting stoniness of the landscape oppressed him. Human striving in such a land seemed futile: "Yet man here harbors . . . like a lizard in a dry well." Regarding Jerusalem and its environs, he wrote: "Here the city old / Fast locked in torpor, fixed in blight, / No hum sent forth, revealed no light: / The valley slept / Obscure, in monitory dream / Oppressive, roofed with awful skies / Whose stars like silver nail-heads gleam / Which stud some lid over lifeless eyes."

Ten years later, in 1867, another notable American author, Mark Twain, traveled to the Middle East. His description of the Holy Land is more entertaining, in places laced with sardonic humor, but no less grim: "Of all the lands there are for dismal scenery, I think Palestine must be the prince. The hills are barren, they are dull of color. . . . The valleys are unsightly deserts fringed with a feeble vegetation that has an expression of being sorrowful and despondent. . . . It is a hopeless, dreary, heart-broken land. . . . Palestine sits in sackcloth and ashes. Over it broods the spell of a curse that has withered its fields and fettered its energies. . . . Renowned Jerusalem itself, the stateliest name in history, has lost all its ancient grandeur, and is become a pauper village. . . . The noted Sea of Galilee was long ago deserted by the devotees of war and commerce, and its borders are a silent wilderness. . . . Palestine is desolate and unlovely."

Most depressing is Twain's description of the Jordan River, which he found crooked, shallow, and puny[7] (hardly the width of New York's

Broadway); and its valley, where "goats and sheep were gratefully eating gravel" ("I only *suppose* they were eating gravel, since there did not appear to be anything else for them to eat.")

The striking contrast between the descriptions of the land in the Bible and by the American writers demands an explanation. Part of the difference, no doubt, was in the eyes of the respective beholders. To the Israelite nomads coming from the desert, the land might indeed have looked more lush than it did to visitors from the densely forested eastern United States. Moreover, the Hebrew writers of the Bible, the American epic poet, and the humorist all had a penchant for hyperbole. Yet, even after we allow for the obvious prejudices and literary exaggerations of the writers, there remain differences in the descriptions that cannot be dismissed. The environment of the country had indeed deteriorated drastically.

Into that ravaged land came the first Zionist pioneers in the 1880s. They left eastern Europe disillusioned with the chimerical promise of the nineteenth century's "Enlightenment," which failed to grant the Jews true equality. Instead of attempting to assimilate in European society or to immigrate to the New World (as many of their people had done), they chose to affirm their Jewish identity by reconstituting their nation's homeland in the ancient Land of Israel. Once there, however, they found the task extremely hard. The tracts of land they were able to purchase[8] were barren or stony or swampy. They had to contend with the harsh climate and the general insecurity of what was then a practically lawless country. Above all, these sons and daughters of town-dwelling merchants had to adjust to the ways of the land, to relearn the long-forgotten traditions and practices of farming.

Paradoxically, however, that initial disadvantage turned out subsequently to be an advantage. Starting from scratch, these new farmers, unfettered by ancient farming methods, could innovate. What they lacked in long experience they made up in ingenuity and improvisation. Soon they began to develop new crop varieties, new cultivation techniques, and—most important—new methods of water management and irrigation.[9]

The first immigrants were followed after the turn of the century by a second wave. The new pioneers were fervent idealists who sought not merely personal and national salvation, but—beyond that—aspired to build an exemplary society based on social justice and cooperation. One of their most notable achievements was a novel form of agricultural settlement and communal life called a *kibbutz*—a collective farm in which all property was owned jointly, all tasks and income were shared, and all decisions were made democratically. Other immigrants chose to establish cooperative but not collective villages, and still others preferred to operate individually in privately owned farms. And there were many more who chose to live in towns and cities and to engage in industry, commerce, or the professions.

Notwithstanding their humane idealism and practical sense, the early Zionists deluded themselves in one important respect. Their premise was: "A landless nation is returning to a nationless land." In so thinking, they ignored or failed to understand that their enterprise encroached upon an existing community of Arabs, disrupted its culture, and threatened its own aspirations. That community—though not yet a distinct nation—certainly had the potential to constitute itself as one, and would in fact be driven to do so in reaction, and in direct proportion, to the very success of Zionist nationalism.

The dilemma found expression in an exchange that took place in the early 1930s between David Ben-Gurion, the leader of the Zionist Labor Party who was destined to become first premier of Israel, and Musa Alami, a prominent nationalist leader of the Palestinian Arabs. In a private conversation reported only years later by Ben-Gurion, he enumerated to his Arab counterpart all the benefits that the Zionist venture can bring to the country: modern medicine, industry, agriculture (including improved varieties, machinery, fertilizers, and irrigation), roads, electricity, international trade, and so on and so forth. Why, then, asked Ben-Gurion, should you oppose Zionism? Don't you want all these benefits for your people? To which Alami replied: We do want all these benefits and more, but not from outsiders. We want to attain them by ourselves. But that might take you a century longer, argued Ben-Gurion. Yes, concluded Alami, but we are patient people, accustomed to the slow pace of history. We wish to progress at our own pace, however long it takes.

Ben-Gurion told me this story personally one late evening in 1954 at Sdeh-Boker, the Negev settlement which I had helped to establish and to which he had later retired. And he added: I understood Musa Alami then and respected his pride, for if I were a Palestinian Arab I would have felt the same way.

Musa Alami, incidentally, played a constructive role among his people following the trauma they sustained in the 1948–49 war. Among other good works, he established a school in Jericho to train the children of refugees in such useful skills as farming, carpentry, metal work, and electricity so they can be gainfully employed rather than perpetually dependent on the provision of welfare from abroad. That exemplary school is still operating, under the guidance of my good friend Dr. Shihadeh Dajani.

The Zionist venture, which began under the Ottoman Turks, gained great impetus following World War I and the British issuance of the Balfour Declaration. It was further impelled by the rise of the Nazis in Germany. With successive waves of immigrants, the Zionists gradually established an integrated society of farmers and entrepreneurs who eventually formed the nucleus of a state.

Still, on the eve of World War II, the Jewish community of Palestine constituted only one-third of the country's total population. In the meanwhile, the Arabs of Palestine were mobilizing to oppose the Zionist venture, which threatened to turn them into a minority where they had long been a majority. The most militant among them initiated a series of violent attacks, first in 1920, again in 1929, and then during the years 1936–39. When the approach of World War II compelled many Jews to leave Europe and strive to enter Palestine, Arab resentment intensified. As the Arabs saw it, they were being forced to sacrifice their birthright for the sins of European anti-Semitism, for which they were not responsible. On the other hand, the Zionists pointed out that the employment opportunities they had created in Palestine attracted thousands of Arab workers from neighboring countries, but these were not counted officially as immigrants.

When World War II ended, the Jews of Palestine demanded the right to receive and resettle the survivors of the Holocaust who had no other place to go. In this, they were supported by public opinion in the United States and elsewhere. The Arabs of Palestine, supported by the region's Arab states, objected to any further Jewish immigration. Violence began to mount once again. Exhausted by the war and by Palestine's continuing turmoil, the British went before the United Nations in February 1947 and admitted they could no longer keep the mandate. Their Palestine policy, which had vacillated between support for the Jews and support for the Arabs, had finally failed.

A special committee of the U.N. then recommended that Palestine be divided into equal parts, to allow the establishment of an Arab state and a Jewish state, which—it was hoped—would form an economic union. In November 1947 the General Assembly approved the plan. A Solomonic compromise had been achieved, or so it seemed. Unfortunately, it did not prove to be any more viable than King Solomon's original suggestion that the baby be divided between the contending mothers. Though the Zionists accepted the partition plan, the Arabs rejected it and an all-out war began.

In that war—which lasted, on and off, until early 1949—the Israelis managed to prevail and to control even more territory than had been allotted them under the U.N. partition plan. Included in their holdings were the densely populated coastal plain, the northern mountain province called Galilee, and the southern desert province called Negev. The war uprooted more than half a million Arabs from their homes. They became refugees, seeking shelter in parts of the country under Arab control, or in neighboring Arab states. In the wake of the same war, equal numbers of Jews living in the Arab countries were uprooted as well. After their property was confiscated, they were allowed to immigrate to Israel.

Following that war, the Arab-held part of the country did not constitute itself as a separate state. The southern coastal strip of Gaza was controlled by Egypt, while the central highlands district was taken by the Arab Legion of Transjordan. The latter district was then annexed by King Abdallah, who called it the "West Bank" of his renamed country, the Hashemite Kingdom of Jordan. Ever since, the Palestinians have been the majority and dominant element in that kingdom. Although King Abdallah himself was assassinated by a Palestinian nationalist in 1951, his kingdom continued to hold sway in that territory, under his grandson King Hussein, until 1967.

In the war of 1967, Israel wrested control of the West Bank (including eastern Jerusalem) from the Jordanians, of Gaza and Sinai from Egypt, and of the Golan Heights from Syria. Six years later, in 1973, Egypt and Syria attacked Israel in an effort to erase that humiliation. Egypt's initial military success in that war enabled President Sadat to conclude a dignified peace treaty with Israel (mediated by American President Jimmy Carter) in 1979, under the terms of which Egypt regained control over Sinai. A peace settlement between Israel and its other neighbors, however, was delayed for more than a decade, and only began to take shape in the early 1990s.

As Israel's occupation of Gaza and the West Bank continued, the local Palestinian population grew increasingly restive. Their resentment finally exploded in 1987 in a popular rebellion, called the *Intifada* ("uprising"). A crescendo of violence and counterviolence marred the relations between the Israelis and the Palestinians. In 1991, however, in the wake of the Gulf War, a peace initiative by the United States brought the contending sides (along with Syria, Jordan, and Lebanon) to the negotiating table. A slow, painstaking, but nonetheless hopeful, "peace process" thus began at last. In September 1993 that process culminated in the mutual recognition of Israel and the Palestine Liberation Organization, leading to the establishment of Palestinian autonomy.

And now back to the simpler subject of geography. Imagine an elongated sea intruding itself like a wedge into the midst of a large land mass, and call that sea the Mediterranean. Now notice a country abutting the southeastern corner of the sea, with half its land area lying above the southern edge of the sea (hence being in the path of rain-bearing winds coming from the sea), and the other half lying below that edge, within a desert belt. If the topography were flat, we may surmise that rainfall in that country would be maximal near the sea (i.e., in the north and west) and would diminish toward the east and south (away from the sea and into the desert).

Now assume that the land rises as we move inland, from an elevation of zero at sea level to an elevation of, say, 1,000 meters at a distance of 50 kilometers from the sea, after which it levels out to form a plateau. Such a topography would produce an "orographic effect," forcing the clouds coming from the sea to rise over the land and to precipitate greater rainfall on the higher ground. In summary, then: elevation, proximity to the sea, and location on the north-south axis would determine the distribution of rainfall in our hypothetical country. In such a country, all streams conveying runoff would tend to flow from the plateau and the west-facing slopes toward the sea.

Now let us boldly assume the improbable: the plateau had been torn asunder in relatively recent geological history, forming a deep, narrow trench on a north-south line parallel to the coast and some 70 kilometers from it. Suppose, even more improbably, that the trench forms the deepest valley on the continental surface of the earth, a chasm whose lowest point lies 400 meters below sea level. The presence of such a gash in the earth's crust would divide the country in two and would likely change the entire water regime of the dissected country. As the moisture-laden air from the sea would be forced upward by the rising west-facing slopes, its vapor would condense and precipitate as rain. Then, as the dried air would descend into the rift valley, it would become still drier and warmer. As a result, the east-facing slope of the highlands together with the rift valley itself, being in the rain-shadow of the cloud-bearing winds, would likely be a desert. And the streams from both sides of the divided plateau and from the mountains north of it would surely rush into the deep valley, creating a river through it and a saline lake at its landlocked lowest point.

All that remains for us to do is to call that river Jordan, and that lake the Dead Sea. By so doing, we have divided our hypothetical country into two geographical entities: the western part, consisting of modern Israel together with the West Bank; and the eastern part, consisting of the Kingdom of Jordan.

The River Jordan (named *Nahar ha-Yarden,* meaning "The River That Descends" in Hebrew; and *Nahr al-Urdunn* or *ash-Shariah* in Arabic, the latter meaning "The Watering Place") is the lowest river in the world. Sunk in its incomparably deep and spectacular trough, it is a wonder of nature and of human history. One cannot but ponder its mysterious hold on the imagination. How can anyone explain why Judaism, Christianity, and Islam developed along its banks and in adjacent lands? The river's hothouse valley—within which lie the ruins of the most ancient of cities, Jericho—is the site where the earliest evidence of irrigated farming was found—dating back perhaps eight millennia.

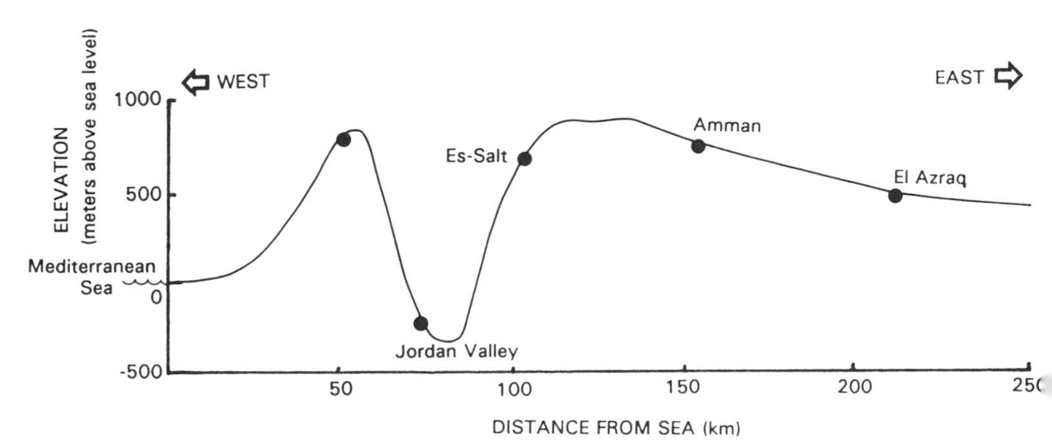

Figure 7.2 A west–east topographic transect from the Mediterranean Sea to the desert, showing the deep trough of the Jordan Valley.

Figure 7.3 The oasis and *tel* of Jericho, one of the most ancient cities and possibly the site of the earliest irrigation.

The Jordan Valley is a long narrow trough averaging about 10 kilometers in width, though it becomes narrower in some places and wider in other places. (The valley's greatest width occurs near Jericho, just north of the Dead Sea, where it is some 25 kilometers wide.) The entire valley is but a section of the gigantic rift system that extends from southern Anatolia southward via Lebanon's Beqaa Valley, the Jordan Valley, the Red Sea, and into East Africa.

The Jordan River begins its course at the foot of towering Mount Hermon, over 2,800 meters high and often snow-clad even in summer. The headwaters of the Jordan are fed by the copious precipitation (some 1,300 millimeters per year) received by that prominence. A fraction of the precipitation seeps into the fissured limestone bedrock and emerges as karstic springs at the foot of the mountain.

The course of the Jordan above the Sea of Galilee is generally designated upper Jordan, and the course below that lake, lower Jordan. Upper Jordan has three principal sources, all rising at the foot of Mount Hermon: (1) the spring of Dan (named for the Hebrew tribe of Dan that settled here) gushes out within Israel proper, near its northern boundary, and delivers a steady 245 million cubic meters per year; (2) the Hasbani (named for the Lebanese village Hasbeya) originates at the western foothills of Mount Hermon in southeast Lebanon and delivers a variable flow ranging between 115 and 140 MCM/Y; (3) the Banias (named for Pan, the Greek

Figure 7.4 Mount Hermon: source of the upper Jordan (J. M. W. Turner, R. A., after a sketch by the Rev. R. Master.)

god of meadows and flocks, whose temple stood here) originates at a site that was within pre-1967 Syria (since held by Israel), and it delivers some 120 MCM/Y.

The three streams converge within Israel, six kilometers south of its northern tip, to form the united Jordan (elevation 78 meters above sea level), which flows into the Huleh Basin. That shallow pan was formerly occupied by a lake and marshes with a dense vegetation of papyrus reeds and water lilies, and was the habitat of a rich fauna. In the early 1950s, the Huleh was drained for the purpose of reclaiming farmland. However, part of the Huleh will now be reflooded in an effort to restore the wetland and its unique habitat.

South of the Huleh, the river plunges swiftly toward the Sea of Galilee, descending 280 meters over a distance of less than 20 kilometers. En route, the river cuts a gorge through a basaltic barrier consisting of congealed lava layers spewed in eons past by ancient volcanic eruptions.

After traversing the Sea of Galilee (known as Lake Kinneret in Israel), the river twists its way in an incredibly sinuous manner through the gray chalky marls of the lower valley. Squirming madly as if trying to escape its fate, the Jordan finally completes the journey from its cool crystal-clear origin to its literally bitter end, where it dies a tired death in the warm, murky brine of the Dead Sea. The straight distance between Lake Kinneret and the Dead Sea is about 100 kilometers, but the actual meandering course of the lower Jordan is at least twice as long.

The southern section of the Jordan Valley is flanked by the Judean plateau (elevation up to 1,000 meters above sea level) in the west, and the Gilead plateau (elevation about 1,200 meters) in the east. The walls of the valley are steep, sheer fault lines, broken only by the gorges of streams descending from the plateaus to the valley, perennial streams on the east side of the valley and primarily intermittent wadis on the west side.

A few kilometers south of its exit from the Sea of Galilee, lower Jordan receives its principal tributary, the Yarmouk, which flows from the east along the boundary between Syria and the Kingdom of Jordan. Farther south, the Jordan is joined by two more tributaries, the Harod on the right bank and the Yabis on the left. Still farther downstream, other tributaries flow in, including Fariah from the west and Zarqa and Nusayrat from the east.

The floor of the valley through which the lower Jordan wends its way is called the *Ghor* ("depression") in Arabic and *Kikar ha-Yarden* ("Flat of the Jordan") in Hebrew. It is blanketed by thick deposits of "Lisan" marl, a friable mixture of clay and chalk. Those deposits had accumulated during earlier geological eras when the valley was covered by a lake, known

Figure 7.5 The Sea of Galilee and the ancient town of Tiberias (J. M. W. Turner, R. A., from a sketch by the Rev. R. Master.)

as ancient Lake Lisan (named for the Dead Sea's peninsula, *Lisan*—meaning "tongue" in Arabic). The horizontal beds of the marl are sculpted by zigzagging rivulets and dry stream beds forming a crazy quilt pattern of soft rocky towers, pinnacles, and badlands on which practically nothing grows.

At the bottom of the valley floor, the river has cut itself a narrow floodplain, several hundred meters in width and some 20 to 60 meters below the level of the adjacent valley floor. That floodplain—overgrown with a jungle of reeds, thistles, tamarisks, willows, and poplars—is called the *Zor* ("thicket") in Arabic and *Geon ha-Yarden* ("Pride of the Jordan") in Hebrew. Within its floodplain, the river zigzags and changes course occasionally, leaving old curving channels stranded.

Along the greater part of the lower Jordan Valley, the amount of rainfall is very low, being no more than 100 millimeters per year on average. Winters are mild here, but the summers are torrid, with temperatures in the lower valley, especially at the Dead Sea, often exceeding 40 degrees Celsius in midday.

Popular lore has magnified the Jordan in the minds of millions out of all proportion to its actual size. Those familiar with it only in legend and song may imagine it to be a mighty river as broad and long as America's "Ol' Man River." In actual fact, the Jordan would scarcely qualify as a rivulet to a tributary of the Mississippi. Along its lower course, the channel of the Jordan is only about 30 meters broad, and only 1 to 3 meters deep. Before its waters were diverted elsewhere for irrigation, its natural annual discharge was only about 1.2 billion cubic meters, of which approximately 660 million cubic meters came from the upper Jordan, 475 million cubic meters from the Yarmouk, and a variable amount from the other downstream tributaries. The Jordan's total natural flow is thus but a tiny fraction (less than 1.5 percent) of the normal discharge of the Nile.

In the context of the local environment, however, the importance of the Jordan cannot be exaggerated. Its role in history, too, has been great beyond rational measure. For that matter, all of Palestine (or Israel) is no larger than New Hampshire or Vermont. But it bears no comparison to any other country or province of comparable size. It combines every environment from the temperate to the tropical, from high hills to the lowest valley in the world, from the snows of Mount Hermon, which spawn the Jordan river, to the steamy Dead Sea in which the river ends. It is a land of contrasts and extremes, all compressed within an incredibly small space saturated with drama and history.

Although the amount of water at issue is small, the rivalry over the Jordan is even more intense than that over the region's much larger rivers. Four states share the catchment: Syria, Lebanon, Israel, and the Kingdom of Jordan. Another entity that is part of the same catchment and a contender for its waters is the West Bank with its Palestinian population. All of these riparians, except Lebanon, suffer aridity. About 85 percent of the area of

Figure 7.6 The lower Jordan River, meandering on its course toward the Dead Sea.

the Jordanian kingdom is desert, as is 60 percent of Israel and nearly 70 percent of Syria. (The latter, however, has access to a much larger river—the Euphrates—and even to a section of the Tigris.)

The relationships among the Jordan River's riparian states are exceedingly complex hydrologically as well as politically. Of the river's total catchment, about 54 percent lies in the Hashemite Kingdom, nearly 30 percent in Syria, about 14 percent in Israel, and less than 2 percent in Lebanon. Of the river's total flow, about 27 percent is generated in the territory of the Jordanian kingdom, about 32 percent in Israel, 10 percent in Lebanon, and about 31 percent (including the Banias[10]) in Syria. However, possession of the Banias (with its 120 MCM/Y) is now in dispute between Israel

Figure 7.7 The Jordan River discharging into the Dead Sea, with the oasis of Jericho in the left foreground and the mountains of Moab in the background. (J. M. W. Turner, R. A., from a sketch by the Rev. R. Master.)

and Syria. That dispute is one of the issues that need to be resolved in the peace negotiations between Syria and Israel.

The prolonged impasse among the riparians of the Jordan River basin is a consequence of their decades-long rivalry and their failure to cooperate in the rational development and equitable sharing of the basin's meager water resources.

During the British Mandate, numerous plans were formulated for developing the Jordan basin's water resources. The basin's development potential was intimately related to the politically sensitive issue of the country's presumed "absorptive capacity," used by the British as a criterion or rationale for limiting Jewish immigration.[11] Several of the plans envisaged constructing canals to convey water from the Jordan and Yarmouk along one or both sides of the lower Jordan Valley from the Sea of Galilee southward for the purpose of irrigating large tracts of land there. A secondary purpose was to exploit the river's energy for generating electricity.

Irrigation of lands just south of the Sea of Galilee was initiated by several kibbutzim in the 1920s and 1930s, using both Jordan and Yarmouk waters. In 1926, the Electric Company, founded under the auspices of the Zionist Organization, diverted the Yarmouk and the Jordan into an artificial lake and built a hydroelectric power station near the confluence of those rivers. That project was rendered defunct by the 1948–49 war. However, those early works provided Israel with the rationale to claim historical rights to a portion of the waters flowing in both the Yarmouk and the Jordan.

In the year 1939, Dr. Walter Clay Lowdermilk, an eminent agricultural engineer and hydrologist then serving as deputy chief of the U.S. Soil Conservation Service, visited Palestine in the course of his survey of soil erosion and conservation around the world. Impressed by the reclamation efforts carried out in Palestine, Dr. Lowdermilk conceived a bold plan for water, land, and energy development. He described that vision in a book, published in 1944 under the title *Palestine: Land of Promise.* It called for the establishment of a Jordan Valley Authority, comparable in scope to the Tennessee Valley Authority in the United States.

Lowdermilk realized that the country had two primary needs: water and energy. While on an aerial reconnaissance of the country, he peered into the deep rift valley, which drops 400 meters below the sea and lies only a short distance from the Mediterranean, and realized that this extraordinary difference in altitude offered an opportunity for a great electric power project.[12] His plan called for the diversion of the Yarmouk River and other tributaries into canals running along the edge of the Jordan Valley for the purpose of irrigating the floor of that valley. Simultaneously, the waters of the upper Jordan would be taken out of the valley to irrigate the Jezreel Valley as well as other lands in the coastal plain and the northern Negev.

Lowdermilk's energy development program called for the conveyance of seawater from the Mediterranean into the Jordan Valley for the double purpose of generating electric power and of compensating the Dead Sea for the loss of the diverted sweet waters. The seawater would be led part way in an open canal and part way in a tunnel from the vicinity of Haifa to the edge of the rift valley, where a hydroelectric power plant would be built. The seawater thus supplied to the Dead Sea would maintain that lake at a constant level. To prevent saline damage to the lands along the course of the seawater aqueduct, concrete-lined canals or closed conduits (pipes) would be needed.

As Lowdermilk perceived the project, it would encompass more than water and power development. Its ultimate objective would be the development of the land, water, and mineral resources of both Palestine and Transjordan (today's Kingdom of Jordan). Water conservation, flood control, irrigation, drainage, soil reclamation and conservation, rangeland improvement, reforestation, and extraction of minerals from the Dead Sea would all be included in the comprehensive development program. He also envisaged the reclamation of lands in the Negev desert, which he recognized as a potentially fertile province. Finally, Lowdermilk stressed that his plan could benefit all the people of the country and greatly increase its absorptive capacity for the immigration of the Jewish victims of Nazism who would need to be resettled after the end of the Second World War.

The program outlined by Lowdermilk was elaborated in detail by James Hays in a report issued in 1948 entitled "T.V.A. on the Jordan." His conception inspired the gradual evolution of a comprehensive plan of action by Israeli engineers to transfer water out of the Jordan basin via a National Water Carrier (NWC) to the coastal plain and the northern

Negev. (The northern part of Israel provides 80 percent of Israel's water resources, whereas most [some 65 percent] of the agricultural land lies in the south.) At the same time, the Jordanians were planning their own water development project in the lower Jordan Valley. Before either plan could be implemented, however, some tacit understanding was necessary regarding each country's appropriate share of the Jordan basin's waters.

For some years following the cease-fire, truce, and armistice agreements ending the war of 1948–49, no movement toward permanent peace between Israel and its Arab neighbors took place, and that despite numerous attempts at political arbitration and conciliation by well-meaning intermediaries. In only one area did any of the dozens of conciliatory efforts come close to bearing fruit. In a region where water means life, the potential of the Jordan River system had long been recognized but not yet actualized. Although coordination of water development and utilization among the riparian states would have been to mutual benefit, direct contacts to achieve such coordination seemed impossible in view of the persistent hostility and distrust.

In 1953, U.S. Secretary of State John Foster Dulles visited the region and reported his impressions in a State Department Bulletin dated 1 June 1953: "The Near East possesses great strategic importance as the bridge between Europe, Asia, and Africa. . . . This area contains important resources vital to our welfare. . . . About 60 percent of the proven oil reserves of the world are in the Near East. Most important of all, the Near East is the source of three great religions—the Jewish, Christian, and Muslim—which have for centuries exerted an immense influence throughout the world. Surely we cannot ignore the fate of the peoples who first received and then passed on to us the great spiritual truths from which our own society derives its inner strength. . . . Throughout the area the cry is for water for irrigation. U.N. contributions and other funds are available to help the refugees, and . . . they can well be spent in large part on a coordinated use of the rivers which run through the Arab countries and Israel."

As he toured the entire region, Dulles observed that the site of the original Garden of Eden can be restored to its former productivity with the revenues from the oil production. Then he continued: "Today the Arab peoples are afraid that the United States will back the new State of Israel in aggressive expansion. . . . On the other hand, the Israelis fear that ultimately the Arabs may try to push them into the sea. . . . There is need for peace in the Near East. . . . The area is enfeebled by fear and by wasteful measures which are inspired by fear and hate. . . . To achieve [peace] will require concessions on the part of both sides. But the gains to both will far outweigh the concessions required to win those gains. . . . These peoples we visited are proud peoples who have a great tradition and, I believe, a great future. . . . It profits nothing merely to be critical of others."

Consequently, Dulles became interested in the idea of a Jordan Valley

Authority. The United Nations Relief and Works Agency had been investigating the possibility but making little progress. At Dulles's recommendation, President Eisenhower appointed Eric Johnston to be special ambassador to the region to negotiate the issue of Jordan basin water development.

Despite the tense state of Arab-Israeli relations in 1953 and 1954, Eric Johnston was able to carry out substantive negotiations over the water issue. Restricted largely to technical personnel, the meetings foresaw a non-political agreement that would produce the maximum amount of water, to be equitably divided among the four nations. Johnston's initial proposal was based on a plan prepared by Charles Main of the T.V.A. Alternative plans were offered by the Arabs and the Israelis. Through patience, persistence, and astuteness, Johnston brought the negotiations to an apparently satisfactory conclusion. His final plan, then called the Unified Plan, was a compromise among the differing proposals, and it suggested a formula for the equitable division of the waters.

The Johnston formula allocated the following quantities: (1) *To Lebanon:* 35 MCM/Y from the Hasbani. (2) *To Syria:* 20, 22, and 90 MCM/Y from the Banias, upper Jordan, and Yarmouk headwaters, respectively. (3) *To the Hashemite Kingdom:* 100, 377, and 243 MCM/Y from the lower Jordan, Yarmouk, and tributary wadis, respectively. (4) *To Israel:* 375 and 25 MCM/Y from the upper Jordan and the downstream Yarmouk, respectively. An important—albeit ambiguous—proviso of the Johnston compromise allowed Israel to avail itself of any excess seasonal flows (over and above the amounts specified) that might become available after the other coriparians had withdrawn their shares. That principle has since been interpreted variously by the contending parties. An equally important lacuna in the Johnston formula was its failure to include groundwater in the water budget.

In 1954 the plan was made public, but because it would benefit Israel as well as the Arab states, the Arab politicians then opposed it. Nonetheless, Johnston had come closer than anyone had before him to creating common ground between the Israelis and the Arabs. His success, albeit incomplete, was due to the technical nature of the deliberations, which were conducted away from the glare and distraction of publicity, and to the obvious potential advantages for all the parties involved. The Johnston plan succeeded de facto in setting the principles of water allocation that were followed, however grudgingly and approximately, by both Jordan and Israel most of the time since. Interestingly, the Arab states later contended that Israel was violating the same agreement that their political leaders had once rejected.

Subsequent developments—including wars, growing populations, shifts in water use patterns, and changed political alignments such as the separation between the Hashemite Kingdom and the West Bank—have long since rendered the Johnston plan somewhat anachronistic. (The Johnston Plan, as well as other plans put forth in the 1950s and 1960s did not regard the Palestinians as a distinct political entity.) For its time, however, that plan was an important advance and a hopeful harbinger of future negotiations that might lead to more comprehensive, detailed, and binding agreements.

The Johnston formula, notwithstanding its shortcomings and the lack of formal ratification, enabled the Israelis and Jordanians—with American assistance—to proceed with their separate development programs, based in principle on the amounts of water allocated to each.

The first step in the Israeli plan was the diversion of water from the upper Jordan and its conveyance through the hilly district of lower Galilee to the coastal plain and the northern Negev.[13] The original plan located the point of diversion above the Huleh, but that site was found to be too sensitive vis-à-vis Syria, so the diversion was relocated to a site below the Huleh. However, that site also proved to be unsuitable for political-military reasons: although entirely inside Israel, it was within an area declared to be a demilitarized zone between Syria and Israel in the wake of the war of 1948–49. The Syrians objected to any Israeli works in that zone. Several violent clashes, and mediation by the United States, persuaded Israel to abandon its planned diversion from the river directly and, instead, to draw its water from the Sea of Galilee.

That change had far-reaching implications for Israel. First, it involved the loss of energy. The earlier diversion sites (about 300 meters above the level of the Sea of Galilee) would have allowed the diverted water to flow by gravity to the planned reservoir of Beit Netofa. It would also have allowed the production of electricity en route by dropping a portion of the water into the Sea of Galilee via a hydropower generator. Instead, under the modified scheme, pumping water out of the lake (at 200 meters below sea level) and lifting it to a height of 270 meters above sea level is an expensive task that annually consumes about 15 percent of Israel's total electric power and 40 percent of the system's operating costs.

An equally serious problem is the increase in water salinity. Whereas the water of the upper Jordan is nearly pure (about 70 ppm), the water of the Sea of Galilee contains a considerable amount of salt (250–400 ppm). The concentration is not so high as to preclude the use of the water for most domestic, industrial, and agricultural purposes, but its prolonged use may eventually contribute to the accumulation of salts in soils and groundwaters.

The change in Israel's diversion site did not assuage the Arab states' fundamental objections to the withdrawal of Jordan water for use outside the river's natural basin. To thwart that program, they decided to implement a counterdiversion of two upper Jordan tributaries then subject to their control, namely, the Hasbani and the Banias. First, in 1959, they considered diverting both streams to the Litani in Lebanon, but such a diversion would have deprived the Kingdom of Jordan as well as Israel. So they changed the direction of their planned diversion from westward to southward. Their modified plan was to lead the Hasbani to the Banias in Syria via a tunnel, and from there to convey the combined waters in a canal over the Golan plateau to Wadi Raqqad, a tributary to the Yarmouk. The augmented waters of the Yarmouk were to be impounded in a dam at Muheiba and used by the Jordanians in the lower Jordan Valley. A summit meeting of the Arab states in 1964 authorized the Syrians to proceed with that plan.

The attempt by the Syrians to implement that diversion plan precipitated a series of armed confrontations with Israel, which contributed indirectly to the war of 1967. The Syrian diversion installations were bombed by Israel in April 1967, only two months before the outbreak of the Six-Day War. The Soviet Union, then Syria's protector, alleged that Israel was preparing a large-scale attack against their client state over the water issue. Israel denied any such intention, but Egypt was persuaded by the Soviets to thwart Israel's presumed action. Egyptian President Nasser moved his army into Sinai toward the Israeli border, ordered the U.N. peacekeeping units out of the Tiran Straits, and blockaded Israeli shipping through the Gulf of Aqaba toward the Indian Ocean. Israel (whose shipping was also prevented by the Egyptians from passing through the Suez Canal) had earlier announced that the closing of Tiran would be *casus belli,* and the so-called Six Day War followed.

In consequence of that war, Israel took control of the Golan plateau and of the Banias headwaters, thereby removing the threat of the Syrian diversion project. Israel's control of the Golan also gave it access to the intended site of the Jordanian dam on the Yarmouk, and hence effective veto power over implementation of that project, which the Israelis feared might deny them their share of the Yarmouk's waters.

Later, incursions and rocket attacks by Palestinian guerrillas into Israel from their bases in southern Lebanon provided Israel with the rationale to seize control of a strip of Lebanese land (the so-called security zone), thereby also obviating the threat of any potential diversion of the Hasbani to the Litani. The Lebanese, on their part, have long suspected that Israel might harbor ambitions toward the waters of its Litani River itself. In the 1970s, however, they diverted the bulk of that river's upstream waters into the Awali River (which discharges into the Mediterranean some 30 kilometers north of the Litani's outlet), mainly for the purpose of generating electricity. Thus, practically none of the intricate steps and countersteps in the regional conflicts of the last few decades can be understood completely without considering the water issue.

A further problem for Israel, largely unforeseen, resulted from its decision to drain the marshes and lake of the Huleh in the 1950s. The exposed deposits of peat, dried and aerated as a result of that drainage, began to subside by shrinkage and to decompose. The decomposition, in turn, released nutrients such as nitrates and phosphates into the water draining toward the Sea of Galilee. Those nutrients can affect the quality of the lake's water by posing the threat of eutrophication.[14]

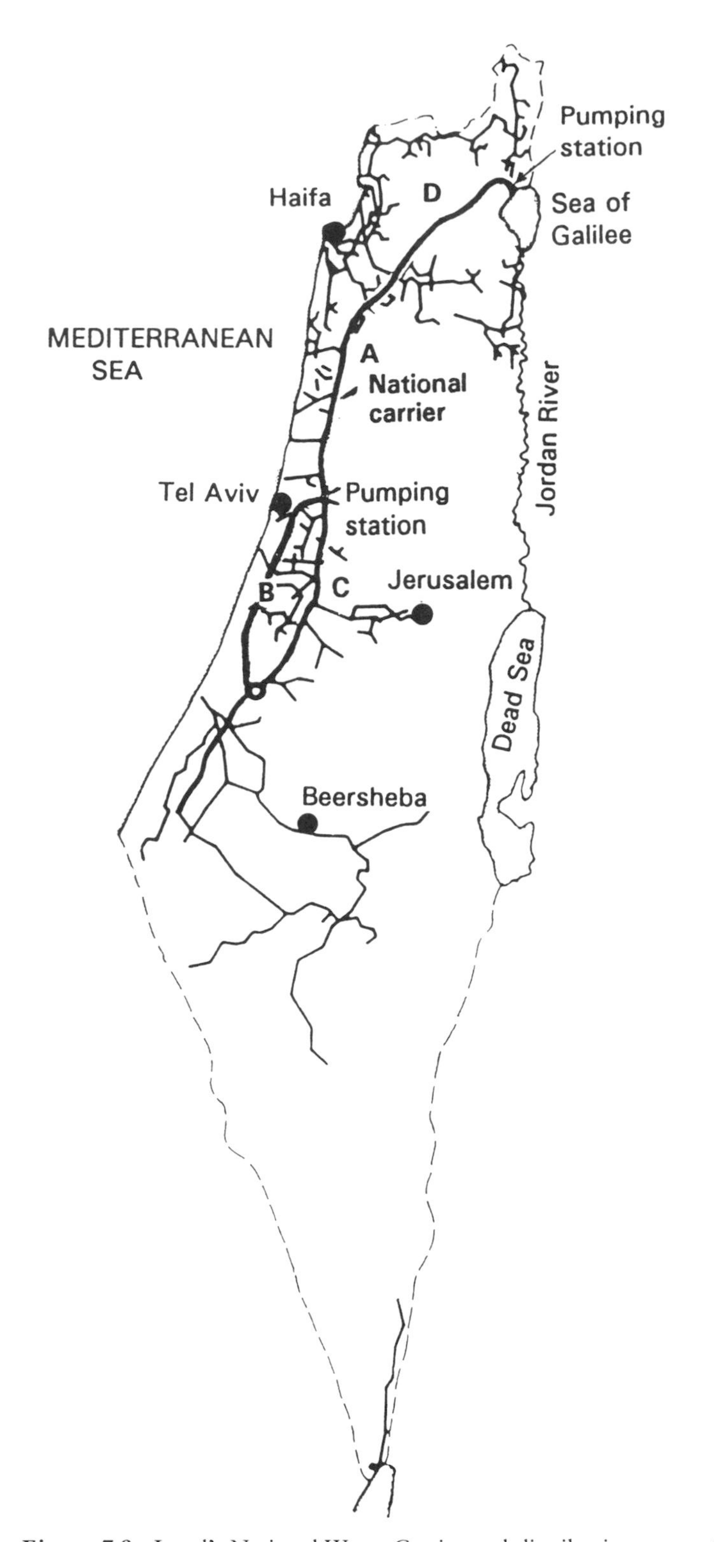

Figure 7.8 Israel's National Water Carrier and distribution network.

Notwithstanding all these problems, Israel's National Water Carrier has functioned continuously and effectively for some 30 years. It has helped Israel to overcome geographical imbalances in water availability. All of the country's fresh water supplies are now joined in an integrated grid, whose arteries deliver quality-regulated water through a system of canals, tunnels, pipes, reservoirs, and pumping stations to cities and villages.

The point of origin of the National Water Carrier is the Sapir pumping station at Tabgha on the northwestern shore of the Sea of Galilee, which lifts up to 20 cubic meters per second from the lake (mean water level 211 meters below sea level) to an elevation of 152 meters above sea level. From this point the water flows by gravity across the Jezreel Valley and along the coastal plain to the south, over a distance of 110 kilometers, via a 23-km-long concrete-lined canal and an 87-km-long 2.7-meter diameter pipeline (Schwartz et al., 1993).

From the terminal point of the National Water Carrier near the Yarkon springs (east of Tel Aviv), a system of two pipelines (the eastern and western Yarkon-Negev pipelines) with diameters of 1.7 and 1.8 meters, respectively, convey the water southward over a distance of 95 kilometers to the arid Negev region.

Several regional networks branch off from the NWC. Depending on the varying demands and the state of groundwater levels in local aquifers, the regional schemes either receive water from the NWC for supply to consumers (mainly in the high demand summer months) or contribute water from local aquifers to the NWC (as needed seasonally). Part of the water from the NWC is used in winter for artificial recharge of the aquifers.

A third pipeline has been constructed recently from the Dan region (the greater Tel Aviv metropolitan district) Sewage Reclamation Project (known by its Hebrew acronym, "Shafdan"). This pipeline runs parallel to the western Yarkon conduit and conveys reclaimed effluents that have undergone tertiary treatment. These effluents are then recharged to the coastal aquifer for pumping by batteries of wells. The water thus recycled is allocated for irrigation only, and its supply to numerous villages and settlements in the south has necessitated the construction of a pipeline network separate from the network supplying domestic water.

The NWC enabled Israel to increase the extent of irrigated farmland from about 30,000 hectares in 1948 to over 200,000 hectares in the late 1980s. (At the same time, more efficient irrigation techniques reduced per-crop water application by nearly 50 percent, even while increasing yields.)

The National Water Carrier was initially intended to extract 320 million cubic meters per year from the Jordan basin. The amount actually drawn increased gradually over the years—totaling 380 million cubic meters in 1970 and reaching 420 MCM/Y toward the end of the 1980s. The increased withdrawals, coupled with the reduced inflows due to the droughts of the late 1980s, caused the water level of the Sea of Galilee to fall by the summer of 1991 to the lowest level considered permissible[15] (below which the water quality in the lake would be likely to deteriorate). Pumping from the Sea of Galilee was then curtailed, and—since the country's aquifers were also in danger of depletion—the entire water supply system seemed inadequate to meet the growing demands of an expanding population and economy. Agriculture, the principal water user, was especially hard hit. Fortuitously, the unusually wet winter of 1991–92, followed by the fairly wet winter of 1992–93, brought relief (only to be followed by a drier-than average 1993–94 winter). It was, to be sure, only a temporary relief; the essential problem remains. The ever-rising demand for water by the countries in the Jordan basin exceeds the limited supply.

Israel obtains nearly one-quarter of its water supplies from the Sea of Galilee. The lake is pear-shaped, about 20 kilometers long and 8 kilometers wide on average, and it covers an area of some 166 square kilometers. In addition to the 540 million cubic meters that it receives in an "average" year from the upper Jordan, the lake collects some 70 to 135 MCM/Y from tributaries that flow into it directly from east and west, as well as some 50 MCM/Y from underground springs and 70 MCM/Y from direct rainfall, for a mean annual total intake of about 750 to 850 MCM/Y. The accretions other than the upper Jordan's are approximately balanced, however, by the annual evaporative loss, amounting to about 270 MCM/Y. Because of the variable nature of the rainfall regime, such average figures are not truly dependable.

The lake serves as a natural reservoir, regulating the flow of the lower Jordan. In recent decades, the lake has also served Israel as an artificially regulated reservoir: a dam built at its southern end allows water managers to control the outflow, thus affecting the level of the water in the lake and the amount that can be pumped out of it. The lake's total storage capacity is nearly 4 billion cubic meters, of which only a fraction is considered to be the "operational storage capacity." Israel's water managers have set the highest allowable water level at 208.9 meters below sea level, and the lowest level at 213 meters. A range of only 4.1 meters between the maximum and minimum lake levels thus limits the usable storage to 680 million cubic meters (less than twice the amount extracted annually). On average, each meter of water level increment is thus equivalent to a volume of about 166 million cubic meters. With such a limited operational storage capacity, the Sea of Galilee alone cannot provide Israel with the long-term storage needed to tide the country over a succession of drought years (such as occurred in the late 1980s).

Theoretically, Israel could increase the storage capacity by raising the maximum allowable water level. However, such action would inundate farmland, roads, and other installations, and—not the least—ancient sites of archaeological and religious significance. Building an embankment around the lake to protect such sites would itself be disruptive, and altogether prohibitive. The alternative is to lower the minimum allowable

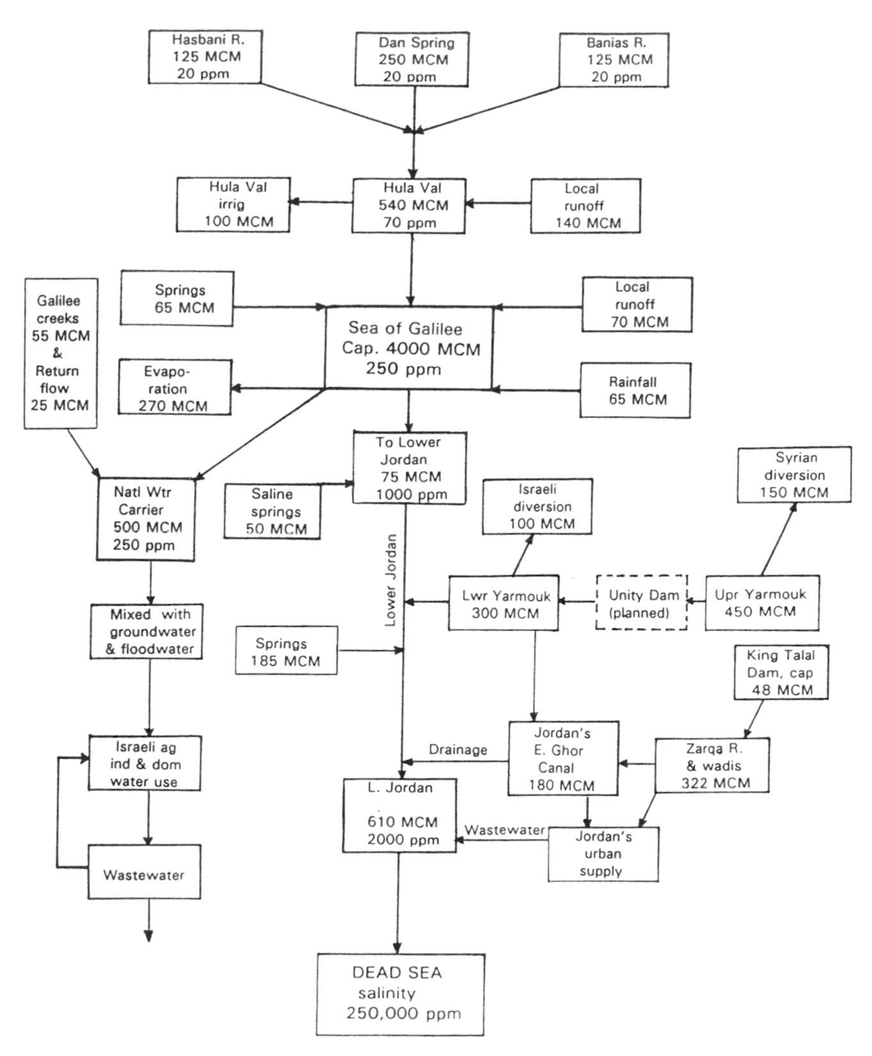

Figure 7.9 Schematic diagram of the Jordan basin's water flows. (The figures given are rough estimates. MCM = million cubic meters; ppm = parts per million soluble salts.)

water level; that is to say, to pump more water out of the lake each summer prior to the onset of the rainy season. Lowering the minimum level by 2 meters would, presumably, increase the operational capacity to about 1 billion cubic meters. The drawback here is that such action would not only reduce the size of the lake each year, but also induce greater salinization, as the saline springs that well up into the lake tend to increase their flow when the counter hydrostatic pressure of the lake is reduced. Moreover, there is danger that depleting the lake in summer might leave the country without sufficient water reserves, unprepared for the possibility that the following winter might be a drought.

In the past, before irrigation was developed in the upper Jordan Valley and before the Sea of Galilee was managed as a reservoir, its natural discharge into the lower Jordan totaled about 650 MCM/Y on average. Historically, that flow was augmented by the lower tributaries (chiefly the Yarmouk), so that the natural inflow of the lower Jordan to the Dead Sea amounted to about 1.2 billion cubic meters per annum. Since the Israelis and Jordanians began diverting the headwaters and tributaries that fed the river, lower Jordan's contribution to the Dead Sea has dwindled to some 200–300 MCM/Y on average. Consequently, with continued evaporation, the area of the Dead Sea has shrunk by more than 30 percent.[16]

In recent decades, Israel has released water from the Sea of Galilee to the lower Jordan only episodically, mostly in unusually wet years. During the rainy winter of 1991–92, for example, there was enough inflow to the lake not only to restore its level but also to necessitate the release of 236 million cubic meters. The winter of 1992–93 was not as rainy as its predecessor, but—because the lake level was higher at the season's outset—the release of water to the lower Jordan was even greater.

As mentioned, the inflow of the Jordan to the Sea of Galilee is of very low salinity. Inside the lake, there is a significant infusion of salts, so the salinity of the water increases more than threefold. Some of the saline springs well up directly under the lake, and some flow into it from the outside. To minimize the lake's salinization, Israeli engineers have developed a system to capture the waters of the latter springs. The brine from those springs (a volume of about 24 MCM/Y) is now diverted from the lake via a special aqueduct and discharged into the lower Jordan. Israeli settlements downstream of the Sea of Galilee dispose of their effluents in the same way. Jordanian villages in the lower Jordan Valley also drain their villages and fields into the old river channel. Sadly, therefore, the lower course of the Sacred River has of late become, in effect, a drainage ditch.

In the late 1970s, the United States government once again tried to promote regional water development as a stepping-stone to regional peace. This time, the issue was the impounding of the Yarmouk River waters behind a dam at Maqarin, to enable the Kingdom of Jordan to utilize the winter flows that had been "going to waste" by running to the Dead Sea. The Carter administration announced that it would fund a large portion of the construction costs if the Kingdom of Jordan could reach an understanding on water allocations with Syria, the upstream riparian, and with Israel, the downstream riparian on the Yarmouk.

Over a three-year period, Philip Habib, then assistant U.S. secretary of state, shuttled among the three riparians to try to forge an agreement. Syria wanted to use the headwaters within its territory above the site of the dam, and Israel wanted to be guaranteed a certain amount of downstream water for its own use, as well as an allocation of water for the West Bank (formerly part of the Kingdom of Jordan but under Israeli administration since the 1967 war). However, the rivals did not display the necessary measure of flexibility to adopt a mutually advantageous compromise, and by 1979 the talks broke down. The Syrians then proceeded to dam the upper tributaries to the Yarmouk for their own purposes, and the Israelis continued to draw water from the lower course of the river for theirs. Consequently, the Kingdom of Jordan has had to curtail its agriculture drastically, as well as its urban water use. And that is substantially the situation today. Unless the rivals to the Yarmouk can renew their negotiations in a spirit of compromise, the situation can only grow worse.

The Yarmouk River, which is the Jordan's principal tributary, can be regarded hydrologically as an independent river. Its catchment (6,800 square kilometers) is only 40 percent as large as the Jordan's (about 17,000 square kilometers), but its mean annual flow (about 475 MCM) exceeds 70 percent of the upper Jordan's. However, since the Yarmouk is fed mainly by seasonal runoff rather than by steady perennial springs, its flow fluctuates widely from year to year, and over the last few decades has varied from less than 300 to more than 800 MCM/Y. The flow also varies between seasons, from less than 150 MCM during summer and autumn to more than 325 MCM during winter and spring. To be used efficiently, therefore, the river's flows must be regulated; that is to say, they must be stored during the flood season of winter for use during the peak demand period of summer. In the absence of such storage and regulation, much of the precious water escapes to the Dead Sea. Unfortunately, the absence of a binding international agreement among the three Yarmouk riparians—Syria, Jordan, and Israel—has thus far precluded the construction of the necessary facilities for harnessing that river's waters.

The Kingdom of Jordan's water supply situation is becoming dire: notwithstanding its rapidly growing population, now nearing 4 million, its total potential fresh water supplies average only about 850 million cubic meters per year. Israel, whose population is not much larger but whose potential fresh-water supplies are nearly twice as great,[17] enjoys the advantage of being the upstream riparian on the Jordan River. The Jordanian kingdom is in a similarly tight situation vis-à-vis Syria on the Yarmouk. Though it is here an upstream riparian relative to Israel, the kingdom is in an inferior position relative to Syria, which controls about 80 percent of the river's watershed.[18]

The Kingdom of Jordan occupies an area of about 90,000 square kilometers. It lies in the transition zone between the semiarid and arid zones, and has a Mediterranean-type climate with a short, rainy winter and a long, dry summer. The annual rainfall varies from a practically negligible 50 millimeters in the desert to a relatively generous 600 in the northwestern highlands, but the distribution of the rain is extremely uneven. More than 90 percent of the area receives less than a meager 200 mm. Only 3 percent of the country enjoys a dependable annual rainfall greater than 300 millimeters, considered the minimum necessary for rainfed cropping.

The total volume of water supplied by rainfall is 7,200 MCM/Y, 85 percent of which is evaporated either directly from the ground surface or from vegetation. (The potential evaporation is very high: more than twice the annual precipitation even in the relatively humid sections and more than 10 times as great as the precipitation in the desert sections.) The remaining 15 percent of the precipitation either runs into surface streams or percolates into groundwater, not all of which is recoverable. Surface water sources account for two-thirds of Jordan's water supply.

Jordan's population growth rate, 3.8 percent per year, is one of the highest in the world. Added to that internal growth was the sudden influx of about 350,000 Palestinians displaced from Kuwait by the Gulf War, which swelled Jordan's population by about 10 percent. About 40 percent of the people reside in the capital city, Amman. The country's population is now projected to reach 5 million by the year 2000.

Hydrologically, the country can be divided into three major water resource sections: (1) the Jordan Valley—which includes the Yarmouk River, local wadis, and groundwater wells in the same area—currently provides some 300 MCM/Y; (2) the wadis draining into the Dead Sea, as well as associated groundwater sources (in Amman, Zarka, Dhuliel, Qastal, and north of Zarka River) provide about 380 MCM/Y; (3) other basins, oases, and aquifers such as those in Wadi Arabah, Jafir Basin, Disi, Azraq, and the Northeast Region, together amount to little over 100 million cubic meters. The total sustainable water supply has variously been estimated to be between 700 and 900 MCM/Y. In actuality, the amount available per year depends on seasonal rainfall, which is highly variable, and on the amount Jordan can withdraw from the Yarmouk, which depends on regional hydropolitics. Damming the winter flows there could augment supplies by 100 to 200 MCM/Y. In addition to its sustainable resources, the country has non-renewable groundwater reserves, mainly located in the southern desert near the boundary of Saudi Arabia.

The Hashemite Kingdom's major water development project was the construction of the East Ghor Canal. This project was originally intended to be part of a comprehensive program that included the Maqarin and Muheiba dams on the Yarmouk, and a series of dams to capture the floods of the other eastern tributaries to the Jordan. As originally planned, there were to be two canals running the length of the lower Jordan Valley, one on the eastern side and one on the western side. Unresolved political tensions between Jordan and Syria delayed the construction of the Yarmouk dams, so the kingdom undertook to build the canals first. Implementation of the project began in 1957–58. The first stage was completed in 1961, and the second stage in 1966, both being on the eastern side of the valley. The war of 1967 prevented implementation of the third stage, which would have transferred a portion of the Yarmouk water across the Jordan River to a "West Ghor Canal" parallel to the eastern one. However, the kingdom did proceed with damming the side wadis within its territory, thus providing a storage capacity of some 125 million cubic meters.

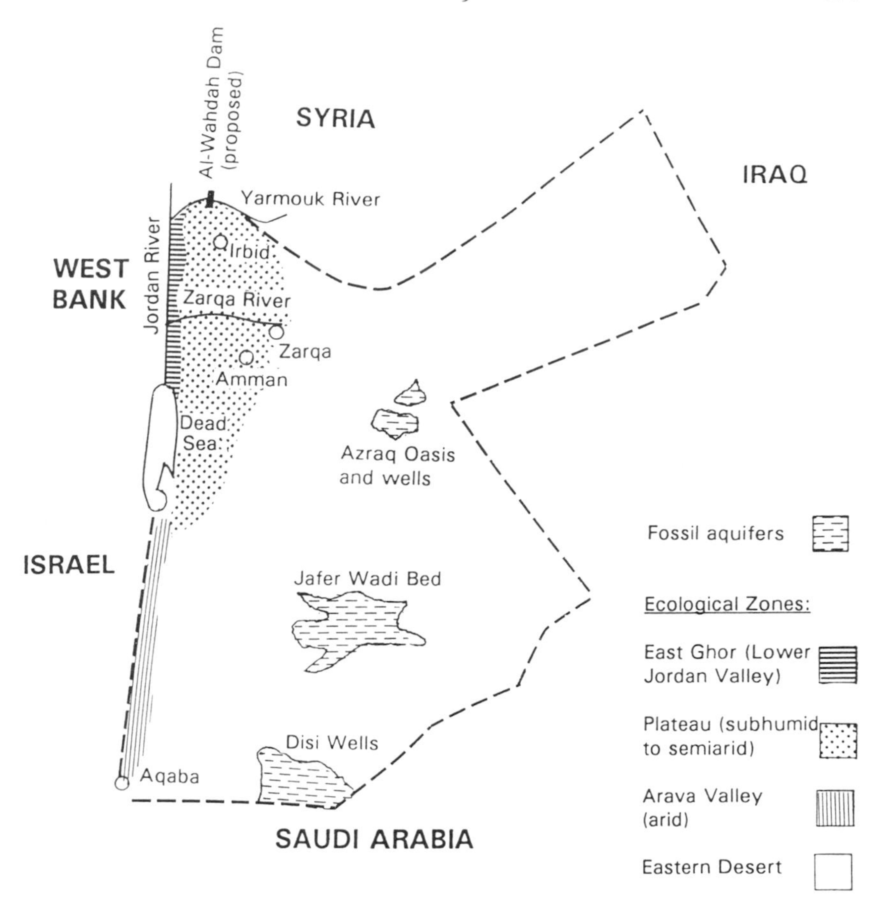

Figure 7.10 Ecological zones and desert aquifers in the Kingdom of Jordan.

Water is diverted from the Yarmouk into the East Ghor Canal by means of a low barrage, designed to direct part of the river flow through a tunnel built near the village of Addasiye. The diverted water then runs south in the East Ghor Canal, from which it is distributed to fields and villages via secondary canals. Drainage waters from the irrigated areas are discharged into the Jordan River. The East Ghor Canal was designed to convey 200 million cubic meters per year, but in practice it has supplied only about 130–150 MCM in most years. Gradually, the canal was extended so that, as of 1991, it runs practically the entire length of the Jordan Valley from the Yarmouk to the Dead Sea.

Prior to the construction of the East Ghor Canal, the lower Jordan Valley was a desolate and sparsely populated province. The project transformed the valley. It included a comprehensive program of land reform, settlement, and infrastructure and community development—complete with educational and medical facilities. The principal crops grown are vegetables and fruits, which are marketed within the kingdom and exported to the Persian Gulf States.

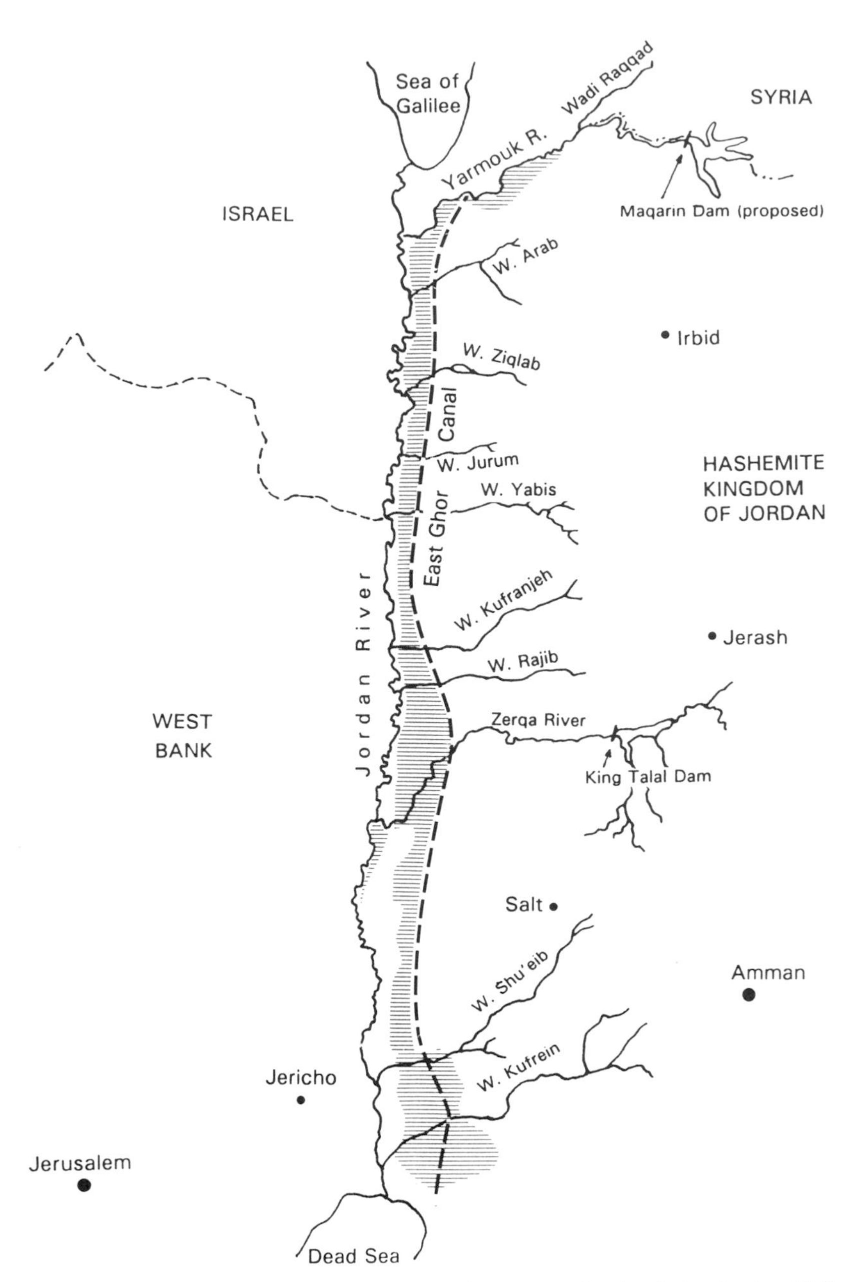

Figure 7.11 Irrigation development on the eastern bank of the Jordan Valley (East Ghor Canal).

The total area of land irrigated in the valley (from the waters of the Yarmouk and of other sources) was reported at one time to be some 30,000 hectares. However, the irrigated area has fluctuated over the years, depending on water availability. The extent of irrigation was drastically reduced in the late 1980s as a consequence of the drought and of the increase in water demand by the urban sector. The Ghor-Amman diversion transfers some 45 MCM/Y to the capital city, whose voracious thirst for additional water is likely to further reduce the amount of water available for irrigation in the valley.

The planned dam on the Yarmouk, near Maqarin, is to be a joint Jordanian-Syrian project, hence it has been given the hopeful name *Wahdah* (meaning "Unity"). The construction of this 100-meter-high dam, with a projected storage capacity of 250 MCM, requires regulation and diversion structures that will divide the Yarmouk flow among Jordan, Syria, and Israel. To date, Israel has opposed the scheme out of concern that it might affect its share of the water supply, and its opposition has blocked the requested financing by the World Bank. Jordan counters the Israeli objection with the contention that the share of the Yarmouk allocated to Israel under the Johnston plan represents less than 3 percent of Israel's water supply, and that the true reason for the Israelis' objection is to protect their own excessive withdrawals of water from the Yarmouk at the expense of the kingdom's rightful share.

The Johnston plan allocated 25 MCM/Y of the Yarmouk's summer flow to Israel, but placed no explicit limits on Israel's use of winter floodwaters. Hence Israel has tended to use as much as it could of the floodwaters left unutilized by Syria and Jordan. The total amount of water used by Israel has varied over the years between 70 and 100 MCM/Y, including 25 million cubic meters of summer water and some 45 to 75 million cubic meters of winter water.

Israel's condition for allowing the construction of the Maqarin dam (which would effectively eliminate downstream winter floods) is to recognize its historical right to a larger share of the river's waters, the exact amount to be determined by negotiation. In addition, Israel demands that the kingdom allocate about 140–150 MCM/Y for the use of the West Bank—the same amount originally intended to be delivered to the West Ghor Canal, which was never built. That, however, is a moot issue, since the kingdom does not have any extra water to deliver.

Despite the prolonged tension between Jordan and Israel over the projected Maqarin dam, there has been a measure of de facto, albeit quiet, cooperation in recent years. For instance, the Israelis have helped the Jordanians to dredge the Yarmouk at the water intake to the East Ghor Canal, where sediments (silt, gravel, and boulders) tend to accumulate during the flood season each year.

Annual water demand in the kingdom has fluctuated in recent years, but the overall trend is inexorably upward. In 1988 the total water demand was 733 million cubic meters. In 1990 it increased to 767 million cubic meters. In 1991, however, the shortage of water was so acute (following a succession of droughts) that the government was forced to impose strict water rationing in Amman and in other cities, as well as in farming districts. Water supplies to residents of the cities were restricted to no more than 48 hours per week. Even so, the total water use could not be reduced below 730 million cubic meters.

Current projections suggest that the annual water requirements are likely to reach 1,000 MCM in 1995, 1,120 MCM in the year 2000, and about 1,230 MCM by year 2005. The maximal potential annual supply of water from all sources (including unconventional sources such as recycled water and the judicious withdrawal of water from fossil aquifers) is estimated at 1,100 million cubic meters. Hence the requirements are projected to exceed supplies well before 2000. These projections are based on the assumption that the annual rate of increase in municipal water demand will be between 7 percent and 10 percent during the coming few years, and that there will be no increase in the water demand for irrigation. The growing water shortage is likely not only to stop new development of irrigated agriculture but also to reduce the extent of current agriculture. This necessity will undoubtedly have an adverse effect on the kingdom's economy, of which agriculture has long been the mainstay.

During the last several years, agricultural water use[19] constituted some 73 percent of the total demand (with 70 percent of the irrigation carried out in the Jordan Valley), whereas domestic and industrial uses amounted to 22 percent and 5 percent of the total, respectively. The irrigation sector has already suffered the most drastic reduction. In 1990 farmers were given less than one-third of their normal water allotments. This reduction in quantity, along with the deterioration in quality of the water supplied, resulted in damages to crops estimated to cost some $100 million in the spring of 1991. Following that season of extreme adversity, there was great anxiety in the country over the prospects for future supplies. Fortunately, there followed the unusually wet season of 1991–92, which helped to restore water supplies. However, that season may well have been an anomaly, bringing temporary respite but not affecting the long-term problem nor diminishing the likelihood of crisis in the future.

Jordan has in fact been overexploiting some of its aquifers. The overdraft ("mining") of non-renewable groundwater was estimated to total 190 million cubic meters in 1990. Most of the groundwater reserves of Jordan are in effect non-renewable, and occur in deep aquifers (150 meters or more below the ground surface) that are already expensive to tap, and will become more so as the water-table falls ever deeper. Moreover, the water in several of the deep aquifers tends to be brackish. As a result of these drawbacks, the costs of pumping, treating, and transferring water from these basins to where it is needed are very high.[20] Even under the best circumstances, these sources cannot supply much more than about 125 MCM/Y.

Jordan's water distribution system is still grossly inefficient, and there is much to be done to prevent leakages and other losses and to ensure that all water is properly metered and that the charges for water are realistic.

The annual per capita supply of fresh water in Jordan is now well below 300 cubic meters. The corresponding supply in Israel is about 360 CM/Y. These amounts are low in comparison with potential supplies in other countries. In Egypt, Lebanon, Iraq, and Syria, the per capita supply exceeds 1,000 CM/Y. (In Canada, incidentally, the potential per capita supplies exceed 10,000 CM/Y.)

The Kingdom of Jordan has already made great investments in the development of water resources, totaling over $3 billion prior to 1990. The projects have included construction of reservoirs and dams, water supply and treatment systems, water distribution networks, and wastewater disposal and recovery facilities.[21] One major project is the King Talal Dam on the Zarka River (named after the father of the present King Hussein), completed in 1977. Its original storage capacity of 56 million cubic meters was increased to 90 million cubic meters by raising the dam 10 meters in 1987.

A major difficulty in supplying water for domestic use in Jordan is the distance from most sources to the principal population centers. This problem increases not only the cost of supplying water but also the hazard of system failure and disruption of supply due to leaky pipes and pump breakdowns. Another problem is the deterioration in water quality. For instance, the reservoir of King Talal Dam, which receives treated effluents from Amman, has shown worrisome signs of water pollution.

From the foregoing it is clear that the Kingdom of Jordan will not be able to meet its future water demands unless it reaches an agreement with its neighbors and institutes drastic changes in the mode of water supply and utilization. In fact, the crisis has already begun. The shortage of water for normal municipal use had reached 50 million cubic meters by the end of 1991. That shortfall will at least triple by the year 2005.

To alleviate its water shortage, the Kingdom of Jordan must undertake several types of action. First, it must increase the supplies that can be made available within the country from conventional and non-conventional sources. Second, the country should improve the efficiency with which water is used in all sectors and allocate water preferentially to the most efficient sectors.[22] Third, it needs to capture, treat, and reuse wastewaters (especially in the agricultural sector, which, in the future, may depend primarily on recycled wastewater). Fourth, Jordan should enter into regional schemes of water sharing and interstate water transfers. While all these approaches will be necessary, the last one will ultimately be inescapable. However, the sine qua non for entering into such regional arrangements is that the Kingdom of Jordan at long last achieve a modus vivendi with Israel, as well as with Syria, allowing these neighbors and coriparians to cooperate peaceably to mutual advantage.

An ancient parable illustrates the contrast between the Jordan basin's two lakes—the Sea of Galilee and the Dead Sea—and the lesson to be learned from them.

An old woman resided in a village located alongside the Jordan River. As she neared the end of her days, she overheard her covetous children quarreling over their expected inheritance, each claiming the lion's share and refusing to concede the rights of the others. The woman then called them to her bedside and said:

> Notice, my children, that this valley has two lakes. The Sea of Galilee brims with fresh water and teams with fish and fowl; its shores are luxuriant with a profusion of plants whose roots sip the lake's healing waters and whose branches bear fragrant flowers and sweet fruits. That lake is nestled among the mountains like a jewel, beautiful to behold. It gladdens the hearts of humans and beasts, it is beloved of all life. The angels sing its praise.
>
> Quite different is the other lake. It is filled with a heavy brine, caustic and poisonous. No fish can frolic in its murky waters, no bird will hover over its fetid air. No plants grace its barren shores. It lies like an evil monster, hostile and lifeless. It is, indeed, the Dead Sea.
>
> What is the reason for that difference? It is because the one lake accepts and releases water in equal measure, it gives as much as it takes. It is generous, hence it is alive and joyful. The other lake is greedy and miserly; it only takes water but gives none. It exists only for itself. Therefore it is lonely and bitter, sterile and dismal.
>
> Be like the generous lake, my children, and not like the greedy one. Live together in joy, not apart in sorrow.

8

The Flowing Streams of Lebanon

A fountain of gardens,
A well of living waters, and
Flowing streams from Lebanon.
The Song of Songs 4:16

Lebanon is a land of rugged coasts and lofty mountains endowed with scenic beauty and a dramatic history. The record of land and water management there is as instructive as it is in Israel, Lebanon's southern neighbor on the eastern Mediterranean littoral. Lebanon was the home of the fabled Phoenicians, who established themselves along the Lebanese seashore during the second millennium B.C.E. and built the port cities of Tyre, Sidon, and Berytus (Beirut). During the period from 1000 to 500 B.C.E., they were the Mediterranean world's foremost navigators and traders. As disseminators of culture, as well as of commercial goods, they introduced the Semitic alphabet to the Greeks.

Lebanon receives enough rain to produce good yields of grain, grapes, olives, and many other crops. But the amount of tillable land in the narrow coastal plain is limited by the proximity of the mountain range to the sea. The highlands are blanketed by snow every winter—an unusual sight throughout most of the Middle East. Hence these highlands were named Mount Lebanon by the ancient Semites, whose word *laban* or *lavan* meant "white" (as it still does in Aramaic and Hebrew). The slopes of the mountains were originally covered by dense forests, the most striking among them being the majestic and renowned cedars of Lebanon.

The high quality timber obtained from the hardwood cedars was used in construction (especially of temples and palaces), and it was essential for shipbuilding. Lebanon's timber was a prized commodity of international trade. So coveted was cedar wood in antiquity that Akkadian King Sargon (circa 2300 B.C.E., even long before the time of the Phoenicians) dispatched special military expeditions northwest from his lowland country to obtain it. So did the Egyptian Pharaohs. When the enterprising Phoenicians took over, they began systematic logging of the cedars, which they sold to treeless Mesopotamia and Egypt.

The biblical Book of 1 Kings reports that Israel's King Solomon, with the agreement of Tyre's King Hiram, sent 30,000 laborers to Lebanon to bring timber for building the Great Temple and the royal palace in Jerusalem. The story is related in Chapter 5: "And Solomon sent to Hiram, saying: 'Behold, I propose to build a house for the Lord my God. . . . Now, therefore, command that thy servants hew me cedar trees out of Lebanon, and my servants shall be with them, and I shall give thee hire for thy servants according to all that thou shalt say, for thou knowest that there is not among us any that hath skill to hew timber like the Sidonians.' . . . And Hiram replied to Solomon: 'I will do all thy desire concerning timber of cedar, and timber of cypress. My servants shall bring them down from Lebanon unto the sea, and I will make them into rafts to go by sea unto the place that thou shalt appoint, and will cause them to be broken up there, and thou shalt receive them.' . . . And Solomon gave Hiram twenty thousand measures of wheat for food for his household, and twenty measures of beaten oil . . . year by year."

Chapter 9 also reports that "King Solomon gave Hiram twenty cities in the land of Galilee . . . but they pleased him not." Nevertheless, when "King Solomon made a navy of ships . . . near Eilat on the shore of the Red Sea . . . Hiram sent his servants in the fleet, shipmen that had knowledge of the sea, with the servants of Solomon. And they came to Ophir, and fetched from thence gold." Here's proof enough that give-and-take bargaining and the quest of mutual advantage has always been the way of the Middle East.

With the prosperity of the Phoenicians came an increase in the population of ancient Lebanon. Following the deforestation, cultivated fields began to creep up the hillslopes, and erosion ensued. By the ninth century B.C.E., the inhabitants found that their degraded farmland and their dwindling lumber supplies were inadequate to support their country's growing population. For a while they were sustained by their extensive trade and their industries, including their invention of the processes for making glass from sand and for extracting the distinctive reddish-purple dye from the murex sea snail. Because the color of the murex dye reminded the Greeks of the mythical red bird, the phoenix, they named the purveyors of that dye "Phoenicians."

Eventually, the Phoenicians could no longer subsist on their own despoiled land alone. They then set out to establish distant colonies, including Carthage on the North African coast and a series of other colonies in Malta, Sardinia, Sicily, and Iberia. (The name of the seaport island Malta means "haven" or "refuge" and the names Carthage and Cartagena—in today's Tunisia and Spain—meant "city" and "new city" in the

Semitic language of the Phoenicians, as do the words *karta* in Aramaic and *keret* in Hebrew.) These faraway colonies supplied food for the home country and accepted its exports.

As the environment of the Phoenician homeland gradually deteriorated, so did the strength of the nation. Finally, it was defeated by the Greeks, whose king, Alexander of Macedon, destroyed the city of Tyre in 332 B.C.E. To overcome the defenses of the city, which was on an island, Alexander ordered his army to fill in the strait separating Tyre from the mainland, thus turning it into a peninsula. Carthage, established on new lands in North Africa, continued on its own for two centuries more. But after its own hinterland had begun to deteriorate, too, Carthage succumbed to the Romans in 146 B.C.E. (Carthage's most famous military leader in its wars with Rome was Hannibal, who was named after the Canaanite god of rain, Baal.)

Modern Lebanon is a narrow strip of land, some 220 kilometers long from north to south and only 35 to 55 kilometers wide from east to west. With a total area only slightly greater than 10,000 square kilometers, Lebanon is one of the world's smallest sovereign states. Its population, estimated to be little over 3 million, is an unblended mixture of diverse religious and ethnic groups.

The country consists of four distinct physiographic landforms:

1. A narrow and discontinuous coastal plain, formed of riverine and marine sediments (clays and sands), fertile enough to permit the produc-

Figure 8.1 The denuded and eroded hillslopes of Lebanon. Remnants of the forest that once covered these hills are evident in the foreground; efforts at reforestation, in the background. Rain-fed farming is practiced in the intermontane valleys.

tion of a variety of crops but not extensive enough for large-scale irrigation. Lebanon's major cities—Beirut, Tripoli (locally called Tarabulus), Sidon (Saida), and Tyre (Sur)—are built in the coastal plain.

2. A continuous range of limestone and sandstone mountains known collectively as Mount Lebanon, which rises steeply from the coast and reaches elevations of 2,000 to 3,000 meters above sea level. The range is incised by streams that form narrow and deep gorges as they carry storm runoff and snowmelt seaward. The range is about 160 kilometers long and between 10 and 55 kilometers wide. Toward the south, the range is of lower elevation, as it blends into the hills of Galilee (which are in Israel). The lower slopes of Lebanon's mountains are still cultivated and grazed, with remnants of the ancient terraces in evidence.

3. The Beqaa (meaning "cleavage" or "depression") is a valley between the Lebanon range on the west and its eastern counterpart, the Anti-Lebanon. This valley, about 175 kilometers long and 10 to 30 kilometers wide, is a segment of the great Syrian-African Rift. In the south, the Beqaa is separated from the much deeper Jordan Valley by a hump of rugged low hills, contiguous with the foothills of the commanding Mount Hermon. The Beqaa, with its fertile alluvial soils formed of sediment from the mountains on either side, is Lebanon's main agricultural district.

4. The Anti-Lebanon mountain range, which runs parallel to, and resembles, the Lebanon range. Geologically the two ranges constituted a continuous plateau until they were cleaved apart by the great rift. The watershed divide of the Anti-Lebanon range forms the political boundary between the Lebanese and Syrian states. Both ranges are largely denuded of their once-luxuriant forests. Only a few isolated relic groves of the original cedars of Lebanon have survived the repeated cutting, burning, and grazing carried out over many centuries. Most of the slopes are now covered by shrubs and low trees (including oaks, pines, cypresses, pistacias, and carobs).

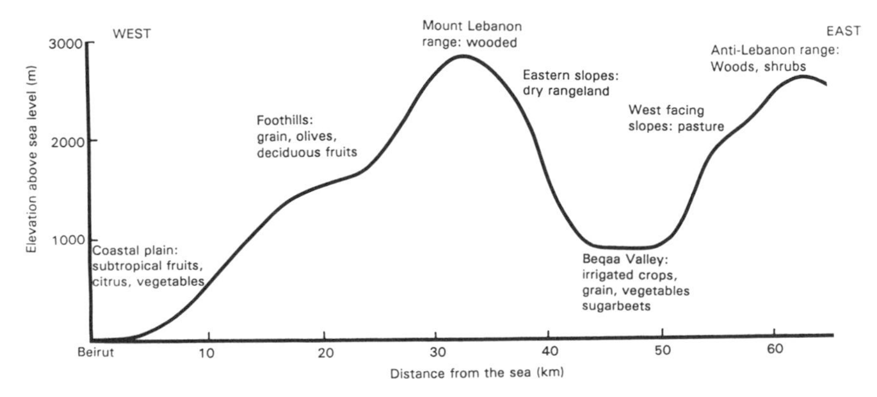

Figure 8.2 Topographic transect of Mount Lebanon, Beqaa Valley, and the Anti-Lebanon range.

The climate of Lebanon is typically Mediterranean, with a cool, humid winter season between late November and early April and a warm, dry summer. The coastal plain is characterized by moderate temperatures and an annual rainfall averaging between 750 and 1,000 millimeters. The mountains are cooler, and receive between 1,200 and 1,500 millimeters of precipitation (much of it as snow that melts during spring and early summer, though the peaks may retain a perennial snow cover). The Beqaa Valley is the warmest and driest area, its annual rainfall varying from less than 400 to more than 600 millimeters.

The streams draining the mountain ranges are mostly seasonal torrents, conveying floodwaters following winter rainstorms and snowmelt in the spring. Only the streams that are fed by perpetual springs (generally derived from karstic limestone) flow year-round. The two principal perennial rivers are the Litani and the Orontes.

The Litani River (160 kilometers long) begins its course in the Beqaa Valley near Baalbek—a site famed for its Roman ruins—and runs southward the length of that valley as if intending to spill into the northern Jordan Valley. Just six kilometers from Israel's northernmost tip, however, it appears to change its mind as it makes an abrupt 90-degree turn westward through a gorge to empty into the Mediterranean near Tyre. (At the bend of the river, on a high prominence, stand the remains of a magnificent Crusader castle named Beaufort, "the beautiful fort," called Qal'at Shakif in Arabic.) The entire length of the Litani thus lies within Lebanese territory.

The Orontes River also rises in the Beqaa, but unlike the Litani, it flows northward and crosses the border into Syrian territory. Prominent among the seasonal streams that drain the Lebanon range are the Barid ("the cold one"), the Abu Ali, and the Awali.

Although arable land is not extensive in Lebanon, except in the Beqaa Valley, the abundance of water supplies provides possibilities for irrigation development. To date, however, irrigated agriculture has lagged far behind its full potential. In the early 1970s, the government of Lebanon intended to increase the area under irrigation from less than 50,000 hectares to over 70,000—mostly in the Beqaa, as well as along the lower course of the Litani and in selected coastal areas—but the civil unrest delayed the full implementation of those plans.

Modern Lebanon was created as an entity by the French when they took over the territory of what is today Syria and Lebanon following the Allied victory over the Turks in the First World War. Their purpose was to provide the Christian community with an independent enclave in a predominantly Muslim region. However, because the areas of Christian concentration in Lebanon were disjointed, the task of carving out a state with a clear Christian majority was difficult. Moreover, in their desire to make the new state economically viable, the French included within it more territory, and therefore more Muslims, than would allow the Christians to maintain a stable majority.

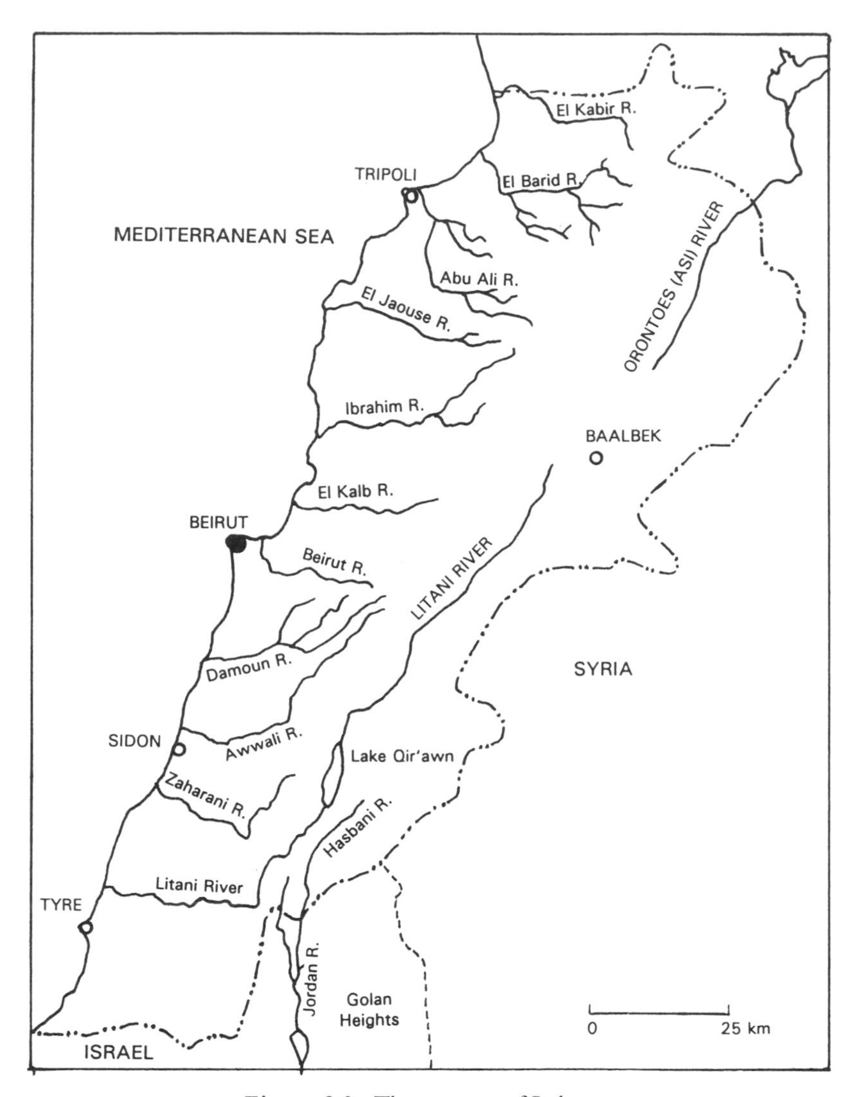

Figure 8.3 The streams of Lebanon.

The Lebanese Republic that emerged when the country gained its independence from France toward the end of the Second World War (in 1943) was based on a National Pact that installed the Maronite Christians in a dominant position. Despite its small size and heterogeneous population, Lebanon did manage for a few decades to become a lively and prosperous commercial center for the Middle East. The Lebanese economy was based primarily on free international trade, financial and other services, and tourism. To promote free trade, the government invested in building a modern infrastructure of seaports, airports, warehouses, and communication networks, while neglecting rural development. The guiding policy was based on minimal government interference with business, no income or profit taxes, bank secrecy laws, and a free foreign exchange market. That policy favored the entrepreneurial class, consisting mainly of urban Westernized Christians. Other sects and regions were mostly left out of the framework and the prosperity it engendered. The very success of that policy in enriching the trading and bureaucratic sectors of the economy while neglecting the village-based rural sector bred resentment, which further exacerbated the ancient religious hostilities of the disparate communities.

When Lebanon first attained independence, the Christians were in the majority. In the decades that followed, however, that tenuous majority dissipated. The birth rate of the mainly rural Muslims far exceeded that of the urban Christians, so that by the early 1980s the proportion of Christians in Lebanon fell from more than 50 percent to less than 40 percent. Moreover, the least prosperous segment within the Muslim community, the Shiites, experienced the greatest rate of growth so their percentage of the population increased. Supported by the Islamic fundamentalist movement led by Iran, the Shiites of Lebanon have in recent years become increasingly assertive. Other communities, such as the Druzes, also demanded their rights in the multicommunal state of Lebanon. The Palestinian refugees, though not indigenous to Lebanon, were numerous and militant enough to constitute a force of their own, particularly in southern Lebanon and in the Beirut area.

In the 1970s, the precarious social balance that had been established among the various communities at the state's inception collapsed in a series of internecine clashes. The superposition of economic grievances upon religious differences caused the eruption of a civil war. The violence was further fueled by rivalries among several militias, and by interventions (ostensibly to bring peace) of external forces such as the United States, Syria, Israel, France, and lately Iran. The strife that engulfed Lebanon for several years resulted in grievous human misery, destruction of infrastructure, and loss of productive capital. Consequently the old order was destroyed. In postcivil war Lebanon now attempting to reconstitute itself, the Muslim community demands greater status, commensurate with its larger proportion of the country's population.

Much of Lebanon is now under direct Syrian military supervision, while a strip of land in the south is under Israeli control. As of the early 1990s, however, there is renewed hope for restoration of order and for economic and social recovery. Nearly 40 percent of the population still lives in villages and depends on agriculture for subsistence. To compensate that community for its deprivations, and for the ravages of the civil war and the various external encroachments, the government is now called upon to promote the economic progress of the underdeveloped provinces. In the rural south, specifically, economic development will require greater use of the Litani, which is the principal water resource.

Over the years, the part of the Lebanese labor force engaged in agriculture has decreased from about 49 percent in 1959 to 12 percent in the mid 1970s. Correspondingly, the contribution of agriculture to the gross national product (GNP) fell from 20 percent to 9 percent. By 1975 some 40 percent of Lebanon's rural population had left or been driven off the land, and most of those displaced had settled around Beirut.[1]

Because the population of Lebanon's hinterlands consists largely of Shiite Muslims, they have been the most affected by the declining importance of agriculture in the national economy, and by the labor-saving technologies that have been introduced into Lebanon. Added to that is the degradation of soil and vegetation resulting from the intensive cultivation and grazing of lands vulnerable to erosion. A wounded environment and a disintegrating old order have combined to create conditions of deprivation leading to disaffection and violence. The only possible cure is a comprehensive program of peaceful economic and social development.

In the early 1970s, the Shiite community of southern Lebanon erupted in violent protests against a plan to divert the Litani to quench the thirst of the rapidly expanding city of Beirut. The demands of the rural people to improve their lives can best be met by harnessing Lebanon's rivers to irrigate their arid but potentially productive lands and to electrify their villages and towns.

Irrigation holds the promise of bringing economic stability to farming districts, to raise the standard of living, and to save the country valuable foreign currency. As the development of water resources can be done in conjunction with hydroelectric power production, there are opportunities to link irrigation to industrialization. Such developments will enable Lebanon to utilize its comparative advantage in water resources, as well as the availability of labor and access to markets, toward the achievement of economic and social progress. It should be clear, however, that the development of the southern province of Lebanon is predicated on peace with neighboring Israel. Reciprocally, the security and well-being of northern Israel depend on the attainment of peace with Lebanon.

The Litani River is a perennial stream, but its discharge fluctuates from season to season and from year to year. Its most copious flow (some 60 percent of the annual discharge) occurs between January and April, during the winter rains and subsequent springtime snowmelt. The annual discharge can vary from a low volume of less than 200 million cubic meters to a high volume of more than 1,000. The long-term average is about 700 MCM/Y. The Litani's low salinity (20–40 ppm) makes it an especially attractive water source.

A Litani Development Authority, established in 1954, has planned the utilization of the river for hydropower and irrigation. The program called for three storage and power dams, one each on the upper, middle, and lower sections of the river. An important aspect of the program was the diversion of a portion of Litani's waters to the Awali stream, mainly for hydropower generation.

The most important of those dams is the Qir'aun Dam, completed in

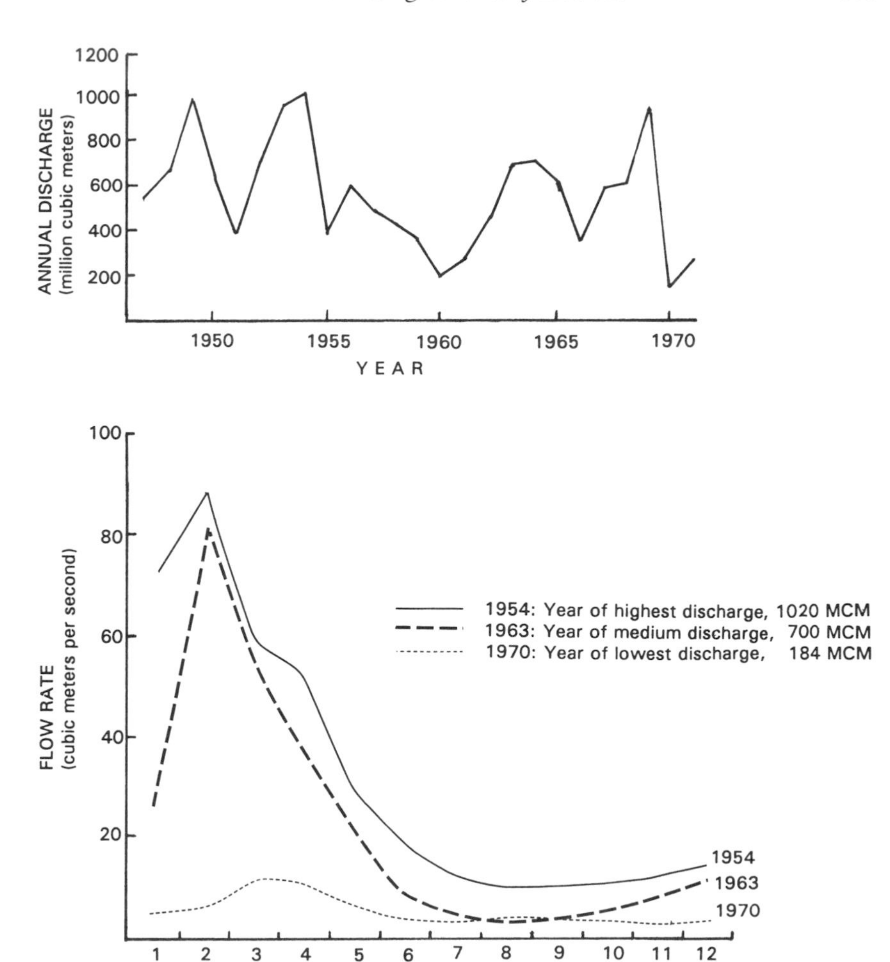

Figure 8.4 Interannual and annual fluctuations of the Litani River.

1966, with a storage capacity of 220 million cubic meters. This dam regulates the downstream discharge. Its outflow is conveyed through a six-kilometer tunnel to the Marqaba power station, where a vertical drop of 140 meters permits the generation of 34 megawatts of power. From there the bulk of the water is diverted via a second tunnel, 17 kilometers long, to the west-flowing Awali stream, where a hydroelectric station utilizes the 420-meter drop to produce 108 megawatts of power. Still another tunnel leads the water to the lowermost power station, where a drop of 150 meters generates an additional 48 megawatts. The system was designed to convey a total volume of 520 MCM/Y, of which nearly 500 were to be diverted from the Litani and the remainder to be collected from springs along the course of the diversion. The 190 megawatts of electricity generated by the system supplies power to the coastal cities of Beirut, Sidon, and Tyre, as well as to the smaller towns and villages of southern Lebanon.

As a consequence of the diversion from the Litani to the Awali, the flow of the lower Litani (called Qasimiah) has diminished drastically.

The early water utilization plans called for irrigating an additional 25,000 hectares in southern Lebanon as well as in the Beqaa Valley. These are the least developed and most impoverished of the country's provinces. Owing to the general turmoil within Lebanon and to the tensions along the border with Israel, the plans pertaining to southern Lebanon are yet to be implemented. At present, a considerable volume of water is drawn from the Qir'aun reservoir for crop irrigation in the Beqaa. However, much of the Litani's water (an amount variously estimated to total between 50 and 200 MCM/Y) still flows through the original channel to the coast, where only a part of it is used effectively. It appears that there is still water enough in the lower Litani to permit significant irrigation development.

As stated, the Litani River is incontrovertibly and exclusively a Lebanese river, its entire course and catchment being within Lebanon proper. There was a time, however, before the map of the Middle East was fixed by the French and British in the wake of the First World War, when the fate of the Litani hung in the balance. The early proponents of the Zionist movement expressed much interest in that river. The lower course of the Litani River seemed to them to be the natural boundary of the district of Galilee in what was to become Mandatory Palestine. Aware of the region's endemic scarcity of water and its economic implications for the future, the Zionist representatives appealed to the French and British governments to set the northern and northeastern borders of Palestine (then being negotiated) so as to include the entire catchment of the Jordan River and access to the Litani.

These requests were made explicit in several letters from Dr. Chaim Weizmann, then head of the World Zionist Organization who would eventually become the first president of Israel, to various British government officials. In one such letter to Prime Minister David Lloyd George, Weizmann argued that Lebanon is a "well watered" region, thus the Litani River's water is "valueless to the territory north of the proposed frontiers. It can be used beneficially in the country much further south." Therefore, the Zionist Organization considered "the valley of the Litani, for a distance of 25 miles above the bend" of the river to be important to the future of the national home promised to the Jewish people by the Balfour Declaration of November 1917.

In another letter to Britain's new foreign secretary, written on 30 October 1920, Weizmann stated: "If Palestine were cut off from the Litani, Upper Jordan and Yarmouk [rivers], to say nothing of the western shore of the [Sea of] Galilee, she could not be economically independent. And a poor and impoverished Palestine would be of no advantage to any power." These suggestions regarding the Litani were turned down when the British, acceding to French demands, agreed to set the border between Palestine and Lebanon through Upper Galilee so as to include the entire length of the Litani within the new state of Lebanon.

Having access to the Litani River was still on the minds of some of Israel's leaders during the state's formative years. The diaries of Moshe Sharett, Israel's second prime minister (in the early 1950s) speculate on the possibility of Israeli occupation of southern Lebanon up to the Litani River.[2] The Lebanese attack on Israel in 1948 could have provided Israel with the rationale to counterattack in the later phases of the 1948–49 war, and possibly to occupy and annex southern Lebanon. The fact is, however, that Israel refrained from doing so, and that, wishing to establish an alliance with the Christians of Lebanon, Israel for many years respected Lebanese territorial integrity. Some Israelis still harbor the idea that water-rich Lebanon might agree, in the context of peace, to a reciprocal arrangement whereby some water from the Litani would be diverted into the upper Jordan Valley. In return, Lebanon could receive the electricity to be generated by the scheme, as well as financial compensation for the water. However, that idea has never become a topic for official negotiation between the governments and now seems highly improbable.

In the wake of the 1967 war, the armed struggle of the Palestinian guerrillas against Israel intensified. Based at first in the eastern Jordan Valley, those guerrillas were expelled by the army of Jordan's King Hussein in September 1970 (a month that is still called "Black September" by the Palestinians). They then established a new base in southern Lebanon, from which they conducted incursions against Israel. In response, the Israeli army entered Lebanon and set up a "security zone" there, to be patrolled by a mostly Christian "South Lebanese Army" that was trained and equipped by Israel.

Over the years, recurrent charges have been made by Lebanese sources, and echoed by a few Russian and even American observers, that Israel is conspiring—or has actually begun—to appropriate the waters of the Litani.[3] Some have opined that Israel's occupation of a strip of land in southern Lebanon is motivated by the desire of the Israelis to access that river for the purpose of diverting it. Israeli authorities have categorically denied all such allegations, fully acknowledging that the Litani is an entirely Lebanese river. Israelis have learned the hard way that an unstable Lebanon unable to develop its resources, with a restive population seething with resentment, is not in anyone's best long-term interest.

The relationship between Israel and Lebanon has changed radically following the loss of Christian dominance in Lebanon. For a time, Israel tried to prevent that loss. The Israeli incursion into Lebanon in 1982 apparently had as one of its principal aims (apart from expelling the Palestinian guerrillas who had threatened its security) to reestablish the Maronite Christians as the dominant element in Lebanon and to make peace with a Maronite-led government. Most outside observers and many Israelis consider that intervention to have been a tragic mistake. Rather than bring peace any closer, it actually inflamed the area by exacerbating the enmity of the Shiite inhabitants of southern Lebanon toward Israel, so that tensions in the border region persist.

In the meanwhile, the idea that Lebanon might agree to sell Litani water to Israel has become moot. Much of the river's water has been diverted to the Awali, upstream of the Israeli security zone, so the quantity of water flowing in the lower river has been reduced markedly, and no excess is likely to be available beyond the needs of the local populace.

Lebanon's other major river is the Orontes, also known as the Asi. It rises near Baalbek in the Beqaa, just a few kilometers north of the source of the Litani, and is similarly fed by runoff and seepage from the rainfall and snowfall occurring over the Lebanon and Anti-Lebanon ranges. Paradoxically, however, the two rivers—so alike and so close at the outset—flow in opposite directions. The Orontes flows northward through the Beqaa Valley, staying within Lebanon for a distance of some 35 kilometers before entering Syria. In Syria, the river traverses the basaltic plain of Homs, runs through that city and the city of Hama, then turns west to enter the valley of Ghab. There it flows northward once again for some 60 kilometers before leaving Syria and crossing into Turkey. The Orontes wends its way for about 50 kilometers in Turkey, making a loop and turning southwestward to spill into the Mediterranean not far from the city of Antioch.

Within Lebanon, the mean annual flow rate of the Orontes is about 380 million cubic meters. Tributaries coming from Syria south of Hama add some 80 MCM/Y to that flow. The river is further augmented in the Ghab Valley by some 700 MCM/Y from tributaries coming from the hilly Ansaria district, where rainfall reaches 1,500 millimeters per year. Along its final leg in Turkey, the river receives still more tributaries, so that its total natural discharge approaches 1.5 billion cubic meters per year.

Unlike the lower Litani, which plunges through a gorge in a rugged landscape that is not readily irrigable, the lower Orontes traverses fertile valleys where it has been utilized for irrigation throughout the course of history. Evidence of its long utilization still exists in the form of Roman-built waterworks (including aqueducts and a dam, as well as numerous ancient waterwheels) in the Homs district. Reservoirs built in that district allow the irrigation of some 20,000 hectares in the area between Homs and Hama.

Irrigation was evidently practiced in the Ghab Valley in ancient times, but the neglect of drainage works there apparently resulted in the inundation of the land. Consequently, the Ghab Valley—over 60 kilometers long and over 10 kilometers wide—remained a sparsely inhabited morass for many centuries. In the 1920s and 1930s, the French administration of Syria made an effort to drain the land, and the task was continued by the Syrians after independence. The river bed was deepened and widened, and a storage dam was built to control flooding and to regulate the supply of water for the irrigation of the reclaimed land. That project was completed in 1968. Today this valley supports a dense population of well over a million and is an important center of agricultural production, particularly of cotton. In all, the various dams built on the Orontes and its tributaries have a storage capacity of about 600 MCM, capable of irrigating about 68,000 hectares. Two of the dams on the Orontes also generate hydroelectric power.

The increasing density of population and intensity of water use along the Orontes have had an adverse effect on water quality. Sewage flowing into the river has caused the spread of water-borne diseases such as typhus, dysentery, and cholera. Consequently the water of the lower Orontes has become unfit for drinking.

Although the Orontes is an international river, flowing in three countries (Lebanon, Syria, and Turkey), it has so far not been the cause of overt conflict. Lebanon does not utilize the Orontes to any significant degree, and is at present constrained from doing so by the Syrians, who hold de facto control over the river's headwaters in northeastern Lebanon. Nor does Turkey, the downstream riparian, depend much on the Orontes.

Lebanon's share of the Orontes watershed is about 12 percent, Syria's is 63 percent, and Turkey's is 25 percent. Of the river's total length (580 kilometers), 5.5 percent is in Lebanon, 81 percent in Syria, 8 percent in Turkey, and 5.5 along the border between Syria and Turkey. The relative contributions of the riparians to the river's flow is difficult to assess, as some of the tributaries cross the political boundaries. Clearly, however, Syria contributes the greater volume to the river's flow. Finally, an assessment of the relative needs of the three states supports Syria's right to be the major user of that river, since both Lebanon and Turkey have relatively abundant alternative water sources.

Nevertheless, an old territorial dispute still seethes under the surface between Turkey and Syria over the district of Hatay, where the lower Orontes discharges into the sea. That district, including the coastal cities of Antioch (Antakya) and Alexandretta (Iskenderun), was separated from Syria toward the end of the French mandate, in 1939, and ceded to Turkey. Independent Syria does not recognize that territorial transfer and demands the return of the district to its sovereignty. Although the territorial dispute is not directly related to the water issue, it may well become a bargaining chip in future negotiations between Turkey and Syria over the allocation of the waters of the Euphrates. That is but one example of the fact that all the water resources of the Middle East are interlinked one way or another, politically if not hydrologically.

> Then he showed me the river of the water of life, bright as crystal, flowing from the throne of God. . . . Also, on either side of the river, the tree of life with its twelve kinds of fruit, yielding its fruit each month; and the leaves of the tree were for the healing of the nations. Revelation 22:1–2

9

Fountains of the Deep

As fountains of the deep showed their might . . .
Proverbs 8:28

Thus far we have focused our attention mainly on surface waters—springs, rivers, and lakes—largely ignoring a water source that may be equally important, and in places even more so. Almost everywhere beneath the land surface, though not everywhere equally accessible, exists an awesome natural resource, a vast invisible ocean of water that permeates porous substrata of sediment and rock. Called groundwater (more properly: underground water), it is—contrary to popular misconception—not a stagnant but a moving fluid. That movement may seem slow to us, but only because of our shortsighted perception of environmental processes.

With the infinite patience that only nature knows, great quantities of groundwater wend their way, sluggishly but relentlessly, through a labyrinth of infinitesimal fissures and pores, through cracked rocks and compacted layers of sand, sometimes taking a year to traverse a mere 10 meters, a human lifetime to travel a kilometer. A drop of today's rain that seeps through permeable soil and subsoil strata and finds its way into the groundwater may not see daylight again in our lifetime or that of our great-grandchildren. When it does finally emerge to feed a stream or lake, or to be drawn from a well, it may carry the residues of the past as soluble salts or other water-borne constituents into the far future.

An aquifer (literally, "water carrier") is a water-saturated geological stratum, occurring at some depth, that can accumulate and transmit water in sufficient amounts to serve as a water source for human use. Aquifers vary greatly in their characteristics. Some consist of unconsolidated coarse sediments such as sand and gravel, and some are pervious or fissured bedrock such as sandstone or limestone.

A distinction can be made between confined and unconfined aquifers.

A confined aquifer, called "artesian," is overlain by an impermeable layer. Such an aquifer is not typically recharged through its impervious cap but from areas some distance away where the permeable formation is exposed. An unconfined aquifer (called "phreatic" from the Greek *phrear,* a well) is not capped, so the water-table is free to fluctuate up and down periodically, depending on the relative rates of recharge and discharge. Some aquifers are restricted in volume and extent, underlying an area of only a few hectares or a narrow strip of land along a stream. Other aquifers may underlie whole regions.

Access to groundwater can generally be achieved by digging or drilling a hole through the overlying strata to the saturated zone. Water then tends to seep into the hole, and the level at which the groundwater comes to rest is called the water-table. The depth of the water-table can be highly variable—in places it may be hundreds of meters deep, whereas elsewhere it may rise to the soil surface.

Where the water-table is very high, it may invade the rooting zone of plants and cause it to be waterlogged. The roots of most crops then suffer from restricted aeration. High water-table conditions are more common in humid regions but may also occur in arid regions, particularly in river valleys where rapid percolation occurs from river beds as well as from excessive irrigation. In such regions, the groundwater may be brackish and the rise of the water-table may induce upward capillary seepage of salt-bearing water to the soil surface, where the water evaporates leaving the salts behind. Thus, groundwater of good quality occurring at an optimal depth can serve as a beneficial source of water, whereas groundwater at too shallow a depth can pose a threat of soil degradation.

The amount of groundwater underlying any location (known as the groundwater reservoir) is finite. That reservoir is recharged whenever rainfall percolating through the overlying soil exceeds the amount returned to the atmosphere by evaporation from the soil surface and from plants. On the other hand, extraction of water from the groundwater reservoir by pumping in excess of the annual recharge will cause its depletion. Overdrawing from shallow aquifers along seacoasts can also lead to saltwater intrusion and aquifer degradation. A prime example of this is the Nile Delta, where saltwater intrusion has now reached as far as 20 kilometers inland.

In desert areas, the meager seasonal rains wet only the surface zone of the soil, which soon dries up by direct evaporation and extraction by natural vegetation, leaving practically no water to percolate to greater depth. Thus, over the greater part of the arid zone there is no significant replenishment of groundwater by downward percolation of rainwater. However, there are exceptions. The occurrence of flash floods (albeit infrequently) may result in localized percolation into the permeable gravel or sand of the stream bed. In places where the permeable sediment is underlain by an impervious layer, the percolating water may accumulate to form shallow "perched" water-tables that can be tapped by wells. Such groundwater reservoirs are likely to occur specifically in alluvial fans, where mountain streams emerge onto plains.

In some arid regions, large aquifers are filled with water derived from an earlier, more humid geological era. Such water, called "fossil water," though it may be accessible, is irreplaceable under present climatic conditions. Fossil water aquifers are known to exist under the Sahara and, to a lesser extent, under the deserts of Arabia, Sinai, and even the Negev. However, these ancient bodies of water may lie hundreds of meters deep, emerging to the surface only in isolated oases. Moreover, in many places the water contained in them is brackish, so their utilization is of questionable feasibility. Where such waters are not too deep and are relatively pure, extraction may be of great economic and social value for a period of some decades. However, the withdrawal of water from unreplenished aquifers is analogous to mining mineral ores or extracting unreplenished petroleum, so preparation must be made for the inevitable time when the finite resource is exhausted or its further extraction (as the water-table falls progressively) becomes prohibitive.

The early development of wells in the Middle East arose because of the scarcity of water in dry areas. Human power and animal power, aided by hoists and hand tools, were the basis for well digging. Evidence of early wells, dating back 5,000 years, has been found in many parts of the Middle East. Undoubtedly the greatest achievement in groundwater utilization in the region was the advent, apparently in Persia over 3,000 years ago, of long horizontal wells called *qanats* or *karez,* which collected water from alluvial fans. Herodotus, writing about the wars in Persia, described the way towns could be subdued by filling their wells and plugging their supply tunnels. The ancient wells, however, were mostly shallow, seldom penetrating more than 100 meters.

Great improvements have been made in the methods of drilling for water in the twentieth century, thanks largely to technology developed by the petroleum industry. Hydraulic rotary methods for well drilling and the turbine pump facilitate the tapping and extraction of deep groundwater. Such activities require considerable capital investment initially and energy for pumping subsequently.

Modern motorized pumps offer obvious advantages in convenience and quantity of water supply but also pose a great danger. Ancient traditions in the Middle East, sanctified by religious edicts, have long prohibited the digging of new wells in the proximity of existing wells, lest the yield of water in the latter be adversely affected. Those traditions were effective as long as all wells were shallow and the amount of water drawn was small. Now, however, powerful pumps enable landowners to draw large volumes of water via deep boreholes, thus affecting the water-table beyond the confines of their property and depleting distant wells formerly considered to lie safely beyond the affected zone. In fact, such pumps can draw down an entire aquifer within just a few years. Traditional safeguards are no longer effective and must be supplanted by effective national—and at times international—regulation.

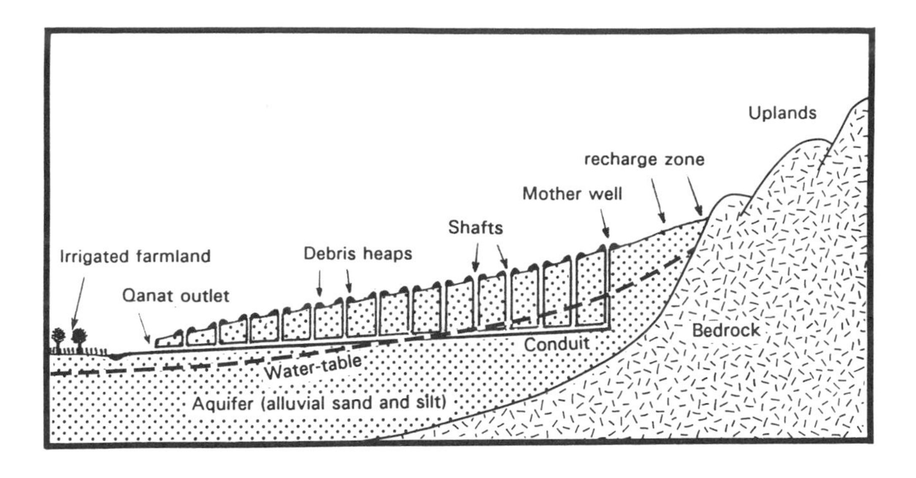

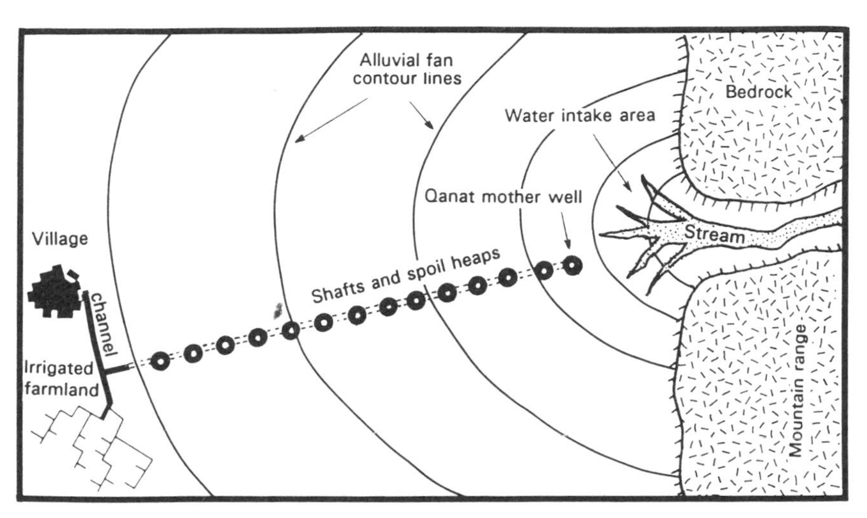

Figure 9.1 Cross-section and top view of a qanat system.

Legislation restricting groundwater extraction has in fact been enacted in virtually every country in the region, but enforcement mechanisms are inadequately developed. In Syria, because of the diversion of fresh water supplies to urban and industrial users, farmers are drilling for alternative sources and are overpumping despite the restrictive legislation. In Yemen's Sanaa basin, some 25,000 wells are overpumping the aquifers, causing an annual decline in water levels of one to seven meters.

Although its extraction is more costly, groundwater is often more desirable than surface water for several reasons. Stored in an aquifer, the water is generally protected against seepage and evaporative losses such as are incurred continuously in surface reservoirs. Groundwater aquifers are less sensitive than are surface reservoirs to short-term droughts. Moreover, groundwater is relatively (but not entirely) free of pathogenic organisms, turbidity, odors, and colors—factors that tend to be filtered by the strata through which groundwater is recharged. Shallow aquifers are, however, vulnerable to contamination by salts as well as by toxic chemicals from industrial, domestic, and agricultural wastes. Aquifer contamination, once caused, can be very difficult to remedy.

Just as neighbors along a river can be rivals, so can users of the same aquifer. Groundwater, like surface water, is a fluid that recognizes no property boundaries. A prime example from the United States is the Ogallala aquifer, which underlies parts of Nebraska, Kansas, Oklahoma, and Texas. In northern Texas, especially, individual farmers have been competing with one another in drawing water from the common source in a quest for immediate profit without regard to the region's future. Here the principle of private ownership of land—and whatever lies beneath it—is carried to its aberrant extreme, as neighboring irrigators "cooperate" only in putting each other out of business. It is now forecast that within two decades or so the water level in large parts of the aquifer will have fallen so far as to make the cost of further pumping entirely prohibitive for the irrigation of crops.

Similar rivalries over common aquifers may occur between nations, and can be even more difficult to resolve than disputes over surface waters. Criteria for establishing rights and equity are particularly ill defined in the case of aquifers. Where groundwater flows naturally from one state to another, the question arises: which state should be granted priority in the use of the resource, and by how much? Even where a resolution can be attained in principle, it may not be applicable in practice simply because the necessary information is not available. The extent of an aquifer's catchment and the quantitative contributions of different areas to it (as well as its rates and directions of flow) are difficult to define even under the best circumstances, and all the more so in a region where there is as yet no coordination in the collection of essential hydrological data.

Examples of rivalries over groundwater abound in the Middle East. One is the case of the Damman aquifer, which underlies parts of the Arabian peninsula along the Persian Gulf. There the Gulf States, in great haste to develop their oil-based economies, have practically depleted shared subterranean reservoirs. Specifically, the water-table in the United Arab Emirates is being drawn down by excessive pumping in Oman. Saudi Arabia, which has developed irrigated agriculture by mining its aquifers, withdraws water that would otherwise flow to Bahrain. A similar rivalry exists between Saudi Arabia and neighboring Hashemite Jordan over the fossil waters of a desert aquifer underlying their border area.

Perhaps even more fateful—though not as immediate—is the rivalry over groundwater that is threatening to develop between Egypt and Libya, in what the Egyptians call the Western Desert. Each of the two countries is pumping water from a fossil aquifer on its side of the border, and each

Figure 9.2 Water pumped from a desert aquifer used for irrigation in Jordan.

is concerned lest the activity of its neighbor eventually affect its own resource. Egypt is drawing groundwater in the Siwah depression, some 50 kilometers from the Libyan boundary, as well as along the Mediterranean coast in the area of Matruh and Sidi Barrani. In addition, it is tapping desert groundwater in such locations as Wahat al-Bahriyah, Wahat al-Farafirah, Wahat ad-Dakhilah, and al-Kharijah, farther east within its own territory. These activities were meant to be part of Egypt's ambitious program to reclaim 1 million hectares of desert land and thus to extend agriculture beyond the Nile floodplain by diverting water from Lake Nasser as well as by developing local groundwater. This program, meant to resettle millions of people, has not so far been implemented on the scale first envisaged. In any case, the amount of groundwater drawn by Egypt is still small relative to that country's overall water use.

A much more grandiose program is now being implemented by Libya. The aquifer system being tapped is contained within the Nubian sandstone formation, an expansive geological structure that underlies huge areas in southern Libya, southwestern Egypt, parts of Chad, and northwestern Sudan (from whose province of Nubia the structure drives its name). It is in fact a complex series of aquifers underlying an area of perhaps 2.5 million square kilometers, and it may contain as much as 50,000 billion cubic meters of fossil water. Extensions of the same formation underlie parts of Sinai, the Negev, the southern region of the Kingdom of Jordan, and Arabia. Water contained in the Nubian sandstone ranges from potable to brackish. In places the aquifer discharges at the surface to form oases, but typically the water-table is hundreds of meters deep. Although hardly replenished under present climatic conditions, the total amount of water actually extractable from this aquifer system is so great as to permit utilization—if carried out in a judicious manner—for many decades.

The origin of this water, paradoxically present in the bowels of the largest and driest desert in the world, is not entirely clear. It seems most probable that it had accumulated during a humid phase of the late Pleistocene, some 30,000 or more years ago, when the Sahara was not the desert it is today but a relatively moist region. Evidence of that earlier climate has been discovered in the Western Desert, where satellite imagery has revealed an array of ancient river beds that had long since been covered by desert sands.

Libya occupies an area that equals half of Europe, but the total amount of rainfall it receives is scarcely equal to that of a single province in France. The greater part of Libya is desert, and even the rainier district, along the Mediterranean coast, lies under the ever-present threat of drought like Damocles under the sword. Renewable sources of fresh water are rare, and desalinization for irrigation is still prohibitively expensive, even in an oil-rich country like Libya.

The first to tap groundwater and introduce modern agriculture in Libya (albeit on a small scale) were the Italians, who held sway there during the early part of the twentieth century until the Second World War. They drilled wells along the coast, mostly in the vicinity of the principal towns of Tripoli and Benghazi. Pumping of groundwater from the shallow coastal aquifer was continued and enhanced after Libya became independent. Soon the water-table fell by several meters, seawater intruded, and many of the wells were rendered unusable. After the revolution of 1969, the new government set as one of its primary goals the achievement of economic independence, including self-sufficiency in food production. That task was severely constrained by the paucity of water in a country that, but for its narrow coastal strip, is entirely a desert.

Back in the 1920s, the Italians explored for oil in the vicinity of the Kufra oasis, deep in Libya's desolate interior. Kufra is a hidden oasis in the midst of the parched Sarir, a section of the Sahara long described as "the world's most impenetrable desert." It lies almost 2,000 kilometers of shifting sands and shimmering mirages away from Tripoli, the country's capital on the coast of the Mediterranean. The dunes, many as high as 100 meters and some as long as 100 kilometers, could then be traversed only on camelback, and the arduous journey took 40 days. To their disappointment, the Italians found only water, a fluid that seemed useless in the remote and unpopulated location where it occurred. On the eve of the Second World War, the Italian prospectors returned to the same area, again in search of oil. And again they found "only" water. After the war came the Americans, also searching for oil, and, to the consternation of Libya's King Senusi, they—like their predecessors—found nothing but water. It was a large quantity of water, and of adequate quality, but from the oil prospectors' point of view it seemed useless.

Some years later, Libya's new leader, Colonel Muammar al-Qaddafi, having deposed the king, learned of the existence of a vast amount of water in the Sahara. In the celebrations of September 1970, marking the first anniversary of his rise to power, Qaddafi declared to the throng gathered in the central square of Tripoli: "Water is more important than oil. Remember this. Water is life. And you, my dear brethren, have the right to search for water everywhere, in Chad as well as, if necessary, in Sicily." The Libyan ruler had grand Pharaonic visions for unifying African and Middle Eastern nations. However, reality intruded and he was forced to forgo most of his dreams. But he persisted in the pursuit of one special dream, notwithstanding his many troubles with Egypt, Chad, France, Britain, and the United States: to bring water to his people.

Yet Qaddafi had no idea how to realize this dream in practice. His vision would have remained just a pipe dream had not another visionary, a more practical one, come on the scene. The man who became the actual godfather of the project was the American billionaire and international entrepreneur Armand Hammer.[1]

Earlier, in 1965, Dr. Hammer had won a concession for his company, Occidental Petroleum, to search for oil in Libya. (His bid stood out from among the numerous others because Hammer had the insight to inscribe it on a sheepskin scroll and wrap it in the colors of the Libyan flag.) To please the Libyans, Hammer also promised to establish an experimental farm at one of Libya's oases. In 1966, in a desert location more than 160 kilometers from the Mediterranean coast, Occidental struck one of the richest deposits of high quality oil in the world. To recover that oil, the company undertook to build a 210-kilometer pipeline across the desert, and did so in less than a year.

Seeing that feat, Qaddafi wished for an even greater one: to convey a huge amount of water from the interior to the coast, so as to make Libya self-sufficient in food and fiber. Fearful that its assets might otherwise be nationalized, Occidental decided to help realize the Libyan's dream. The water was found at Sarir, Kufra, and Tazerbo, sites that are many hundreds of kilometers from the coast of Benghazi in eastern Libya. Other sites with large groundwater reserves were found in western Libya, equally distant from the coast. The Kufra basin, covering an area of 3,500 square kilometers, has been estimated to contain several thousand billion cubic meters of water. Two other basins, Sarir and Murzuk, may contain a similar amount. Though such estimates may be grossly exaggerated, there is no doubt that an enormous quantity of water is hidden under the desert sands. How much of that water can feasibly be extracted and used is, however, an entirely different question.

The choice before Qaddafi was either to transfer people to those remote locations, or to transfer the water to where the people live—along the coast. Fortunately, the topography is such that after being pumped to the surface, the water can flow by gravity from an elevation of some 270 meters above sea level toward the coast—Benghazi, Sirte, Tripoli, and irrigable lands along the way. So the decision was made to leave the people where they were and move the water.

Accordingly, a huge project was undertaken, to be named the Great Man-made River. In 1979, when the project was begun, it was expected to cost $25 billion. The first stage of the project would consist of drilling about 120 wells in Tazerbo and an equal number in Sarir, and laying a pipeline from those sites to Benghazi, and thence westward along the coast to Sirte. The amount of water to be conveyed would total about 750 million cubic meters per year, enough to supply the domestic needs of the towns of eastern Libya and to irrigate about 50,000 hectares of land. The second stage would transfer a similar amount of water from the desert aquifer underlying southwestern Libya via a pipeline to the western coast—including the capital city of Tripoli. The third stage would augment the eastern system with some 2.2 billion cubic meters annually, to be obtained from a series of wells in the even more remote site of al-Kufra. Finally, the pipeline delivering the extra water to the coast of eastern Libya (Cyrenaica) would be linked to the coast of western Libya as well as eastward (toward the Egyptian border) as far as Tobruk. The total volume of water to be piped from the desert to the coast would thus amount to about 3.5 billion cubic meters—three times the total natural flow of the Jordan (along with its tributaries) and nearly one-twentieth the flow of the mighty Nile.

The first phase of the project was accomplished in late 1986, the second in now in progress, and the third and final phase is to be completed in 1995. According to a U.S. Agency for International Development report published in 1993, Libya has already invested $18 billion in laying

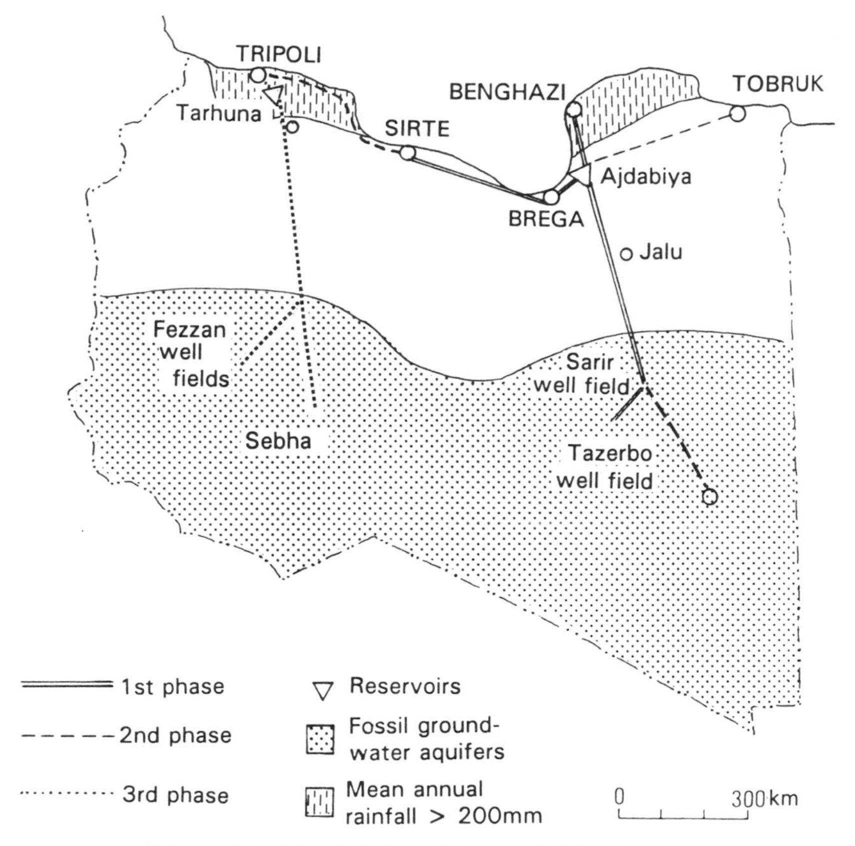

Figure 9.3 Libya's "Great Man-made River" project.

1,800 kilometers of pipes and in building reservoirs, pumping stations, canals, and roads. Though the actual implementation was contracted to a Korean consortium, the planning and technical supervision were carried out by an American firm from Texas. Following the American aerial attack on Libya under the orders of President Reagan in 1986, the supervision was formally transferred to "neutral hands," though it seems that the same Americans continued the work under the guise of being "Canadians."

The Great Man-made River is practically invisible at the surface. The water flows in gigantic buried pipes especially manufactured for the project at Marsa el-Brega. The steel tubes are coated with concrete and wrapped with steel cables for reinforcement. The proud Libyans claim that the amount of concrete used for the pipes would have been sufficient to build 16 of Egypt's largest pyramids, the length of steel cable used could encircle the globe 25 times, and the volume of earth moved in digging the trenches for the pipes would suffice to build 12 Aswan High Dams. Two reservoirs are already in operation, at Benghazi and at Surt. The length of the pipeline from Tazerbo to Benghazi is 800 kilometers, and from Sarir to Sirte 783 kilometers. The pipelines now under construction will lead water from

Kufra to the coast, and from Fezzan to Tripoli (the latter will convey 2 MCM per day, or 730 MCM per year).

The Libyans seem intent on continuing the project despite its rising costs and the reduced price of oil (which has lowered the country's revenues). Whether they will indeed complete the project in its entirety, and whether it will prove to be economically justified, no one can yet tell. In the meanwhile, questions remain regarding the speculation that the aquifer is at least partially replenished from the rainier region south of the Sahara, and regarding the aquifer's rate of depletion under pumpage and its effective lifetime. Finally, it is unclear whether the tapping of this aquifer by the Libyans might affect the groundwater reserves of neighboring Egypt.

The Egyptians are somewhat concerned over the latter possibility, for—while they now rely primarily on the Nile—they regard the groundwater under the desert as a valuable strategic reserve for the future and have already begun to tap into it. It seems doubtful that the radius of influence of the Libyan wells (the maximum distance from the wells subject to a significant lowering of the water-table) will extend more than 100 kilometers even after some years of pumping. However, the situation should be monitored objectively in order for hydrologists to learn more about the utilization potential of the Nubian sandstone aquifer system. What can be predicted with confidence is that, since the source is finite, the real cost of the water will increase over time, as the water-table within the well fields falls, and as the yield of the wells is reduced.

Outside Libya, Qaddafi is distrusted for his provocative gestures and his alleged support for international terrorism. It would be ironic if this erratic man were to enter the history books as the visionary leader who brought water out of the wilderness to his thirsty nation and gave life to a parched region. It is also possible, contrariwise, that Qaddafi's lavish expenditures to squander so precious and non-renewable a resource on the ephemeral greening of a strip of desert land will be seen within a few decades (when the aquifers are effectively depleted and further pumping becomes prohibitive) to have been a 27 billion dollar act of folly.

Of Israel's total potential supplies of fresh water (estimated at roughly 1.6 billion cubic meters per year, on average), about 60 percent is found in the country's two largest aquifers: the coastal plain aquifer and the mountain aquifer. Waters obtained from these major aquifers and from several minor ones,[2] together with the waters drawn from the upper Jordan River basin, are linked by the National Water Carrier. This integrated system allows transfers of water from one location to another in accordance with optimal quantity and quality requirements. Additional water supplies are obtained from reclaimed wastewater and from the interception of the occasional floods of intermittent streams. Much of the latter water, in turn, is stored seasonally in the coastal aquifer by means of artificial recharge.

The coastal aquifer underlies Israel's central and southern coastal plain, from Mount Carmel in the north to the Gaza Strip in the south.

The plain extends over a length of about 120 kilometers and a width varying from 3 kilometers in the north to 20 kilometers in the south, for a total area of some 1,800 square kilometers. The aquifer strata, consisting of sandstones and unconsolidated sands of late Tertiary and Quaternary age (Pliocene-Pleistocene), are about 150 to 180 meters thick at the shoreline, and become progressively thinner inland. The aquifer derives its water naturally from two sources: the excess of winter rainfall (which amounts to 400–600 mm/Y) over evapotranspiration in the coastal plain itself, and the seepage of floodwaters from streams that traverse the plain as they carry runoff from the mountains to the sea.

The mean natural recharge is estimated to be about 300 MCM/Y. In addition, some 90 MCM/Y seep into the aquifer from irrigated lands (the coastal plain being a major agricultural area where most of Israel's citrus is grown), as well as from septic tanks and leaky water supply systems. Since a certain amount of water must be allowed to flow from the aquifer to the sea in order to flush out salts and other contaminants, the amount of water that can be withdrawn safely is of the order of 240–300 MCM/Y. The actual amount pumped from the numerous wells in the coastal plain has exceeded that judicious limit for many years. Efforts have therefore been made to recharge the aquifer artificially with wastewaters from coastal cities and with waters taken from the Sea of Galilee during the winter season, when irrigation and other water requirements are minimal. Recharging is done by spreading the water in open sandy basins and by injecting it into wells. The amount of water thus recharged into the aquifer has varied widely from year to year, and in general has not been sufficient to compensate for the cumulative overdraft. Inevitably, the "mining" of this aquifer has affected both the quantity of water in storage and its quality.

The coastal aquifer is a so-called phreatic, or water-table aquifer. Owing to the excess of withdrawals over replenishment, the water levels in wells have generally been declining. The decline was steep in the 1950s and early 1960s, prior to the initiation of the National Water Carrier. When water was brought by the carrier from the Jordan basin, it replaced the water formerly withdrawn from the coastal aquifer, and was also used to recharge the aquifer. Consequently the water levels there recovered somewhat during the late 1960s and early 1970s. In the 1980s, however, the decline resumed and the water-table became dangerously low. Ideally, the water-table in the aquifer should maintain a 1 percent (0.6 degree) slope toward the sea. Following the long-continued overpumping, the slope diminished to near zero. In some places, the water-table had even fallen below sea level, thus producing a reverse slope. The inevitable result has been a process of seawater intrusion and salinization of wells.[3]

Fresh water, being less dense than seawater, tends to float over the latter with a rather sharp interface. Along a coast, that interface assumes a wedgelike shape. As the overburden of fresh water is reduced by pumping, seawater invades the aquifer from below, and the tip of its wedge moves

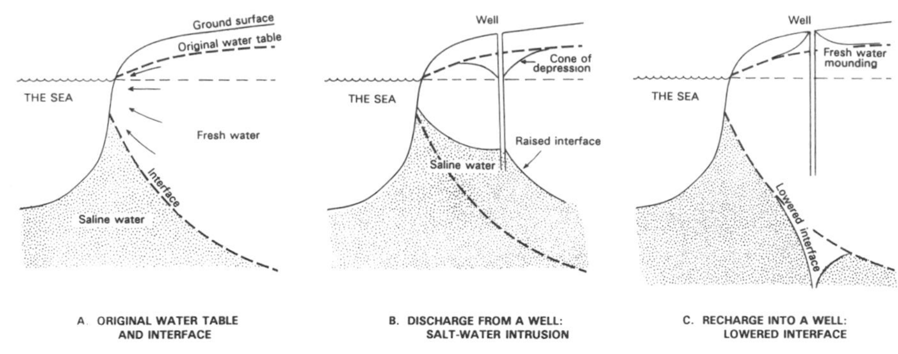

Figure 9.4 The dynamic salt water–fresh water interface as affected by groundwater extraction and recharge.

inland. Once that intrusion has taken place, it is very difficult to push it back and to flush out its residual salts. Even where flushing is possible, there might be permanent damage to the aquifer, as the salt may cause the lime in the calcareous sandstone to reprecipitate and thus partially clog the aquifer's porosity. Therefore, efforts are made to maintain the wedge of salt water underlying the coastal plain at an average distance of no more than 1.5 kilometer inland from the seashore.

The total fresh water storage volume of the coastal aquifer is several billion cubic meters. Only a fraction of that volume, however, is usable. Being the largest reservoir in the three-basin National Water System (consisting of two major aquifers and the Sea of Galilee), the coastal aquifers serves as a principal regulatory "water bank" for the short-term storage of winter surpluses and the long-term balancing of supply and demand throughout the country. Continued overuse of this aquifer is bound to endanger its capacity to fulfill such a vital regulatory function. For this reason, Israel's water managers have been monitoring the aquifer—like doctors monitoring the vital functions of a patient—using a network of 300 observation wells on a two-kilometer grid to observe the groundwater level, and a series of multipipe observation wells along the western strip of the coastal plain to monitor seawater intrusion.

In early 1991 the state comptroller of Israel, Miriam Ben Porat, submitted a scathing critique of the former water commissioner's policies, especially as regards the overdrawing of groundwater reserves and the too-liberal allocation of water to farmers. By then, the coastal aquifer was estimated to have accumulated a deficit of nearly 1 billion cubic meters. The comptroller concluded that, as a consequence of the commissioner's policy, "In practice Israel has no more reserve water stored in its reservoirs" and that "there is real danger that the country will not be able to maintain the quantity and quality of its water supplies even in the short term." The heavy rains that fell during the winters of 1991–92 and 1992–93 brought temporary relief but did not fundamentally change the problem.

The coastal aquifer extends southward into the Gaza Strip. There, a crisis situation already exists, as population growth has overtaxed water supplies at an alarming rate. The aquifer underlying this district is capable of providing up to 65 million cubic meters of sustainable supplies annually, but the Gaza Strip's 750,000 inhabitants already pump more than 120 million cubic meters. The inevitable result has been a drastic lowering of the water-table and severe intrusion of seawater into the local wells.[4] Continuation of the present rate of overpumping invites an economic disaster. There can be no doubt that the Gaza Strip's severe water shortage is one reason for the district's extreme poverty.

When peace is achieved and economic development takes place in earnest, Gaza—given its unique location—may well become a major center of international trade and industry, indeed the Hong Kong or Singapore

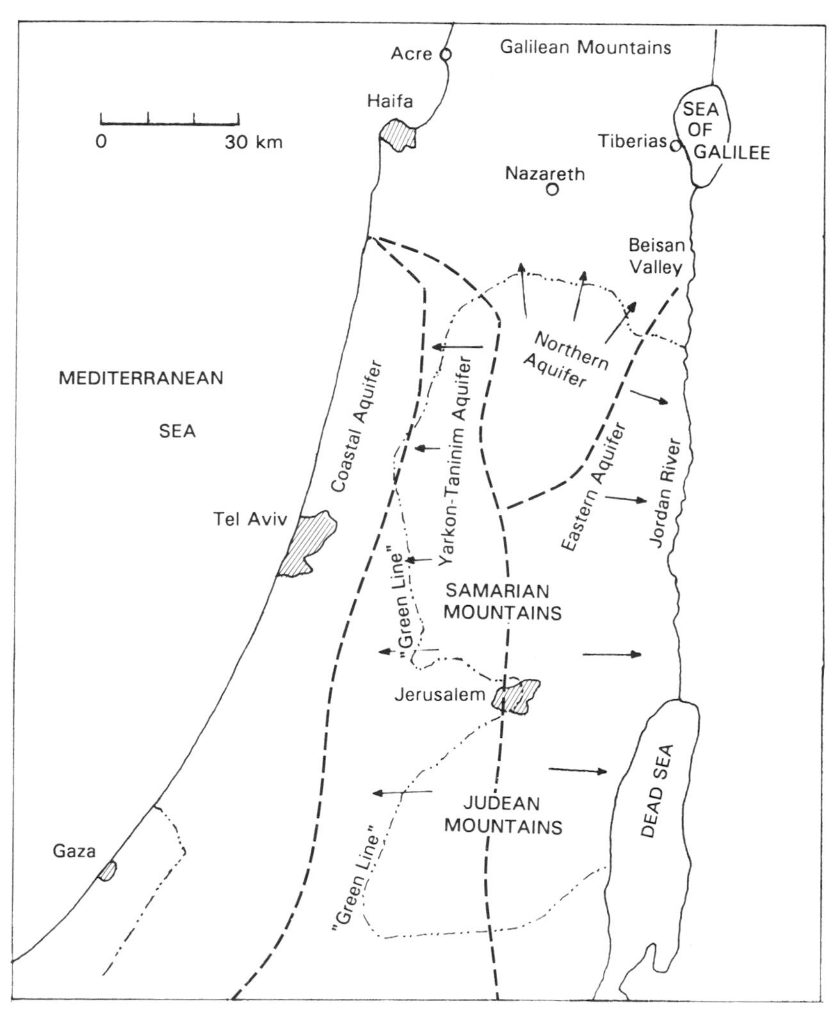

Figure 9.5 Aquifers underlying Israel and the West Bank.

of the Middle East. Water will then be shifted from low-return agricultural use to more remunerative uses. Even if irrigation is drastically curtailed, however, Gaza is clearly in urgent need of additional supplies, either by importation or by desalinization.

Whereas the coastal (sandstone) aquifer lies entirely within Israel, the mountain (limestone) aquifer extends eastward beyond the country's pre-1967 boundary. The greater part of the catchment of this aquifer lies, in fact, in the mountainous territory of the West Bank (called Judea and Samaria by the Israelis).[5] It is therefore a source of considerable tension between the Palestinians and the Israelis.

Hydrologists refer to the mountain aquifer as the "Yarkon-Taninim aquifer," after its two main natural outlets, where the groundwater breaks to the surface on the eastern edge of the coastal plain within the territory of Israel proper. The one outlet is the Yarkon springs, which rise at Rosh Haayin (meaning "Spring Head"), about 15 kilometers east of Tel Aviv, and feed the River Yarkon, which runs through that city. The other natural outlet of the aquifer is the Taninim springs, emerging 60 kilometers farther north and feeding the River Taninim (the name meaning "Crocodiles" for the aquatic reptiles that once abounded there).

The aquifer itself consists of limestone and dolomite formations[6] of the Cenomanian-Turonian age. These anticlinal formations are exposed in the central highlands, and it is from rainfall over those highlands that the aquifer receives its natural replenishment. Westward, the water-bearing strata dip toward the sea. In the foothills they are covered by layers of Eocene chalks and marls, a few tens to a few hundred meters thick; and in the coastal plain they are covered additionally by relatively young Pliocene-Pleistocene-Holocene deposits.

The entire aquifer is shaped like a slanted slab, some 150 kilometers long from north to south. In addition to the springs that discharge at the surface of the coastal plain, this aquifer may also have an outlet to the Mediterranean Sea itself, but the configuration of such an outlet and the quantity of water flowing through it are yet uncertain.

Extraction of water from the limestone aquifer was begun by Jewish settlers in the 1920s. Substantial pumpage was instituted in Israel in the early 1950s from wells located along the eastern fringe of the coastal plain. Some of the water thus extracted was piped to the northern Negev for the purpose of irrigating newly settled lands there. By 1963 the groundwater levels had fallen below the elevation of the Yarkon springs (16.5 meters above sea level), so those springs ceased to flow. This cessation was considered a favorable occurrence, as it prevented the "wasteful" outflow of water through the Yarkon River to the sea and made part of the aquifer volume available for artificial recharge. To ensure against occasional spontaneous discharges from the aquifer following wet seasons, the spring's outlet crest was raised artificially by 5 meters.[7]

The Taninim springs (the aquifer's other natural outlet), emerge at an

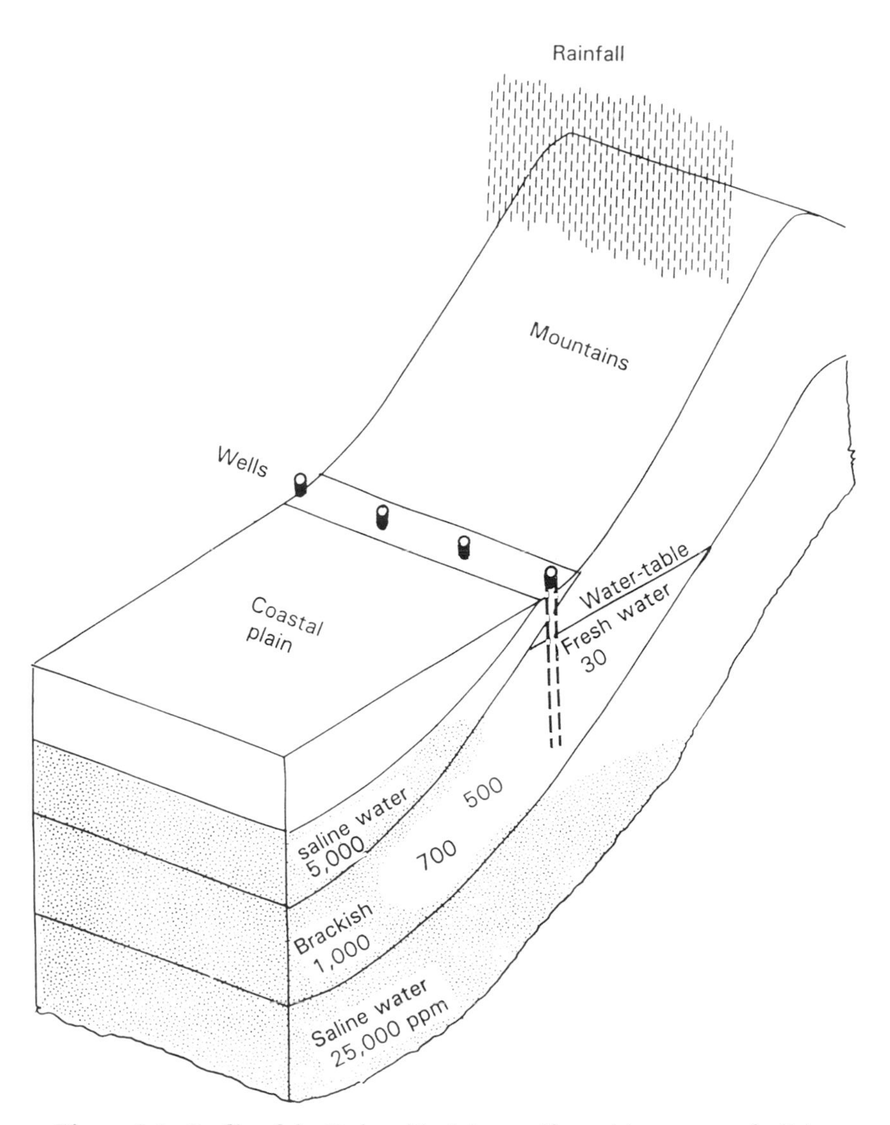

Figure 9.6 Profile of the Yarkon-Taninim aquifer and its sources of salinity.

elevation of 3.5 meters above sea level. These springs continue to discharge at the rate of 30 to 50 million cubic meters per year. Their flow is somewhat brackish, owing to mixing with intruding seawater or with connate bodies of saline groundwater. Hence the water is used mainly for fishponds rather than for crop irrigation. Experience has shown that an outflow of some 40 MCM/Y is necessary to prevent the progressive salinization of the aquifer's groundwater. The delicate task for Israeli hydrologists, therefore, is to manage the aquifer in a such a manner as to ensure the necessary discharge from Taninim even while preventing wasteful discharge from Yarkon.

Apart from the discharge of the brackish springs, the Yarkon-Taninim aquifer's sustainable natural yield of fresh water is estimated to be about 300 MCM/Y.[8] The aquifer is tapped by some 300 wells, whose total annual pumpage averages 375 million cubic meters. As this rate of withdrawal exceeds the average natural recharge, efforts have been made to replenish the aquifer artificially via specialized and dual-purpose wells, with water piped from the Sea of Galilee. Nonetheless, the aquifer suffered depletion during the late 1980s. As in the case of the coastal aquifer, the situation was redressed by the abundant rains of the 1991–92 and 1992–93 winters, yet the aquifer must be protected against overpumping in the future. (The euphoria over those two wet seasons was ended abruptly by the recurrence of less than normal rainfall in 1993–94.)

Israel's utilization of the mountain aquifer's entire safe yield is resented by the Palestinians of the West Bank. Nearly one million of them live in this mountainous district of only 5,800 square kilometers. It consists of two subdistricts, which the Israelis call by their biblical names: Samaria—with the principal towns of Nablus, Jenin, and Ramallah—in the north; and Judea—with the principal towns of Bethlehem and Hebron—in the south. The city of Jerusalem is at the juncture of the two subdistricts. The east-facing slopes of the West Bank, constituting nearly one-third of the total area, are extremely arid, and much of the west-facing slopes—though relatively humid—consists of shallow and rocky soils. Of the 200,000 hectares of arable land, only an area of about 15,000 is irrigated—mainly along the Fariah valley, the vicinity of Jericho, and alongside the few perennial springs in the other valleys of the district.

Prior to 1967, agriculture in the West Bank—then under Jordanian jurisdiction—was almost entirely rainfed, with just a limited amount of irrigation that depended on local springs. Only a few wells were in operation, and those were primarily shallow wells used for domestic consumption. On the relatively rainy west-facing slopes, rainfed agriculture was practiced much as it had been since biblical times. It was mainly subsistence farming with limited marketing to local towns and practically no exporting of produce abroad.

The Israeli occupation changed local agriculture profoundly. It introduced modern technology, including mechanization, precision tillage, pest control, plastic covering of crops for temperature control, high-yielding varieties, postharvest processing of produce, marketing, and export outlets. It also introduced efficient methods of irrigation, including sprinkler and especially drip irrigation. Consequently, output increased greatly, and farming was transformed from a subsistence enterprise to a commercial industry. Some West Bank products were exported via Israel to Europe, and some—more importantly—were exported directly to the Kingdom of Jordan and through it to the lucrative markets of the oil-rich Gulf States. Ironically, this improvement in productivity and profitability, in itself a positive result of the Israeli-Palestinian interaction, also increased the demand to expand water use and thus created a growing rivalry over limited water resources.

In 1967 the potential fresh water supply in the district was estimated to be about 400 MCM/Y. At that time, the total water use from springs and wells was under 100 MCM/Y, mostly for agriculture and the rest for domestic purposes. Practically no water was used by modern industry, of which there was very little then. The output of all wells on the western slopes of the West Bank and in the Jordan Valley was no more than 38 MCM/Y, mainly drawn from shallow local aquifers.[9] The larger mountain aquifer, requiring deep drilling, was hardly tapped at all by the Palestinians. Its waters flowed naturally toward the coastal plain, where they had been used for some decades by Israel. Therefore, the Israelis claim historical rights to this aquifer.

After 1967 about 100,000 Israelis settled in the West Bank, on what the Israeli government considered "state lands." Projections suggest that by the year 2000 the population of the West Bank may increase by 20 percent or more to 1.2 million Arabs and perhaps 150,000 Jews, and that water demand will rise accordingly.

In the mid 1980s the water supply potential of the district was reappraised at about 600 MCM/Y. Consumption by that time had risen to about 165 MCM/Y, of which 125 were used by the Arab population and about 40 by the Israeli settlements. The remainder of the potential yield of the mountain aquifer continued to flow to Israel, where the extractable portion of it has been fully utilized, and in some years overutilized.

To protect its water supplies, Israel has restricted the right of the Palestinian residents of the West Bank to drill into the aquifer underlying their district. Stringent measures have been enforced to control water use there. The permission of the Civil Administration was made mandatory for all new well drilling. Approvals have been granted, albeit sparingly, for domestic use but not for the expansion of irrigation. Moreover, restrictions were placed on how much water can be pumped annually from each well. The Palestinians complained about these regulations, but the Israelis replied that the same restrictions on drilling and pumping apply in Israel itself. Clearly, the Israelis are concerned that the drawing of water from the limestone aquifer in the West Bank is likely to reduce the water yield of wells in Israel. In the words of Meir Ben Meir, a former water commissioner: "If the demand is for drinking water, we must say yes. . . . But we are not going to stop irrigating our preexisting orchards so they can plant new ones." Israel is also concerned lest the development of industry in the West Bank entail the disposal of wastes that could percolate into and pollute the aquifer.

At present, water allocation to Arab agriculture, estimated to be 100 MCM/Y, is only marginally greater than in years past. That amount of water allows for the irrigation of about 15,000 hectares, roughly 7.5 percent of the cultivated area of the West Bank. This figure is in marked contrast to Israel, where about 40 percent of the cultivated land is irrigated.

The discrepancy is not simply due to Israeli policies toward the West Bank; it reflects the difference in topography as well as in the degree of intensification of agriculture.

The West Bankers also contend that they pay more for each cubic meter of water delivered to them, since they are not granted the same subsidies that Israeli settlers receive. Those settlers have had deep wells drilled for them, and they tend to consume much more water per capita than their Arab neighbors. The former Israeli government, controlled by the nationalist Likud party, maintained a policy of encouraging Jewish settlement in the West Bank and of opposing any possibility of severing the link between that territory and Israel. The intention was to allow limited autonomy to the Palestinians eventually, but not full sovereignty. That policy has clearly been changed in principle since the return of the Labor Party to power in 1992, but it remains to be seen how the new relationship will evolve in practice.

The Israelis claim that the Kingdom of Jordan "owes" the West Bank about 150 MCM/Y that were slated to be diverted from the East Ghor Canal to the planned (but never built) West Ghor Canal, but by now the kingdom is experiencing an acute water shortage of its own and cannot possibly comply with the original plan. The kingdom counterclaims that Israel is taking more than its rightful share of water from the lower Yarmouk. The West Bank Palestinians thus suffer from both sides, and are especially resentful of the disproportionate allocation of water to the Israeli settlements "planted" in their domain. The Palestinians contend that the Israeli settlers in the West Bank are using an inordinate share of the aquifer's waters, that is, that 100,000 settlers are allowed to draw some 40 million cubic meters while nearly 1 million Arabs are restricted to 125 million cubic meters. The Israelis counterclaim that a major portion of the water allocated to Israeli settlers is used in the irrigation of newly established farms in the Jordan Valley, an extremely arid subregion that had been largely uninhabited before, and where farming cannot be done without intensive irrigation.

Security-conscious Israelis believe that even after the Palestinians are granted autonomy or sovereignty, Israel should retain control over the Jordan Valley, as a buffer zone against the possible entry of hostile forces from the east. Apart from military security, there is the issue of hydrological security. Occupation of the West Bank as a whole guaranteed Israel control over the waters of the mountain aquifer, which provides about one-quarter of the state's fresh water supplies. While not the only reason, of course, this was certainly one of the reasons why Israel was reluctant to relinquish control over the West Bank.

The simplistic formula "land for peace" ignores the fact that "land" (however important in itself) is more than mere territory—it implies control over water. Should an independent state on the West Bank begin to drill for and utilize the water of the limestone aquifer that lies beneath it, and—additionally—should it pollute that aquifer, Israel would be denied nearly 400 MCM/Y, an amount on which it has come to depend. The contention between the Palestinians and the Israelis regarding the waters of the mountain aquifer, boiled down to its essentials, is a classical clash between two principles: sovereignty over the source of the water (claimed by the Palestinians) versus the right of prior use and the natural course of the water (claimed by the Israelis). As such, it is analogous to the competing claims of Turkey versus Syria and Iraq over the Euphrates, and of Ethiopia versus Sudan and Egypt over the Blue Nile. As far as the Israelis are concerned, considering the groundwater in the mountain aquifer to be "Palestinian water" would be like telling the Egyptians that the Nile is "Ethiopian water." On the other hand, considering that groundwater to be "Israeli water" would be like telling the Turks that the Euphrates is "Iraqi water."

The rivalry over water threatens to demolish any political arrangement regarding the future status of the West Bank and Gaza that does not allocate water in a manner satisfactory to all sides. The polemics, back and forth, can be endless. Much of it will fade into irrelevance once peace is firmly established. At that time, there is likely to be a great deal of economic cooperation and many joint ventures among the Palestinians, Jordanians, and Israelis, to the great benefit of all three economies. To achieve a workable and enduring peace, and to promote economic progress, the erstwhile rivals must achieve a compromise that will ensure the equitable sharing of all available groundwaters and surface waters, while strictly protecting their quality and sustainable yields, as well as their efficient utilization.

10

The River of Waste

... as water spilt on the ground, which cannot be gathered up again.
II Samuel 14:14

A great river runs through the entire Middle East, more ample than any of the rivers we have thus far described. It is a diffuse and subtle river, as invisible as groundwater yet even more pervasive; a multinational river, flowing all too freely in and out of every one of the region's countries. And though it is a potentially abundant water resource and relatively easy to tap, it is unmarked on any map and is practically ignored by geographers and hydrologists alike. For want of a better name, we call it the River of Waste.

Although the international rivalries over water seem most dramatic and attract the greatest attention, in fact the water crisis relates as much to the nature of water allocation and utilization within each state as it does to the water allocation between states. The tendency of political analysts has been to make alarmist predictions based on extrapolations of competing demands for water in the face of dwindling supplies while ignoring the great potential for managing and moderating those demands internally.

Many observers and commentators concerned over the voracious thirst for water in the Middle East regard the problem entirely as a shortage of supply. They are apt to say that there is simply not enough water in most of the region's countries, and that if only the supply could be increased the problem would be solved. And how? Perhaps by damming another river, tapping another aquifer, building another aqueduct or pipeline, engineering another interbasin transfer, "milking" more rain from passing clouds or more runoff from barren slopes, towing icebergs from the Arctic or Antarctic, or—as a final resort—desalinizing brackish water or even seawater.

What the "supply siders" miss is the need to manage the demand side of the ledger. The demand for water by a society is not a preordained, unquestionable quantity that must be satisfied at all costs simply because it exists or is claimed. The perceived or purported need is not necessarily the real or rational need. To go on investing in ever more elaborate schemes to meet the already excessive and ever-expanding demands without examining and attempting to moderate those demands will inevitably lead to economic, ecological, and political crises.

The prevailing supply orientation treats demand as a "given" that must be satisfied by ever-greater supply. Needed is a more balanced approach that regards demand as a variable quantity to be controlled; and that recognizes conservation, efficient use, and protection of sustainability and quality as primary goals of rational water resource management. The imperative to adopt a new approach is inescapable in view of the rising costs of supplying additional fresh water and disposing of wastewater, and the environmental damage inflicted in the process.

Promoting efficiency, conservation, and resource protection is a challenge no less crucial than augmenting supplies. It is a task faced by many nations around the world, the United States not excepted. But in the Middle East it is most urgent. At present there is much latitude for improving the way water is used, and doing so seems the most immediate, practical, and economical approach to the water shortage. Opportunities exist in every sector: agricultural, industrial, and domestic.[1] But adoption of water-efficient practices requires vigorous promotion. Unfortunately, many societies, especially tradition-bound societies, are extremely conservative and tend to cling to anachronistic and wasteful practices.

The task calls for a fundamental change in perception and attitude. Water must be recognized as the precious and vulnerable resource that it is, no longer to be taken for granted as an inexhaustible free good. To effect the necessary change, each state must institute a program of public education and a system of incentives and rewards for conservation. Such a program might include appropriate pricing policies for water[2] and encouragement of private and public investment in water-saving technologies and strategies, along with administrative mechanisms designed to promote and enforce such a program.

A useful parallel can be drawn between the demand for water and the post-1973 demand for energy in the industrial countries. Both water and energy have long been priced below their true costs of supply and without regard to the ultimate environmental consequences of profligate use. Both have been governed by institutions committed to augment supply rather than manage demand, and both have been used so freely and for so long that most people came to doubt the very possibility of conservation. However, in the case of energy, the externally imposed price rises beginning in the 1970s induced pro-conservation government policies, and these have in fact resulted in significant improvements in efficiency. Although there is still much room for further improvement, the trend of energy use can already serve as an indication of what can and should be done with water.

Since in most of the region's countries agriculture accounts for more than 80 percent of the water demand, the greatest opportunities for conservation exist in the irrigation sector. Conservation strategies can do more than save water; they can—equally importantly—produce more food and fiber while avoiding the worst ecological damage resulting from present-day practices.

Despite the declining importance of agriculture to national income and employment in nearly every country, most of the region's governments are continuing to subsidize water for agriculture in an effort to promote self-sufficiency in food production. Attempts to change that policy where it is clearly uneconomical often encounter the opposition of sectors with vested interests in maintaining the status quo, however futile such efforts may be in the long run, given the widening gap between the profligate demand for water and the dwindling supply. The extent to which the policy of food security can drain the limited water budget is exemplified by Libya and Saudi Arabia, both of which are engaged in the extraction of irreplaceable groundwater to produce wheat at costs that may be ten times greater than the cost of wheat available in the world market.

Another essential principle of sound water management is to devote as much effort to conserving water quality as to conserving quantity. Too often the discussion regarding water in the Middle East focuses exclusively on quantity while ignoring quality. Hydrologists or economists who consider only the water balance tend to forget that water-borne pollutants (e.g., salts) have their own mass balance and tend to concentrate where water evaporates. Water-borne pollutants may migrate into domestic sources of supply, including reservoirs and wells. Wastewaters, including toxic chemicals and pathogenic organisms, are commonly dumped into streams, through which they flow into surface reservoirs and seep into underground aquifers. Consequently, much of the region's water is already polluted by industrial, agricultural, and domestic wastes. Pollution is bad enough for surface waters. It is even worse for groundwater, which is more sluggish and less readily flushed out and renewed than surface water in streams.[3]

An extremely important aspect of conserving both the quantity and quality no water is the collection of wastewater (sewage) from urban centers as well as from irrigated areas. After appropriate treatment to remove toxins and pathogens, such water can be reused to irrigate many types of crops and to serve in industry. Wastewater can also be applied, with certain safeguards, to groundwater recharge. Thus, a potential source of pollution can be turned into a beneficial resource.

There is a relation between the quantity of water use and the effect of use on water quality: efficient irrigation systems can reduce soil salinization; and recovery processes in industry and low-flow household appliances can cut discharges into sewers. The same policies that promote economically efficient use can also help to protect water quality and the environment.

The most prevalent method of irrigation in the region is still the surface flooding of basins or furrows, a method that has, at best, a 50 percent efficiency. That is to say, it involves the needless (and in fact harmful) percolation or runoff of half or more of the water delivered. (Sprinkler irrigation can be operated at an efficiency of 70 percent or more, while drip irrigation can attain an efficiency as high as 90 percent if competently managed.) Moreover, in much of the Middle East, cropping intensities (average number of crops grown per year on the same plot of land) are about 1.4, whereas in East Asia they surpass 2.4.

The perceived scarcity of water is strongly tied to the pattern of allocation of water and to policies that either discourage or encourage its efficient utilization. Direct or indirect subsidies that lower the price of water below real cost, especially to farmers, generally result in excessive use. Similarly, the provision of water on a fixed schedule without monitoring the amount taken invites careless overuse. On the other hand, strict monitoring of the water volumes used and imposing realistic prices in proportion to the amount taken generally cause the recipients to use the water sparingly and to practice water conservation. Most effective are policies that charge for water progressively, in increasing proportion to the amount used. For example, if prior research establishes the optimal water requirements of a given crop, then growers of that crop should be charged a progressively higher price per cubic meter as they use more and more in excess of the optimum. In some cases, the imposition of fines on overuse of water might be justified, though positive incentives and pricing mechanisms are in principle preferable to punitive measures.

Traditionally, water has been regarded as a free gift of nature. In most countries, notably in Egypt, water is provided virtually without charge. Water pricing as a mechanism for recovering the cost of providing it, and as an inducement for promoting conservation, is resisted vehemently and all too often flouted. Moreover, the pricing of water per unit volume requires the installation and maintenance of metering devices, which can easily be sabotaged or bypassed; as well as the imposition of penalties for excessive use by administrative personnel, who may not always be incorruptible.

A distinction can be made between absolute and relative scarcity. Absolute scarcity refers to a supply insufficient to satisfy the indispensable minimal needs of a populace for drinking, sanitation, and basic self-supporting economic activity. That amount is estimated to be roughly 100 cubic meters per person per year, on average. Relative scarcity, on the other hand, refers to an insufficiency of water for maintaining activities that may be desirable or traditional but are not essential to the life of the community.

Certain types of water use should be discouraged. Among such are the irrigated production in arid regions of water-guzzling crops that can readily be imported from water-rich regions; and the continued reliance on inefficient irrigation methods that, incidentally, tend to waterlog and salinize the soil. Another wasteful practice, especially in cities, is the irrigation of lush ornamental plants where much water could be saved by replacing such plants with drought-resistant species. The latter can be just as beautiful, and ecologically more authentic in an arid region. In general, better economic utilization of water, encouraged by appropriate pricing and allocation policies, can go a long way toward alleviating current and future water shortages, and therefore conflicts.

Current water policies in the Middle East have engendered a culture of waste. In most cases, water is priced far below its marginal cost of production, leading to overirrigation and the production of water-intensive crops in a region endemically short of water.

It is a disconcerting fact that irrigated farming in many areas falls far short of achieving its potential yields. Even more disconcerting is the fact that extensive irrigated areas are losing productivity because of waterlogging and salinization. The question arises whether land degradation is an inevitable consequence of prolonged irrigation, or merely of its careless practice. In other words: is irrigation necessarily self-destructive sooner or later, or can it be sustained in the long run?

Experience shows that the problem is due primarily to mismanagement. At fault is the unmeasured and generally excessive application of water to the land, with little regard either for the real cost of the water (in contrast with its arbitrarily set price, which is typically too low), or for the potentially destructive processes that result. Another frequent and closely

Figure 10.1 The still prevalent practice of conveying water in unlined channels and applying unmeasured (often excessive) amounts of water to the land, generally without provision for drainage.

related fault is the neglect to provide for drainage and to manage the salts as well as the water so as to prevent the process of soil salinization.

A universal fallacy of humans is to assume that if a little of something is good, then more must be better. In irrigation (as indeed in many other activities) just enough is best, a controlled quantity of water just sufficient to meet the requirements of the crop and to prevent the accumulation of salts, no less and certainly no more. Applying an insufficient amount of water is an obvious waste, as it fails to produce the desired benefit; on the other hand, applying an excess amount can be still more harmful, as it tends to impede aeration, leach nutrients, induce greater evaporation, raise the water-table, cause salinization, and greatly increase the need for and the cost of drainage.

The term "efficiency" is generally understood to be a measure of the output obtainable from a given input. Irrigation and water-use efficiency can be defined in various ways, depending on the nature of the inputs and outputs considered. For example, one may attempt to define as an economic criterion of efficiency the financial return in relation to the investment in water supply. One problem is that costs and prices fluctuate from year to year and vary widely from place to place. Another problem is that some of the costs of irrigation, and certainly some of the benefits, cannot easily be quantified in tangible economic or financial terms, especially in places where a market economy is not yet fully developed. Often, only the short-term costs and immediate benefits are discernible, whereas the long-term advantages or disadvantages of alternative actions are unknown a priori.

Quite different from the strictly economic criterion of efficiency is the physiological efficiency. The criterion here is the amount of total plant growth, or of the harvestable yield, per unit volume of water taken up by the crop from the soil. This depends on the choice of crop and variety, on local environmental factors, and on the package of agronomic practices designed to optimize the controllable conditions affecting crop growth.

Still another criterion of efficiency relevant to irrigation is the technical application efficiency, or what irrigation engineers call simply "irrigation efficiency." It is generally defined as the amount of water added to the root zone (or the amount supplied to the crop) as a fraction of the amount taken from some source. The actual level of this efficiency depends on whether it is applied to regional projects or to individual farms. In each case, the difference between the amount withdrawn from the source and the net amount of water supplied to the crop represents the losses incurred in conveyance from source to crop. Those losses consist of seepage and evaporation from canals and ditches en route to the field, and runoff, evaporation, and deep percolation beyond the root zone within the field.

Where open and unlined distribution ditches are used, uncontrolled seepage and evaporation, as well as transpiration by riparian vegetation, can cause major losses of water. Even pipeline distribution systems do not always prevent loss and may leak water from loose or corroded joints or

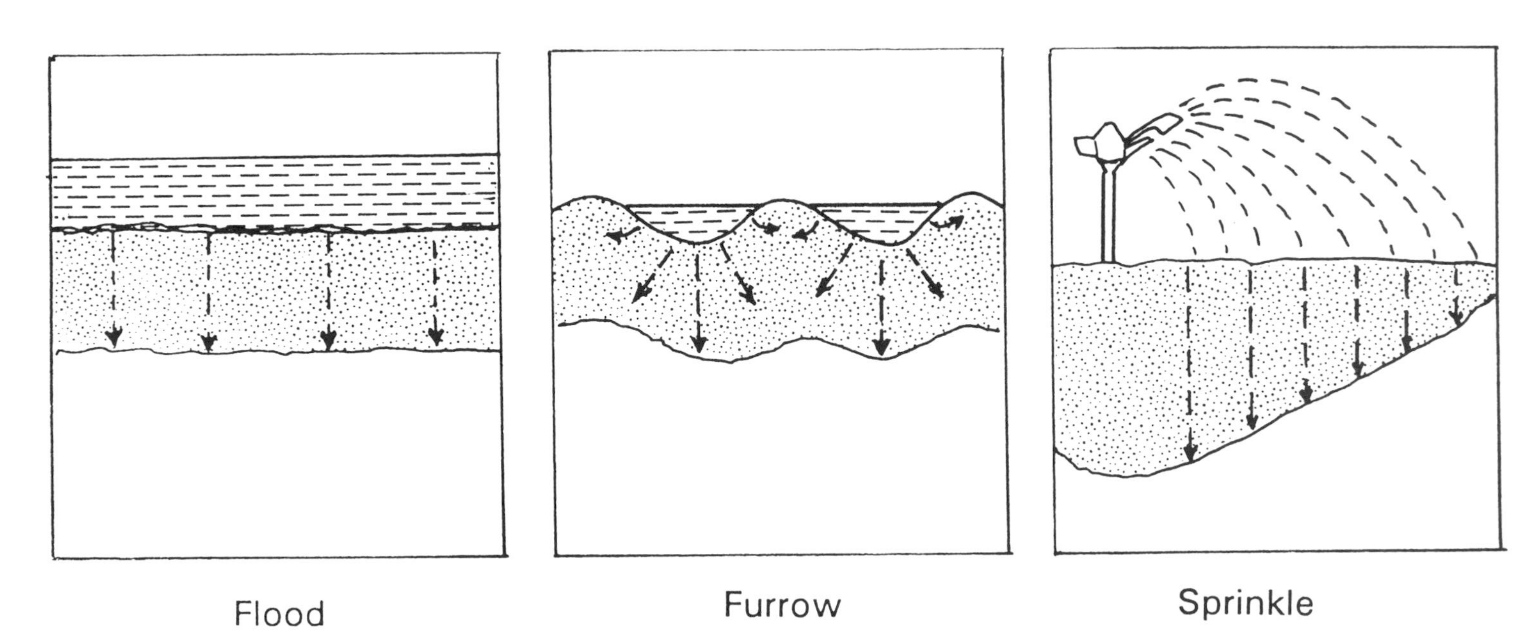

Figure 10.2 The pattern of water distribution in the soil under different irrigation methods.

from ill-maintained valves. Where the pipelines are buried in the soil, such losses of water are not readily apparent and may escape economic valuation entirely.

It is unfortunately true that in most traditional irrigation schemes, based on conveyance of water in unlined earthen canals and on running water freely over the land surface, the application efficiency is generally much less than 50 percent and in many cases even less than 30 percent. This means that most of the water taken from a river or a reservoir not only goes to waste but exacerbates the problems of drainage and land degradation. Since it has been proven that application efficiencies as high as 90 percent can be achieved, there is obviously much room for improvement.

From the point of view of water use, some large-scale irrigation projects operate in an inherently inefficient way. Where water is delivered to the consumer only at fixed times, and charges are imposed per delivery regardless of the actual amount used, customers tend to take as much water as they can while they can. Particularly difficult to change are management practices that lead to deliberate waste not necessarily because of insurmountable technical problems or lack of knowledge but simply because it appears more convenient, or even more economical in the short run, to waste rather than to conserve water. A typical situation is when the price of irrigation water is lower than the cost of the labor or equipment needed to avoid overirrigation. More often than not, the price of water is kept deliberately low, for political reasons, by government subsidy. The cost of water may be distorted even in the absence of governmental subsidy. For example, in the case of users drawing water from an aquifer in excess of the rate of annual recharge, the cost of pumping may be only a small fraction of the cost of replenishing the aquifer after it has been depleted. However, by the time the aquifer is emptied, the overusers may have passed on, leaving the problem to future generations.

New irrigation techniques, by which water is delivered in closed conduits and applied in small volumes at high frequency directly to the root zone of crops at a controlled rate in exact response to their optimal needs, offer the greatest opportunities for water conservation. The opinion is often expressed that these modern methods are too costly[4] for adoption in the less developed countries, so these countries have no choice but to remain with their traditional methods of surface irrigation. That opinion is fallacious. It considers only the direct installation and operating costs, but ignores the hidden (yet real!) costs of water wastage, waterlogging, and land degradation, and the eventual need for, and increased costs of, drainage and land rehabilitation. Although the latter costs are not immediate, they are practically certain to occur eventually, and when they are taken into account the relative costs of modern versus traditional irrigation methods change radically. Furthermore, there is a great and as yet unrealized potential for simplifying and reducing the costs of modern irrigation methods and of adapting them to the needs and the circumstances of the less developed countries, where the relative costs of capital and labor are very different from those that prevail in the industrialized countries.

The principle of economic efficiency calls for allocating scarce resources preferentially to their most productive uses. In Israel, for example, many of the products of the irrigated agricultural sector are still water intensive. A similar or even worse situation exists in the other countries of the region. The policy that encourages wasteful use of water also spurs investments on the supply side to satisfy the "needs" of a thirsty agricultural sector that is made artificially profitable, even though the capital allocated to this effort could contribute more to the overall economy if directed to other activities or sectors.

Optimal use of resources may be promoted by making prices proportional to marginal costs of production. Market prices are often distorted by administrative interventions, subsidies, or taxes. In this connection, it is important that the true scarcity price be determined and used as the decisive criterion. The most economical allocation of water can then be determined by comparing alternative uses. However, while pure fresh water can usually be transferred from one use to another, marginal waters (such as brackish or reclaimed sewage) have restricted uses and cannot be substituted from one sector to another.

Clearly, water use in the Middle East is still very much suboptimal. By shifting water among alternative sectors and activities, governments and economic enterprises have much room to improve efficiency and reduce the severity of perceived shortages. Having said that, however, we must acknowledge that any reallocation of water within (and certainly between) states is likely to come at the expense of, and be painful to, the social and economic sectors that had benefited from the status quo ante. As these might represent politically powerful interests, changes in policy recommended by economists may encounter strong resistance. To overcome the inertial resistance and to effect the necessary changes, there must be strong and farsighted leadership, able to convince the larger constituency of the ultimate benefits of a new water management policy. It seems most likely that the needed atmosphere of change will come in the context of a comprehensive move toward peace and regional economic development in the Middle East, such that will offer the hope of a better life for all the region's people and will lessen nationalistic concerns for the attainment at all costs of "food security" or self-sufficiency in agriculture.

Consider the case of Egypt, where the economy depends most heavily on irrigation. Egyptian water management is still largely characterized by a rigid distribution system and by the flood-irrigation of level basins, and it is threatened by a rising water-table and by saltwater intrusion into the Delta.

Traditional, centuries-old surface irrigation systems are still used in most of the Old Lands of the Nile Valley and Delta. Moreover, the cropping pattern is often wasteful. For example, close to 30 percent of the water supply for irrigation is allocated to rice and sugarcane, both of which are water-extravagant crops. Such crops may seem profitable from the individual farmer's perspective, but they are ultimately very expensive to the country. Water is supplied free of charge to the farmers in the Delta, whereas their counterparts in the New Lands pay substantial sums for the delivery of water. Clearly, there is a need to change the traditional cropping patterns and irrigation systems in the Nile Valley that cause losses of water and require expensive drainage.

There are, however, stubborn obstacles to improving the efficiency of irrigated farming in the Nile Valley. Fragmented lands and disjointed holdings make it particularly difficult to adopt modern methods. The present fixed-schedule water supply pattern is incompatible with modern technologies, for example, sprinkler and drip systems, which require a continuous availability of water. It also constrains the introduction of new crops, which may necessitate altering the frequency and quantity of water application.

The introduction of more efficient irrigation methods and a greater selection of crops into the Old Lands depends on changing the time-honored system of water delivery. That can be done only if the physical infrastructure and the administrative (as well as social) structure of Egyptian farming are modified, a difficult task indeed. Sustainable water management also demands the installation and operation of an efficient drainage system to prevent water-table rise and salinization, and to allow recovery and reuse of drainage effluent.

Most water delivery systems in Egypt are inefficient, so that water loss by seepage and evaporation may be half or more of the amount of water allocated to a given area. Likewise, the delivery of water to the domestic sector (3 BCM/Y) and to the industrial sector (4.7 BCM/Y) involves losses exceeding 50 percent. Some of the seepage is recovered for reuse, but at a lower quality and larger expenditure than the original Nile water. In addition to allocating water for agricultural, domestic, and industrial purposes, Egypt must allocate water (estimated at 1.8 BCM/Y) to maintain navigational levels in the Nile, particularly for tourist cruisers.

Treated wastewater is an important auxiliary source of water for irrigation. However, current treatment procedures, at best, remove only organic and pathogenic components but not the salts. Therefore, drainage water cannot be reused repeatedly without eventually exceeding the threshold of salinity tolerated by some crops, and without increasing the hazard of soil salinization. So provision must be made for the safe disposal of unusable drainage waters. At present, agricultural drainage water in Upper Egypt is discharged into the Nile River, and this increases the river's salinity. Improvement of irrigation efficiency can reduce the volume of drainage but might also increase the salt concentration of the drainage waters. A serious challenge to irrigated farming is the safe and permanent disposal of salts, to prevent their accumulation in the soil. The salt-laden drainage water can be disposed of into the sea, if the area is near the coast, or to the desert. Specially constructed conduits are needed to prevent the seepage of saline waters into the groundwater.

The Egyptian government has made great efforts to expand the area of land under irrigation and to introduce modern methods of irrigation in the New Lands, hoping thereby to meet the needs of a rapidly growing population. The government has decreed that the New Lands be irrigated by modern methods such as drip or sprinkler. Where applied successfully, modern irrigation methods have allowed the New Lands to improve agricultural production and to shift cropping patterns toward high-value crops, including fruits, vegetables, and oilseed.

In the early years of the New Lands program, the imported technology was too often inappropriate. Instead of developing small-scale systems attuned to the indigenous social structure and the needs of individual farmers, the government attempted a short cut to development by promoting large commercial or state-owned farms. This approach has been inefficient: it has robbed farm workers of incentive and made the practice of farming too impersonal and bureaucratic. Consequently, attempts are being made to establish more responsive systems of land ownership, water rights, and water supplies; and to adopt scaled-down variants of modern water control and application technology. This implies making water available on demand to accommodate the variable needs of farmers, rather than according to a rigidly fixed schedule that allows them no individual discretion.

The sustainability of water resources in Egypt is threatened by the declining quality of fresh water supplies. The continuous discharge into the Nile of wastewater from untreated domestic and industrial sources and from agricultural fields is causing the progressive pollution of that water course and of the groundwater underlying the Nile Valley. The use of chemical fertilizers, especially water-soluble compounds of nitrogen, phosphate, and potash, has increased fourfold in the three decades following the construction of the Aswan High Dam (which holds back the fertile silt that had long been Egypt's natural fertilizer). Residues of those fertilizers tend to eutrophy the water in canals and in the river itself. The wholesale use of herbicides to control submergent weeds in irrigation canals and water hyacinths in drainage ditches has caused serious environmental hazards and is likely to be discontinued. This is only one among many cases in which new technologies introduced to solve specific problems have adverse environmental impacts that eventually threaten the sustainability of the farming system as a whole.

Egypt's population, now nearing 60 million, is projected to increase to 70 million by the year 2000. If the demand for water rises proportionately, Egypt will require at least 75 BCM/Y, including some 10 BCM/Y for its new reclamation projects. However, much can be done to conserve water now used wastefully. Greater use is to be made of marginal water resources, notably drainage water and recycled effluent. The wastewater recovery project for Greater Cairo, for example, is designed to treat and release nearly 2 BCM/Y.

The looming water shortage has induced the government to apply various measures to regulate water use in agriculture. Distribution networks are being converted from open channels to closed conduits, such as pressurized pipe systems, and automated water control devices are being introduced. Restrictions are imposed on planting low-value crops that consume too much water, and positive inducements are given for planting high-value, water-thrifty crops. Water allocations to farmers are limited and are dependent on the annual assessment of water availability. Research is encouraged on water-efficient crops and irrigation methods and on optimal cropping patterns. Finally, pricing policies and other incentives encourage farmers to conserve water. Egypt's agricultural and technical cooperation with Israel, made possible by the peace accord of 1979 and pursued with American encouragement but without fanfare, has helped greatly in promoting water use efficiency.

Some 30 years ago, a revolutionary development occurred in the science and art of irrigation. Appropriately, that revolution took place in the Middle East, the region where irrigation has been practiced longer than in any other part of the world. Surprisingly, however, the initiators of this revolution were not the region's most experienced irrigators but newcomers to the practice who were able turn the disadvantage of their initial inexperience into the advantage of a fresh approach, unfettered by ancient traditions. The new irrigation began in Israel within less than two decades of its birth as a state. In retrospect and prospect, this technical (and non-political) development may be Israel's greatest contribution to the welfare and progress of the entire region,[5] and of similar regions throughout the world.

Having witnessed and participated in that extraordinary development, I am able to describe it directly. It was a rare and heady experience for a young scientist to be present at the birth of an innovation that had both fundamental and practical importance and has since proven to be applicable globally. I refer to the advent of high-frequency, low-volume irrigation techniques, specifically of drip and microsprayer irrigation. What began as a questioning of the guiding premise of conventional wisdom gradually evolved into a conceptual change, an inversion of the traditional approach, and the invention of a whole new technology.[6]

During the 1960s and early 1970s, I was engaged in conducting and coordinating research on the efficiency of water utilization by irrigated crops in Israel. We began by taking what was then the accepted approach to irrigation, a time-honored practice presumably bolstered by research carried out during the early decades of this century in California. The California irrigationists formulated a hypothesis that soil moisture remains practically equally available to crops until evaporation and extraction by roots deplete it to some residual low value called the "permanent wilting point."

In practice, this hypothesis justified an infrequent irrigation regimen designed to wet the soil periodically to its maximal field capacity, then to let the crop deplete soil moisture almost to the wilting point before irrigating once again to replenish the soil moisture reservoir. The conventional irrigation cycle thus consisted of a brief episode of irrigation followed by an extended period of extraction of soil moisture by the crop. Accordingly, the topsoil was periodically saturated, a condition that leached nutrients and deprived the roots of aeration; then it was allowed to dry to a degree that desiccated the roots that were near the soil surface. Practical limitations on the frequency of irrigation by conventional methods made it difficult to test alternative methods of irrigation designed to avoid such fluctuations by maintaining an optimal level of soil moisture continuously in a well-aerated (moist but not saturated) root zone.

The low-frequency mode of irrigation seemed to make economic sense because most of the conventional irrigation systems are more expensive to run at a higher frequency than at a lower one. With such systems, irrigators tend to minimize the number of irrigation per season by increasing the time interval between successive irrigations. (For example, where the cost of portable tubes is a major consideration, it pays to make the least amount of tubing serve the greatest area by shifting available tubes successively to as many subplots as possible before returning to the same subplot for the repeat irrigation.) The question was, thus, how dry can the soil become before the crop experiences a "significant" reduction in yield? The hypothesis of undiminished soil moisture availability to crops until they near the wilting point is an interesting example of the human predilection for contriving theoretical justification for what is convenient.

Gradually, the contradictions became too obvious to ignore. Evidence accumulated that soil moisture is not "equally available" but rather less and less readily available to crops as it is progressively depleted. Therefore, it occurred to some of us that we ought to try the opposite approach: to irrigate as frequently as possible but with very small volumes of water. For the traditionalists, our unorthodox approach seemed foolish. Irrigate more frequently in an arid environment? No, they said, that would increase water use per unit area. But smaller water use per unit area never should have been an end in itself. A better criterion is water use per unit of production. And this is where the new approach eventually proved itself. Rather than ask crops to go thirsty without diminished performance, we began to ask crops how much better they could perform if they were grown in a constantly moist soil and prevented from ever suffering thirst.

Irrigation researchers had focused for so long on plant behavior in a drying soil that they seemed oblivious to what might happen to crops grown in a continuously moist soil. When we tried that, we discovered that crops often show a pronounced increase in yield when irrigation is provided in sufficient quantity and frequency that water never becomes a limiting factor,[7] especially when nutrients are applied along with the water.

But how to accomplish that in practice? The optimal condition could not easily be achieved by traditional surface flooding and even by the then-available sprinkler irrigation methods. Fortunately, newer irrigation methods became possible, allowing the application of water in small quantities as often as desirable, without additional cost for the extra number of irrigations. These include permanent installation of low-intensity sprinklers, subirrigation by porous tubes, and especially the techniques of drip, trickle, and microsprayer irrigation.

In the technique of drip irrigation, first tried on a field scale in the Negev in the mid 1960s, water is provided at the base of each plant or cluster of plants drop by drop, at a slow rate that can be regulated precisely to meet the water requirements of each specific crop, no more and no less. The technique had earlier been applied in greenhouses, but what made it practical in the field at last was the advent by that time of low-cost tubing that could be fitted with specially designed drip emitters, all made of weather-resistant plastic materials. Such assemblies could be laid on the ground in any configuration to conform to crop rows and either be left in place perennially or be gathered up again and shifted elsewhere at will. To prevent clogging of the narrow-orifice drippers (by particles, algae, or precipitating chemicals), filters developed earlier by the swimming-pool industry were adapted for use in the field. Later, the application system was supplemented by ancillary equipment such as timing or metering valves to set the exact volume of water for irrigation and then shut off, and by injectors to feed soluble fertilizers into the water supply. These devices lent themselves readily to labor-saving automation.

Within a few seasons, the results of the early trials confirmed the new hypothesis beyond initial expectations. Yields of orchard and field crops, perennial as well as annual, rose dramatically, and the amount of water required fell markedly.

Raising crops by the old method of infrequent, heavy irrigations seems analogous to raising babies by force-feeding them on Sunday, then waiting a full week, until they are practically famished, before gorging them once again. In contrast, drip irrigation is like spoon-feeding a baby with frequent, small portions, in continuous response to need so that the baby is neither overfed nor underfed at any time.

With drip irrigation, it is possible to create favorable moisture conditions even in problematic soils (such as coarse sands, gravels, or heavy clays) that had previously been considered unirrigable. It is also possible to deliver water uniformly to plants growing in fields with variable elevation, slope, wind velocity and direction, and soil characteristics. Drip irrigation can maintain the root zone at a moist yet unsaturated condition, so the roots lack neither water nor oxygen. Finally, and very importantly, the ability to target irrigation precisely to the base of crop plants, thus wetting only a small fraction of the surface while avoiding wetting the inter-row areas, reduces compaction by traffic, evaporation, and weed proliferation. The latter effect further helps to conserve water.

Where salinity is a hazard, as where the irrigation water is somewhat brackish, the continuous supply of water to the soil ensures that the salts do not concentrate progressively in the root zone. Moreover, since drip irrigation is applied underneath the plant canopy, it avoids the hazard of leaf scorch by brackish water and reduces the evaporation of intercepted water as well as the incidence of fungal diseases that thrive on moist foliage (such as occur under overhead sprinkler irrigation).

After the initial success of drip irrigation, it was not long before several variants of this method were adapted to an array of soil and crop conditions. The variants are based on the same principles of high-frequency, low-volume, partial-area irrigation. Among these variants is bubbler irrigation—a low-cost alternative especially adapted to basin and furrow irrigation, in which water is allowed to "bubble up" through vertical tubes of predetermined height. Another variant is microsprayer irrigation, by which water is sprayed over an area of a few square meters rather than dripped at points.

Together, drip and microsprayer irrigation are called "microirrigation." These methods contrast with alternative modern methods of automated high-frequency irrigation such as the center-pivot irrigation machines used in the high plains of the United States and visible as green circles from the air. The latter systems are designed specially for large-scale irrigation and require that water be provided at relatively high pressure. Microirrigation, on the other hand, can be scaled down to any size of field or plot and operated at low pressure, so it is inherently more flexible and adaptable to the needs of small-holders in developing countries.

Above all, the new irrigation methods, when properly applied, can help avoid the most common failing of conventional high-volume irrigation, namely, overirrigation. Precise, high-frequency, low-volume irrigation not only reduces the hazards of waterlogging and salinization but also helps to lower drainage costs. (With conventional irrigation, the installation and operation of drainage is often more expensive than the irrigation system itself.)

But microirrigation in and of itself is no panacea. Meticulous operators can achieve high efficiency with these methods, but careless operators can be just as wasteful with such systems as with conventional ones. Drip emitters have narrow orifices that are easily clogged by materials carried in the water, such as sand particles or algae. Moreover, the fact that only a fraction of the soil volume is wetted, which we extolled as an advantage, can also be a problem. Although even large trees can thrive when water is supplied to less than 50 percent of the soil volume (provided that the water and nutrient supplies within this limited volume are sufficient), a typical crop under drip irrigation becomes extremely sensitive and vulnerable to any disruption in the water supply. If the irrigation system does not perform

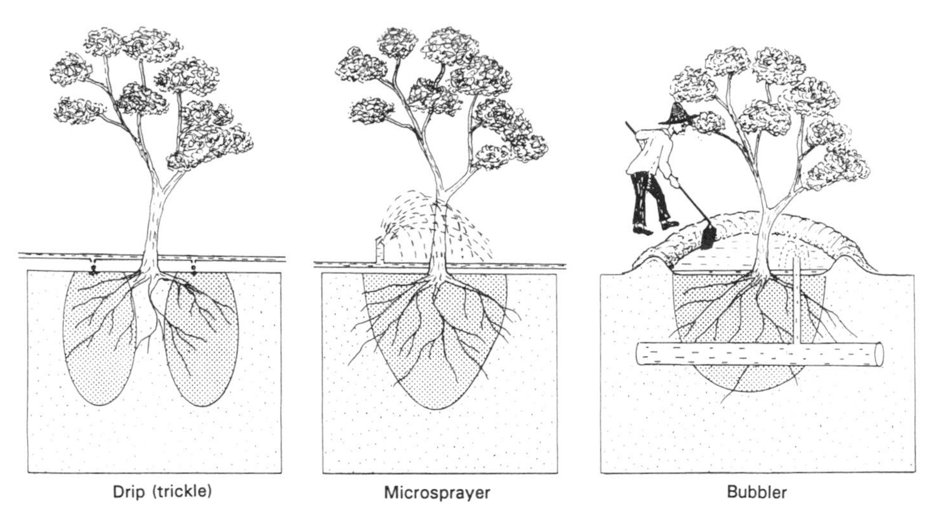

Figure 10.3 Irrigation of single trees by drip, microsprayer, and bubbler methods.

perfectly and continuously, crop failure may follow quickly, since the soil moisture reserve available without replenishment is so small.

Apart from the capital investment and technical problems associated with installing and maintaining the irrigation system in perfect operation without interruption, there is the initial problem of making water available at all times. The necessary infrastructure is seldom provided in countries lacking capital and technical expertise. Also lacking in some countries is research-based guidance to irrigators on the optimal water requirements of specific crops during the successive stages of their growth, as influenced by variable weather conditions.

Owing to its small size (an area of only some 20,000 square kilometers, of which more than half is desert) and the extreme shortage of water, Israel has had to base its development on strict water conservation from the outset. For this reason, Israel has been forced to undertake the role of a pioneer in the management of scarce water resources. In so doing, it has learned lessons that can be of immediate and lasting benefit to its neighbors. Peaceful cooperation could enable the other countries of the region to avail themselves of Israeli technological innovations, especially in irrigation and wastewater reuse.

There are 430,000 hectares of land under cultivation in Israel, of which some 215,000 are irrigated. The average water use per hectare has been reduced by 50 percent with the introduction of sprinkler and (especially) drip irrigation systems. Planners now envisage a one-third reduction of water allocations for agriculture, to a level under 900 MCM/Y, with an increasing proportion of that water to come from reclaimed sewage.

Figure 10.4 Drip irrigation by means of regularly spaced emitters in narrow plastic tubes.

Moreover, the shift from high water-use crops such as cotton and citrus, already begun, will doubtless continue.

Even though Israel has an exemplary record of achievement in water resource development, sewage recycling, and irrigation water use efficiency,[8] it now faces increasingly acute water shortages. The problem is closely associated with the preferential allocation of water to agriculture. If it were possible for the country to subsist without irrigated agriculture, then it could make do with no more than 100 cubic meters per capita per year. For a population of 5 million, that would amount to a total of 500 million cubic meters. Eliminating irrigated agriculture is obviously impossible. However, regarding agriculture merely as an economic enterprise (rather than as serving higher national or ideological goals) would allow shifting to alternative uses part of the billion or more cubic meters of fresh water now provided for irrigation.

In the opinion of some economists, Israel's economy might actually benefit from such a shift. They also claim that it does not make much economic sense for Israel to grow irrigated crops in the outlying Negev or the Golan Heights when the same crops can be raised more efficiently in a climatically more favorable subregion closer to the water source and at lower elevation. It might even be more efficient to import a greater quantity of agricultural produce rather than to continue the effort to achieve agricultural self-sufficiency (and to export irrigated fruits and vegetables) by using expensive water while denying that water to potentially more remunerative non-agricultural enterprises. Such changes, however necessary, are certain to be resisted by Israel's politically powerful farmers, who invoke the ideological foundations of the state to support their stand.

Figure 10.5 A young banana plantation under drip irrigation.

Already, there is internal competition between the rapidly growing needs of Israel's largely urban-based economy and the vested rights of its farming sector. Agriculture, which still uses 70 percent of the country's water, now employs less than 5 percent of Israel's labor force, and the fraction is continuing to diminish. Given the rate of immigration and internal population growth, and the imperative to employ more people in non-farming enterprises, it seems inevitable that Israel will reduce the production of low-value crops such as cotton that are readily obtainable on the world market. Such a change need not occur by coercion, but as a consequence of the necessity to raise the price of water. That rise in price will happen when the state reduces its subsidies for irrigation water, an action it seems likely to undertake sooner or later despite the resistance of the agricultural sector.

Israel's underpricing for water has been purposeful. It originated historically from the state's explicit aim to emphasize the development of agriculture, not merely as a basic industry but, indeed, as an exemplary way of life. The early Zionists exalted farming as the soul of the nation, its spiritual and physical connection to the ancestral land, to nature, to history, to biblical lore, and to the original meaning of the Jewish holidays. There was a practical purpose as well to the encouragement of agricultural settlement. Needing to occupy the outlying, empty parts of the country quickly, the state delivered water everywhere it could and subsidized people willing to live and to farm in the frontier areas of the desert and along the borders. The state also wished to promote economic independence through self-sufficiency in food production. Indeed, Israeli agriculture now provides (either directly or indirectly through income-producing

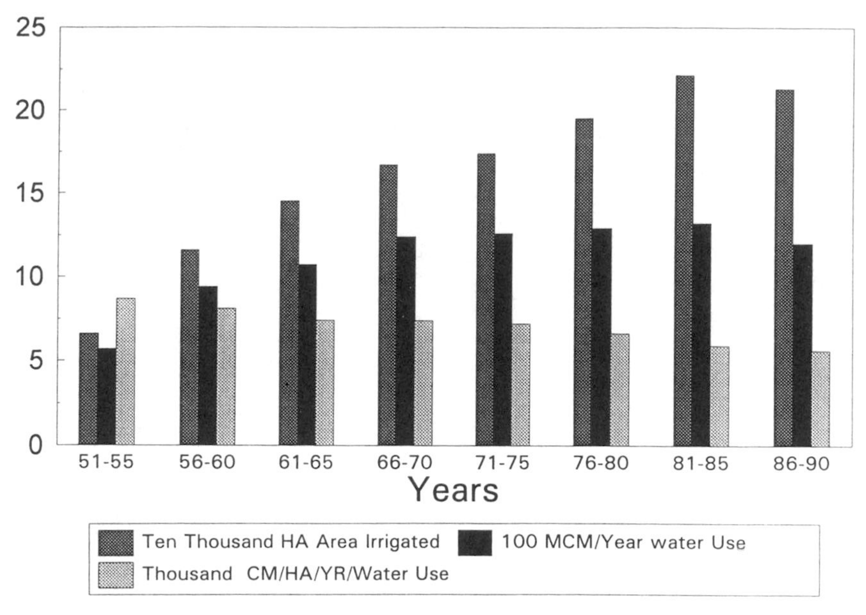

Figure 10.6 Irrigation trends in Israel, 1951–1990. (HA = hectares, CM = cubic meters, MCM = million cubic meters)

exports) substantially all the food requirements of the country. However, the agricultural sector has diminished in relative importance since the 1960s and now accounts for only about 7 percent of the gross national product (employing about 5 percent of the labor force and producing about 4 percent of the total export earnings).

Israel, incidentally, is not unique in subsidizing water for agriculture. The United States does the same in the western states. In practically no country can farmers afford to pay the real cost of irrigation water without drastically raising the prices of their products.

The charges for irrigation water in Israel depend on the quantity used. Each farmer is given an allocation of water yearly, the amount depending on the crop and the area irrigated. For up to 80 percent of the allocation the price is about \$0.125 per cubic meter, for the last 20 percent of the allocation the price is \$0.20/CM, and for excess use the price rises to \$0.26/CM. The price is 40 percent higher during the peak demand period of July and August. The price charged urban users for domestic water is of the order of \$0.50/CM. As the actual average cost to supply water in Israel is currently estimated to be about \$0.36 per cubic meter, one could argue that the subsidy to farmers is low relative to the corresponding subsidies granted by other countries in the region and elsewhere (e.g., in California). However, this preferential treatment of agriculture cannot be long continued at the same rate, as the needs of the domestic and industrial sectors continue to grow and as the share of agriculture in Israel's GNP continues to shrink.

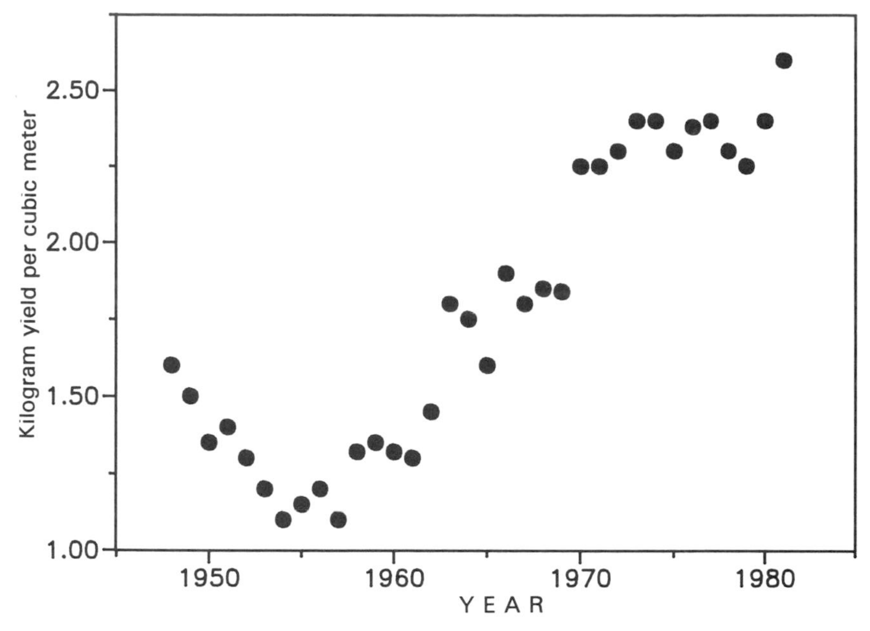

Figure 10.7 The improvement of crop water use efficiency in Israel.

Critics of Israel's current water policy contend that control of the country's water resources and their allocation must be taken from the Ministry of Agriculture and given to an administrative body concerned with the entire national economy rather than with a particular sector within it. To justify its allocation of water, Israeli agriculture must continue to enhance water use efficiency.[9] Hence Israel is now attempting to develop even more intensive forms of agriculture. Future agriculture will be based increasingly on the production of high-value crops in controlled-environment greenhouses where yields can be maximized per unit quantity of water utilized. The development of such new forms of agriculture, already begun, requires large investments in research and in capital.

On the positive side of the ledger, agriculture helps to solve the sewage disposal problem by providing an economically useful means for its recycling. In the process, wastewater is not only utilized but also rendered largely harmless. Some (though not all) of the potential pollutants are detained, retained, and decomposed in the soil. Thus they may serve as nutrients or otherwise be prevented from contaminating streams, lakes, or groundwater. The trend to utilize marginal water resources is well under way in Israel. In 1990, a total of 195 million cubic meters of recycled effluents was used, of which 114 million cubic meters served for irrigation. The amount of waste water to be made available will increase in the course of the decade of the 90s to some 300 MCM/Y. Additional "marginal" water resources to be utilized include ephemeral floodwaters from intermittent streams (40–80 MCM/Y) and brackish groundwater (145–160 MCM/Y) to irrigate salt-tolerant crops.

The amount of water that can be allocated to agriculture is not a constant: it fluctuates over time, depending on the vagaries of climate. Since the needs of the domestic and industrial sectors are relatively inflexible, agriculture must absorb most of the annual fluctuations in water supply. A volume of some 800 million to 900 million cubic meters per year can be supplied to agriculture with a high degree of assurance (say, at a probability of 19 out of 20 years). An additional 300 million cubic meters can be supplied at a lower level of assurance (say, in 8 out of 10 years). The exact probabilities cannot be predicted precisely because the future course of seasonal rainfall is unknown.

In principle, the agricultural sector in Israel must be flexible enough to respond to such fluctuations in the allocation of water. In practice, a cropping pattern based on annual crops is more flexible than one based on perennial crops that must be maintained over a period of years regardless of seasonal fluctuations. Over time, a national strategy will evolve to optimize agricultural water use, based on an optimal mix of annual and perennial crops with diverse water requirement patterns.

Israeli farmers are now being urged to increase production of specialty crops that bring high export earnings rather than go on using up precious water to grow crops that other countries can produce and sell more cheaply. Consequently, some citrus and avocado growers have either reduced their orchards or regrafted new varieties onto them in hopes they will yield better returns. The efficiency of water use (rather than of labor or machinery) is now more than ever the crucial test for Israeli farmers. Crops such as cotton and citrus that were the mainstays of Israeli farming are losing their primacy. In their stead, specialty crops (including out-of-season vegetables, flowers, spices, and medicinal herbs) are gaining importance, particularly for their export potential. Such crops may be grown in smaller areas in highly controlled conditions under plastic cover rather than in the open field over extensive areas.

And so Israel, having so proudly made the desert bloom, must now soberly recalculate the long-term cost of its remarkable development of irrigated agriculture. Painfully, it now finds this symbol of national success withering in the harsh light of evolving economic and environmental realities. That pain will be mitigated in large part when regional peace opens up new markets for specialized agricultural as well as industrial products, and promotes the growth of alternative economic enterprises that will bring greater prosperity to the entire region.

An important lesson from all the foregoing is that perceived or claimed water shortages are relative to particular patterns of demand. Changing those patterns can change the entire future outlook for water. One instructive criterion is the amount of water available per capita. In Israel it is approximately 370 cubic meters per year, already below the amount of 500 CM/Y considered by some[10] to be the threshold of adequacy per person for all uses. Yet the water supply in Israel does provide an adequate standard of living and sanitation for the country's population, including highly developed industrial and agricultural sectors and recreational facilities. And although water use in Israel in some years has tended to exceed the annual supplies, the necessary adjustments can be made to bring demand and supply into balance. As Israel's population continues to grow, greater efforts will be necessary both to augment supplies and to curtail per capita demand by making water use even more efficient than it already is.

In contrast, the per capita annual supply of water in Egypt exceeds 1,200 cubic meters. Of course, agricultural water requirements in Egypt are higher than in Israel, given the warmer and drier climate, the lack of effective rainfall, and the greater proportion of the population engaged in agriculture. Even so, the amount available per capita seems quite adequate, and it is likely to remain so if the Nile's upper riparians do not appropriate too much of the river's headwaters and if Egypt itself improves its efficiency of water use. In Lebanon, Syria, and Iraq the potential per capita water supplies are well above those of Egypt.[11]

In the Kingdom of Jordan, however, the quantity of water available has already fallen below 300 cubic meters per capita per year. Jordan is in a real quandary, as its population increases even as its water supplies are reduced by Syrian appropriation of the Yarmouk's headwaters and by the depletion of fossil aquifers. Such a progressive reduction in per capita supplies will force the desert kingdom to change its pattern of use drastically and to enter into a new set of international arrangements to alleviate what will otherwise be a worsening crisis.

Those who wish to resolve the water issue must go beyond the polemical bickering and concentrate on the real solution. The solution must include measures to augment supplies, to avoid waste, to utilize available supplies more efficiently, and to reuse water by capturing and treating sewage and drainage effluents. All these tasks can be achieved more readily in the context of peaceful cooperation among the nations of the region sharing the same limited supplies and the same challenges.

11

Augmenting Supplies

We send down water from the sky according to due measure, and we cause it to soak in the soil;
And we certainly are able to drain it off with ease.

Koran XXIII:18

The preceding chapter made the case for water conservation: so much water is now wasted or used inefficiently, and so much can be done to save it. So much, however, and no more. Ultimately, the question reverts to the supply side: can the water reserves of the region be augmented? Are there hidden reserves to be tapped within the separate countries, or has each reached the limit of what can possibly be obtained independently? And, if so, can supplies be enhanced cooperatively?

The search for additional supplies continues. And where no more natural supplies of fresh water are readily available locally, ingenious schemes have been offered to summon up additional supplies by unconventional means.

Consider the quandary now facing Israel, Jordan, the West Bank, and Gaza, all of which are almost at a dead end in their separate efforts to find additional local resources, and in all of whose territories aquifers have been exploited beyond their sustainable yields. A more equitable sharing of existing water resources may ease the current tension temporarily but will not provide a solution for the future. Only joint action to develop additional resources (as well as to further enhance the efficiency of water utilization) can resolve the problem in the longer run.

The primary source of fresh water is precipitation, rain and snow. However, in an arid climate precipitation is unstable, and drought is an ever-present threat. Throughout history, severe dry spells have caused great upheavals, including mass migrations, wars, revolutions, and extinctions of cultures. Where no alternative sources of water (rivers, springs, or groundwater) existed, all that people could do to promote more rainfall was to pray for it, as did the ancient Hebrews,[1] or to perform a rain dance, as did the American Indians. Statistically, the success rate of such rituals could not have been greater than 50 percent, since seasons of above-average rainfall are no more prevalent than seasons of below-average rain.[2]

From whence comes the rain? The ancients did not understand clouds as recondensed vapor. Still, they looked to the clouds for rain. And so do we today.

Until quite recently, Mark Twain's quip, "Everyone talks about the weather but nobody does anything about it," was a truism. Nothing could be done about the weather. The way to survive a drought was to store the excess grain or the excess water of wet seasons for subsequent use during dry seasons, and to use water sparingly. In the last few decades, however, there have been serious efforts to do more than that: to actually induce greater rainfall in specific regions.[3] The usual technique to accomplish this is to "seed" clouds with volatile or dispersible materials aimed to "nucleate" raindrops by causing the minute droplets floating in a cloud to coalesce into larger drops, thereby transmuting an ethereal mist into actual falling rain.

The idea was born in the late 1940s, when a scientist named Vincent Schaefer accidentally discovered that he could trigger the precipitation of fog in a laboratory chamber. He was working with a cold chamber fashioned from a deep freezer. Worrying that the ambient room temperature might cause the chamber to warm up, he slid in a chunk of dry ice (carbon dioxide frozen into a solid block). Instantly, ice crystals formed in the cloud around the dry ice, and the dense fog turned into drops of water that trickled down like rain. Shortly after that, Schaefer's colleague Bernard Vonnegut found a chemical salt, silver iodide, that was almost exactly like ice in its molecular structure but could be carried and applied more easily than dry ice.[4]

The discovery was immediately hailed as a great blessing. The most obvious expected benefit was augmentation of water supplies. More rain could promote crop growth in semiarid areas and enhance groundwater recharge, stream flow, and water storage in reservoirs. Moreover, in salinity-prone areas an increased infusion of fresh rainwater or runoff into existing reservoirs could improve water quality.

Over the next few decades, enthusiasts invested millions of dollars into cloud seeding. Commercial entrepreneurs, though they themselves barely understood the scientific basis of the new technology, were quick to peddle it to thirsty communities. The results were mixed, and therefore ambiguous. Attempts at weather modification in too many places failed to produce convincing effects. Moreover, there were those who opposed the entire effort as a violation of ecological principles. To them, it represents an interference in a natural process, with consequences that might be difficult to predict or control. Man as a geologic force has built dams and levees, and has altered the flows of great rivers, thus upsetting ecological balances and destroying entire ecosystems. What further havoc might human meddling in atmospheric processes wreak upon the earth?

Skepticism and resistance were therefore provoked by the excessive enthusiasm of early advocates, who tended to claim more than they were able to deliver. During the early years of weather modification work in the 1950s and 1960s, the shift from scientific research to commercial application was too hasty. Fast-talking promoters sold their service indiscriminately, and that service was soon found to be unreliable and fraught with problems. Even where success was apparent and rain seemed to be enhanced for the operators' clients, there was the danger that others would complain (and sue!) for having had too much rain perpetrated over them. And people living downwind of the target area might—and did—complain that they were deprived of the rain that would otherwise have fallen on their territory. None, however, could prove such contentions conclusively.

Complex legal and public policy questions arose. Who owns the passing clouds? Who is liable for damages from floods or other weather events possibly resulting from (or merely attributed to) weather modification activities? How can the rights of those who want more rain be reconciled with the rights of those who prefer less, or none? What if rainfall is seen to increase in a catchment over which cloud seeding took place but at the same time seems to decrease in another catchment? Has the latter been unfairly deprived of its rightful rainfall? And there are other questions: How can the amount of new water be quantified for credit and distribution? On what basis are the costs and benefits of weather modification to be allocated among the purported beneficiaries? Also not to be neglected are possible environmental problems resulting from weather modification. Local or regional manipulation of climate could affect present plant and animal populations. For example, increased precipitation might mean increased infestation by weeds or by other pests and diseases, as well as accelerated erosion or impeded drainage. Concern has also been expressed about the unknown environmental effects of introducing artificial condensation nuclei (e.g., silver iodide) into the atmosphere.

All these questions, however important they seemed, were premature as long as the uncertainty remained regarding the most basic question, to wit: is weather modification both sound scientifically and feasible in practice? By how much might it increase precipitation over a given area? In other words, does it work?

Eventually, most commercial entrepreneurs withdrew from weather modification and only the hardy souls who were serious scientists persisted with the effort. Now, at last, it appears that their labor has brought the dream somewhat closer to fruition, albeit on a more modest level than originally conceived. By tracking seeding chemicals with high-speed analyzers and computer programs, researchers have been able to follow the path of their chemicals from the time of their release into the storm cloud until they form raindrops. They now have a better idea of the best clouds to seed, as well as the best time and the best way to seed them. And they are more realistic: they claim no dramatic increases in a region's rainfall regime, only marginal enhancement. And because no rain can be made where no clouds exist in the first place, no one should expect weather modification to green up a desert or to end a drought completely.

So, like the ancients, modern atmospheric physicists look to the clouds for rain, but unlike the ancients they complement prayer with action. Using science, they venture to do what only mystics had tried before: to coax rain out of reluctant clouds. The modern rainmakers attempt to do to a cloud formation what farmers normally do to a field: they "seed" it. And like farmers, they hope that their "seeds" will develop and mature to provide a timely harvest. Only the harvest they expect is atmospheric, not terrestrial; physical, not biological. Their intended harvest is water, in the form of enhanced rainfall.

Radiation from the sun supplies the energy to evaporate water from the sea or from wet surfaces on land. The vapor rises into the air, then wind sweeps the vapor-laden air over rising terrain such as occurs in the mountain ranges paralleling the eastern coast of the Mediterranean Sea. Rising higher, the moist air enters the colder layers of the atmosphere. Tiny water drops are then formed through condensation, and a cloud results. A typical cloud is an assemblage of trillions of suspended droplets, too small to fall to the earth. In fact, more than 99 percent of the volume of even the densest clouds is air.

Before they can become raindrops or snowflakes, the droplets must somehow jump across the air spaces separating them and clump together, millions at a time, until they are heavy enough to fall to the ground. A common mechanism by which droplets coalesce is through the formation and growth of ice crystals at cold temperatures. However, this does not necessarily happen as soon as the temperature falls below the normal freezing point. If the cooled droplets do not encounter a nucleating agent about which they can begin to crystallize, they may remain in the dispersed liquid state. The water's purity and the lack of nuclei prevent the droplets from crystallizing into ice. They are called *supercooled droplets,* and form supercooled clouds. A nucleating agent is a collection of particles, generally crystalline, that provide templates to which water molecules can attach themselves, thus initiating the formation of ice crystal. As each crystal grows and is buffeted by air currents, it collides with many more droplets, which, in turn, attach themselves to the crystal and enlarge it. Eventually, the crystal becomes too heavy to remain suspended in the cloud, and it begins to descend. In so doing, it encounters more droplets and is further enlarged. As it becomes heavier, the crystal accelerates its fall toward the ground, along with innumerable others that share its fate. If the air below the cloud base is warmer than freezing, the ice particles tend to melt en route and to fall as rain.

The first condition for the success of rainfall inducement by cloud seeding is, of course, the presence of suitable clouds. But the mere presence of such clouds does not necessarily result in local rain. In the absence of a sufficient concentration of condensing nuclei, the clouds may simply float away from the area over which they hover, or, if they encounter warmer and drier air, they may dissipate by reevaporation. The purpose of cloud seeding is to induce the formation within the cloud of crystals that can act as foci for the formation of raindrops and thus hasten the occurrence of rain. This can be done by means of aircraft that spread either dry ice (frozen carbon dioxide) or silver iodide microcrystals into the atmosphere. The latter material can also be applied from ground-based generators, though less efficiently than from aircraft.

That, at least, is the theory of rainfall enhancement, and the rationale advanced by its advocates. But the theory needed to be validated in practice by careful experiments. The fundamental problem encountered by scientists is the difficulty of proving that any particular rainfall event that may occur following seeding would not have occurred otherwise, or would have given less rain. Direct cause and effect and quantitative relations can be proven under highly controlled conditions in the laboratory but not under the variable and capricious conditions of nature, where all relevant variables are generally not controlled and often not even recognized. Hence the only practical way to study the effects of cloud seeding is by statistical methods, which require much time. Several decades of consistent seeding are needed in the same geographical area to accumulate sufficient statistical data to yield reliable results. Even then, such methods provide only probable indications rather than absolute conclusions. Statistical uncertainties make it difficult to obtain unequivocal proof that the changes in rainfall noted on the ground do in fact result from the treatment and not from naturally occurring random variations.

Despite the fact that the trials have continued for several decades, the technique of cloud seeding has not yet emerged completely from the research phase into full-scale application. The obstacles are that the very process of rain formation in clouds and the physical effect of artificial seeding under varying meteorological conditions are not yet completely understood. Clouds and the generation of rainfall have turned out to be much more complicated than anyone had thought originally. Cloud seeders now must determine whether or not the water in any set of clouds is supercooled; and whether, when, and how to seed those clouds for maximum effect.

The original cloud-seeding experiments dealt primarily with convective cumulus clouds, the most common and widely distributed among cloud forms and the world's most important sources of precipitation. The short life span and the instability of such clouds complicate seeding operations. Orographically enhanced clouds (which form as moist air rises up the windward slopes of mountainous areas) are better for seeding because they last longer and therefore allow weather modification experiments to be implemented and monitored more readily.

In 1989, the World Meteorological Organization counted 118 weather modification projects operating in 32 countries, under a coordinated international program. In the United States, nearly all the western states have had active programs of weather modification research. Nowhere, however, has the work been carried out more painstakingly and consistently than in Israel.

Ever since the 1950s, Israeli meteorologists have been conducting systematic trials to ascertain whether rainfall can be increased by cloud seeding. The purpose has been not only to add moisture to the soil for greater crop growth and to reduce the requirement for irrigation but also to increase the water yields of such natural sources as springs, streams, and aquifers. The experimenters hoped that if the method proves itself on an experimental scale, it can perhaps be applied on a country-wide basis to augment water supplies and thus obviate the need for such inherently expensive alternatives as desalinization and importation.

Most of Israel's rainfall (occurring from early November to late April) is contributed by convective clouds associated with winter extratropical cyclones that form over the relatively warm waters of the eastern Mediterranean. The rain cloud systems occur typically in bands associated with cold fronts and cold air masses behind them. These rainbands generally move from the Mediterranean Sea inland, where the winds sweep the vapor-laden air and the clouds over rising terrain. Thus the clouds are subject to orographic enhancement.

The hypothesis underlying the rain inducement work in Israel, as elsewhere, is that the conversion of cloud water into precipitation is inhibited by an insufficient concentration of nucleating ice crystals in some of the clouds, and that such crystals can be formed by seeding the clouds with nuclei of silver iodide. The desired concentration of the added crystals was thought to be 1 to 10 per liter. Based on these concepts, systematic cloud-seeding experiments were begun in Israel in 1961.

Rain enhancement trials have been held in the northern part of Israel,

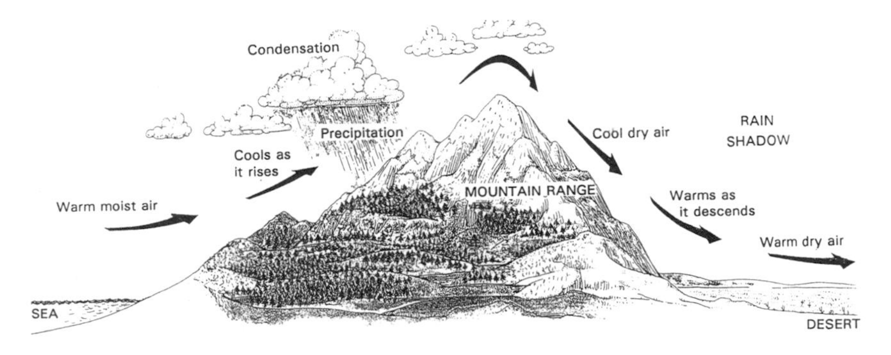

Figure 11.1 The orographic effect on rainfall.

where natural rainfall is 400–1,000 mm/Y, and in the central part of the country where rainfall is 300–600 mm/Y. The procedure is to introduce silver iodide into clouds from aircraft flying along lines parallel to the coast, and to leave "control" sections untreated. The results are then monitored and recorded in terms of daily rainfall, along with stream and spring discharges.

Each winter, whenever storm clouds roll across the southeastern shore of the Mediterranean, a specialized aircraft takes off and flies along a preassigned north-south trajectory parallel to the shore. For maximum effect, the aircraft flies at cloud-base level along a line upwind of target areas. When reaching the base of the clouds, the pilot releases a plume of microscopic particles of silver iodide salt. Simultaneously, operators on the ground, located along a similar transect, begin burning silver iodide–impregnated coal in special furnaces meant to release upward whisks of smoke toward the same clouds. The tiny crystals are meant to disperse within the clouds and initiate the formation of rain.

The first experiment (1961–67) apparently resulted in a 15 percent enhancement of the rainfall in the target area. The second experiment (1969–75) appeared to give a 13 percent increase of rainfall in the northern part of the country, but not in the south. The seeding effect was found to be largest at a distance of 20–50 kilometers downwind from the seeding line, where it evidently increased rainfall by as much as 22 percent. Subsequently, the seeding line in the north was moved inland to shift the area of maximal positive effect to the catchment of the Sea of Galilee.

Following the early trials of the 1950s and 1960s, clouds in northern Israel have been seeded operationally since 1975, while randomized experimental seeding is continuing in the south. In attempting to interpret the differences between the results obtained in the northern and southern parts of the country, Israeli scientists have hypothesized that the prevalence of dust in the air in the south[5] (nearer the desert) reduces the effectiveness of the seeding treatments. It appears that dust blown from the Sahara and the Arabian deserts[6] bordering the target area in the south can serve to nucleate clouds naturally, hence artificial seeding is only beneficial when this dust is absent. On days without dust, the researchers reported that rain was enhanced over the catchment area of the Sea of Galilee by as much as 34 percent. Whether such episodic enhancement has any significant long-term effect remains in doubt, however. In seasons with naturally abundant rainfall, such enhancement, if effective, may even cause excessive rainfall and thus do more harm than good.

All these difficulties notwithstanding, the indications are that the seeding of convective clouds under winter conditions in the eastern Mediterranean probably can enhance rainfall to the extent of about 10–15 percent on average. (The demonstrated augmentation of stream flow, however, is only about 6 percent.) This increment may seem marginal, and by itself it is certainly no panacea for the basic water shortage of the region. But under the circumstances, and in conjunction with other promising approaches, cloud seeding appears to be sufficiently effective to make the program worthwhile, especially during drought years.

So the work continues. The operation itself is fairly inexpensive and flexible: the equipment can be moved from place to place at will, and it requires little in the way of costly installations. The additional water theoretically obtainable by this means may cost less than 1 cent per cubic meter.

There are, however, caveats. The "milking" of rain from a cloud formation in one area may affect the rain that would otherwise fall on another area located some distance downwind. In other words, the gain to one location may involve a loss to another. Because the areal extent of the effect may be unpredictable, and because the distribution of clouds as well as the courses of the winds, streams, and aquifers do not recognize political boundaries, the practice of cloud seeding may itself exacerbate rather than alleviate the rivalry over water. This caveat does not negate the practice but does argue for pursuing it in a context of regional cooperation. A comprehensive international effort may well be of benefit to all sides.

Studies have shown that cloud seeding in Israel enhances—rather than diminishes, as some had feared—rainfall over the parts of Jordan and Syria that lie downwind of the direct target area in Israel. Recent research offers the possibility of enhancing rainfall even further, particularly if cloud seeding is carried out on a more extensive regional scale. That can only be done cooperatively within the framework of a comprehensive peace agreement. Even at best, however, cloud seeding offers only a limited potential for relieving the growing water shortage of the thirstiest countries of the Middle East.

Gentle rains falling onto permeable soils are absorbed entirely, part to be detained in the root zone for subsequent uptake by plants and part to percolate beyond the root zone to the groundwater reservoir below. But intense rainstorms may exceed the ability of the soil to absorb water (called the soil's *infiltrability*), especially where the soil is shallow and relatively impermeable. Whenever the rainfall rate exceeds the rate of absorption, free water tends to accumulate over the soil surface. This water collects in depressions, thus forming puddles, the total volume of which depends on the smoothness or roughness of the surface as well as on the overall slope of the land.

If an intense rain persists over a sloping area, eventually the puddles begin to overflow and water tends to run off the surface of the land. This water, called *surface runoff* or *overland flow*, begins to run downslope as "sheet flow" but eventually collects in channels variously called *rills* or *gullies*. Such channels generally form a treelike pattern of converging branches leading to larger and larger streams.

In agricultural fields, runoff is generally undesirable, since it results in loss of water for the crops and often causes erosion, the amount of which increases with increasing amount and velocity of runoff. Uncontrolled runoff also causes off-site damage, as it results in destructive flooding and sedimentation of downstream areas.

In many arid regions, however, large tracts of land remain unused owing to insufficient or unstable rainfall, non-arable soils (too shallow, stony, or saline), or unsuitable topography. Such tracts can be set aside for the purpose of inducing and collecting runoff to augment the water supply of the region. The collection and utilization of runoff have been practiced by people in semiarid and arid regions since antiquity. In large areas of the Middle East (particularly in the Negev of Israel[7]), it is still possible today to find functional ancient systems of terraced stream beds that were used for farming, with their water supply augmented by runoff water collected from the adjoining hillsides. Similar techniques were used to collect runoff in cisterns for subsequent use by humans and domestic animals.

As described in the chapter on ancient civilizations, the early inhabitants of the Negev were apparently not satisfied with the natural rate of runoff, which generally did not exceed 10 percent of the seasonal rainfall. They actually attempted to enhance the formation of runoff by removing the stones from the surface of the hillsides that served as their catchments. This method of inducing runoff, though ingenious, was laborious and of limited effectiveness. At best, it only doubled the amount of runoff. Today, we can surpass their achievement, using modern machinery and chemical

Figure 11.2 A roadway breached by a flash flood in the Negev Highlands.

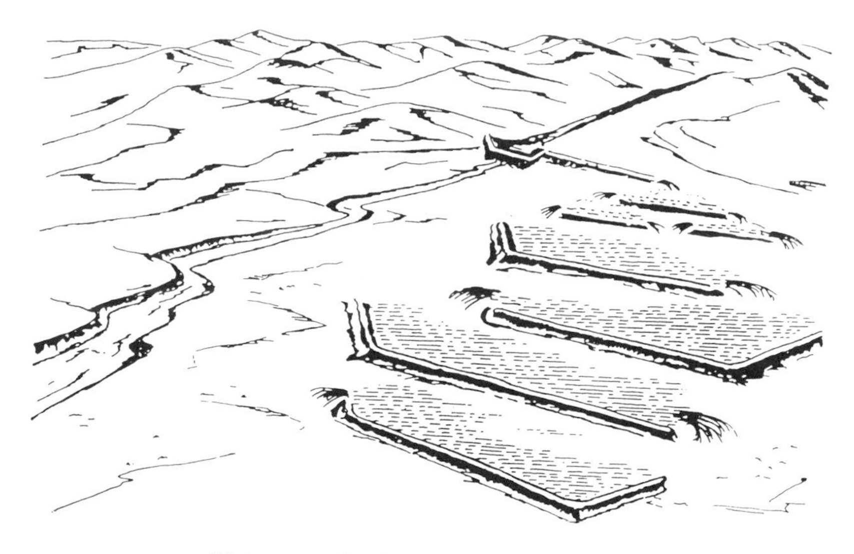

Water spreading by earthen dikes

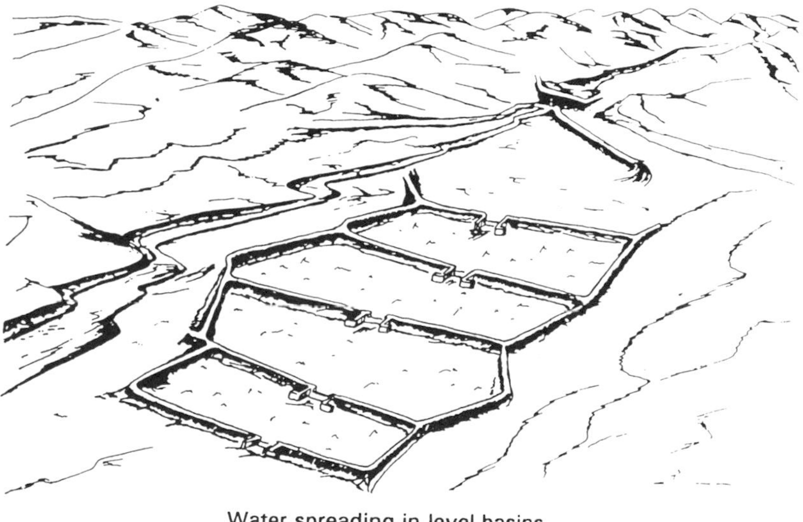

Water spreading in level basins
with stone or concrete spillways

Figure 11.3 Two methods of water spreading by diversion from a wadi.

substances unimagined by our predecessors. With such means, we may be able to induce nearly 100 percent of the rainfall to form runoff from designated areas. Recall that 100 millimeters of rainfall on just 1 square kilometer (0.4 square mile) constitutes 100,000 cubic meters (nearly 30 million gallons) of water of the highest quality!

Figure 11.4 Flood-irrigated orchards in the Negev.

During the 1950s and on to the early 1970s, I conducted extensive research on methods of runoff inducement,[8] also called *water harvesting*. The simplest way is to cover the surface with an impervious apron of plastic, metal, or concrete. The problems encountered with such an approach are how to make the covering materials adhere permanently to the soil surface, which may be irregular and have jagged protruding stones; and whether such materials can resist weathering. An alternative approach is to cause the soil itself to shed, rather than absorb, the rain. In general, we can classify soil treatments for runoff inducement as follows: *mechanical treatments*, such as clearing stones and shrubs, then smoothing and compacting the surface; *dispersive treatments*, to promote the self-sealing tendency of clay at the soil surface; *hydrophobic treatments*, to reduce the wettability of the soil surface with such materials as sprayable silicone solutions, the way fabric raincoats are treated to make them water-repellent; *surface-binding treatments*, to permeate and seal the soil with an adherent material (such as asphalt) that can cement the loose soil into a firm matrix; and, finally, various combinations of such treatments.

In some arid areas, it is possible to raise crops by means of a modern system of runoff farming, where part of the land surface is shaped and treated appropriately for runoff inducement while another portion of the land is arranged to receive the runoff water so produced. Several systems have been tried in respect to size and arrangement of the runoff-contributing area in relation to the runoff-receiving area. For example, parallel contour strips of land can be treated on a slope so as to shed their share of the rainfall to adjacent lower strips in which rows of trees can be planted. Alternatively, the land can be shaped to provide each tree with its own microwatershed.

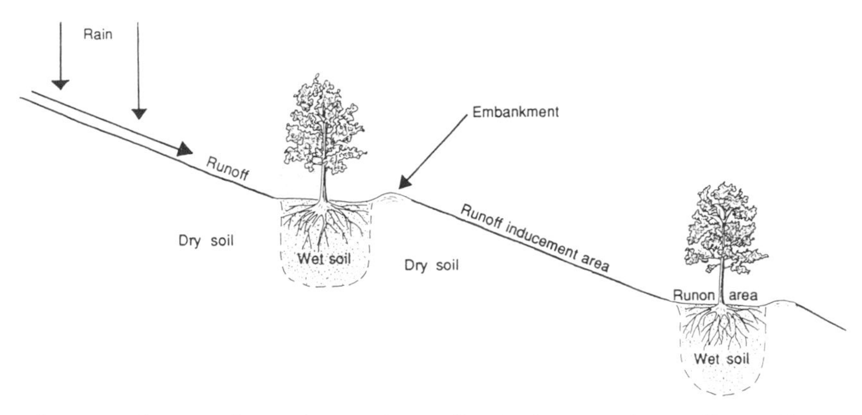

Figure 11.5 A modern scheme of runoff agriculture, with runoff contributing strips treated to enhance water yield.

Our studies showed that such methods can greatly increase the runoff yields obtainable from otherwise barren slopes. Such lands can serve to augment water supplies for reservoirs as well as for crops in areas where water is particularly scarce, especially in highland areas far from alternative sources of supply. However, whether and which of the various water-harvesting methods may be economically feasible will depend in each case on local circumstances.

An intriguing possibility, given much publicity in recent years, is the transfer of water across national boundaries to water-poor countries from water-rich countries that are either in the region or outside it. Of the various imaginative proposals that have been made in this regard, some are so far-fetched as to be dismissed out of hand. Others, however, deserve serious examination. Needed, therefore, are regional studies on the economic and engineering feasibilities and long-term implications (ecological, social, and political) of water-transfer options. The apparent advantages must be weighed against the potential disadvantages of each possibility, in comparison with various alternatives. In the absence of the necessary detailed studies, there is some danger that visionary water-transfer schemes might be promoted simplistically as "quick fixes" to the water shortage of one country or another. That could turn out to be a costly fallacy.

Interbasin and transnational water transfers are certain to require large investments of capital and energy, and to necessitate much time for preparation and implementation. Moreover, they are conditional upon the willing cooperation of several countries that must either supply the water or allow it to be transported across their territories. For some of the proposed schemes, the necessary cooperation cannot be assured until the entire region achieves a comprehensive peace settlement. Even then, transnational schemes make the recipient countries vulnerable to unexpected dis-

Figure 11.6 Microcatchments for runoff-irrigation of trees.

ruptions of supply. Such disruptions may result from technical malfunctions, acts of sabotage, or disavowal of agreements due to changes of regime or policy by one party or another. (Such eventualities are not unlikely in this region.) Hence there is understandable fear on the part of each prospective importing country of handing over control of its water security to outsiders.

The situation may seem similar to, but is really worse than, the dependence of a country on the importation of petroleum, since the disruption of the latter supply from any particular source can always be made up by purchases from another source (given the flexible global nature of the petroleum market). The situation is obviously not so simple in the case of a bilateral or multilateral water-importing arrangement, for the possible abrogation of which the importing country may have no redress.

The vulnerability of recipient nations, at least to temporary interruptions of supply, could be lessened if the imported water were applied to recharging aquifers rather than to immediate consumption. Day-to-day dependency would thus be avoided. However, the costs of recharging aquifers and of subsequently pumping from them, and the danger of water quality deterioration, must be taken into account.

The amounts of additional water needed by the thirstiest nations of the region over the next 25-year planning period can only be estimated roughly. It seems reasonable that the Palestinians will require some 200–300 MCM/Y of water for the West Bank and about 150 MCM/Y for Gaza. Israeli planners have projected the need for some 300–400 MCM/Y of additional water for their state. The requirements for Jordan are estimated to be about 300 MCM/Y over and above the additional water that Jordan expects to obtain from implementation of the Unity Dam on the

Yarmouk. The three figures add up to at least 1,000 MCM/Y of additional supplies to be imported or desalinized by the year 2020.

Among the numerous visionary schemes for long-distance interbasin water transfer are the following: pipelines from Turkey to Arabia and the Gulf States; an aqueduct from the Euphrates in Iraq to Amman, Jordan; a line from Iraq to Kuwait; a canal or pipeline from the Nile across northern Sinai to Gaza; a line from the Karun River in Iran to Bahrain; even a line from the Indus River in Pakistan to Saudi Arabia; and—more recently—a line from Turkey to the newly independent Central Asian Republics of the former Soviet Union. All these schemes are in addition to older ideas of towing an iceberg from Antarctica to Arabia (the Red Sea or the Persian Gulf), and diverting Litani River water from Lebanon to Israel and Jordan. (To skeptics, it might appear as if the fanciful idea of solving the Middle East's problems of water shortage by artificially reshuffling regional water resources has run away with itself.)

We shall describe a few of the possibilities in turn.

Turkey's "Peace Pipeline"

The most highly publicized proposal for transnational water transfer was made in 1987 by the late Turgut Ozal, then prime minister of Turkey who soon afterward became his country's president. Turkey, with its abundance of water resources, is in a unique position to serve as a water supplier and regulator in the Middle East. Its highlands receive copious amounts of rain and snow, feeding several large rivers that are largely unutilized. Turkey's potential water supplies have been estimated at some 180,000 MCM/Y,[9]

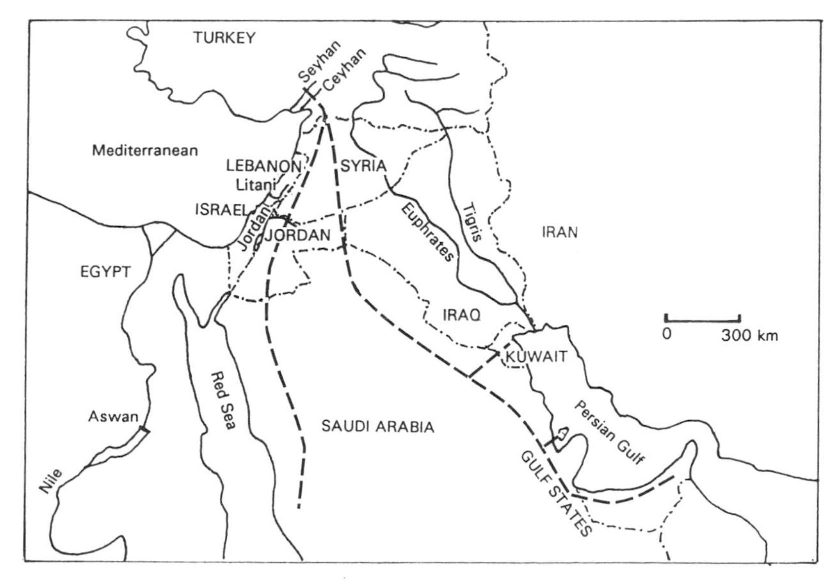

Figure 11.7 The proposed Turkish "Peace Pipeline."

only about 15 percent of which is currently utilized. Even when full economic and agricultural development is achieved according to present plans, Turkey is still likely to have significant amounts of unutilized water reserves.

President Ozal's plan, grandly named the "Peace Pipeline," actually envisioned two pipelines to carry water from two rivers in southern Turkey (the Seyhan and the Ceyhan) all the way to the Persian Gulf States in the east, and to Saudi Arabia's Red Sea coast in the west. En route, the pipelines could supply water to Syria, Iraq, Jordan, and perhaps Israel as well.[10] Altogether, the scheme could bring 2.2 billion cubic meters of potable water each year to more than 15 million people at a construction cost of more than $20 billion. Of course, so heavy a financial commitment can only be undertaken in the context of a stable political atmosphere and with a binding water-sharing agreement. Certainly, the financial resources are available in the Middle East, and when peace comes, all the region's nations (especially the oil-rich states) should be willing to divert some funds from the largely sterile pursuit of military security to ensuring water supplies. Even if such an agreement were reached today, however, the Peace Pipeline would require at least 10 years to construct, and may not necessarily be the best investment under the circumstances.

Critics have raised the suspicion that Ozal's plan may have been a public relations ploy, motivated in part by a desire to counter the unfavorable publicity surrounding Turkey's unilateral action in damming the headwaters of the Tigris and Euphrates as part of its Southeast Anatolia Project. When fully implemented, that project may deprive Syria and Iraq of half the flow of the upper Euphrates (and a substantial portion of the upper Tigris as well). Be that as it may, Ozal's idea excited new interest in the possibility of regional water-sharing arrangements, including international trading in water as an exportable commodity. Because Turkey sits atop the region's greatest water resources, its strategic importance in the post–cold war Middle East has increased.[11]

The Peace Pipeline proposal was not greeted with universal enthusiasm. Even some of the intended beneficiaries expressed skepticism and a reluctance to place themselves in a state of dependency upon the continuing goodwill of an outside power with which their nations had a long and not entirely happy relationship. Some economists doubt that the price of water to be finally delivered by the envisaged pipelines would be lower than the alternative cost of local desalinization. In any case, the entire plan lost its impetus following the elections held in late 1991, in which Ozal's party lost power, and especially following his death in early 1993.

The victor in the 1991 election, Prime Minister Suleiman Demirel (Ozal's opponent), soon disavowed the Peace Pipeline scheme as "not in Turkey's best interest." So, for the time being, the Peace Pipeline was reduced to a pipe dream. In the longer run, however, that exercise in constructive imagination may be a harbinger of things to come. What is now perceived as impractical may yet spring to life in the future.

Water Transfer from Turkey to Jordan and the West Bank

While Turkish President Ozal's concept of a Peace Pipeline to all the water-short states of the Middle East seems too grand to be implemented at this stage, a less ambitious plan, dubbed the "Mini-Peace Pipeline," might be more practical. It calls for the conveyance of some 600 MCM/Y from Turkey through Syria to Jordan and the West Bank. The Sea of Galilee and the Unity Dam (yet to be built) could be used for seasonal storage to supply water throughout the summer peak demand period to Jordan and the Palestinians.

Still another alternative to the conveyance of water from Turkey to Jordan would be to deliver the water only as far as southern Syria. This could induce the Syrians to release the flow of the Yarmouk tributaries that they have been damming in recent years, and thus enable the Jordanians to utilize the full flow of that river. Accordingly, the Turkish pipeline could be made shorter and less expensive, and its operation less complicated, since it would involve only two countries.

Nile to Gaza Water Transfer

The idea of a pipeline to convey water from the Nile through El Arish to the Gaza Strip was originally proposed by the late President Anwar Sadat of Egypt following his peace agreement with Israel in 1979. The Gaza district was even then in the throes of a severe water shortage. Sadat's idea was never implemented, and the water shortage in Gaza has worsened.[12] It is one reason, along with the population explosion and the lack of opportunities (especially for young people), for the dire plight of that district's people.

Since that time, Egypt's own water prospects have deteriorated as the water surpluses enjoyed during Sadat's time have dwindled. Egypt itself may eventually face water shortages, especially if its population continues to grow and if its upstream coriparians appropriate significant quantities of water from the upper Nile.

Nevertheless, the Nile to Gaza transfer scheme might yet be revived. Egypt still releases considerable amounts of water to the sea to maintain its hydrological balance and to flush out the salts that would otherwise accumulate in the soil. The water so released is of better quality than the groundwater now available in Gaza. Moreover, the amount of water needed by the Gazans (estimated to be 150 MCM/Y), while supremely important to the prospective recipients, is not significant when considered in terms of Egypt's water budget (being about one-quarter of 1 percent of it). Nor would the Egyptians need to commit themselves to supply the water indefinitely. The normal life of a project of this sort is about 30 years. If it should outlive its economic utility or its desirability to any of the parties, it could—and probably will eventually—be replaced with desalinization. For the Nile to Gaza transfer scheme to be worthwhile for the

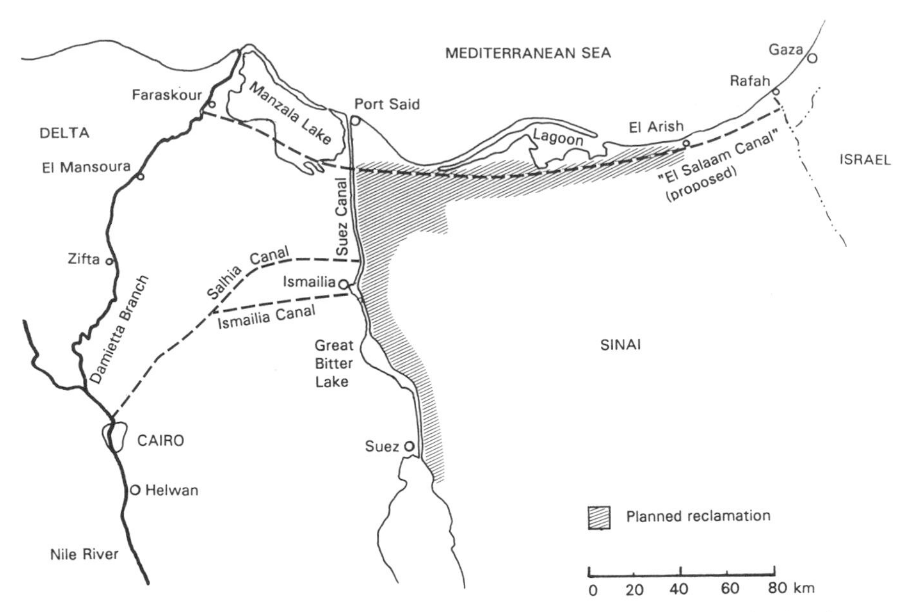

Figure 11.8 The late Egyptian President Anwar Sadat's proposed El Salaam (peace) Canal across northern Sinai toward Gaza.

Egyptians, they should, of course, be compensated adequately; but that is a matter to be negotiated between the parties at the appropriate moment. The moves toward peace between the Palestinians and the Israelis may well bring that moment soon.

In the meanwhile, the Egyptian government has reacted negatively to a similar suggestion made recently by some Israeli academics.[13] They claimed that if Egypt avoids wastage in storage and conveyance, improves its irrigation system, and makes necessary repairs of its facilities at Aswan and elsewhere, it will in fact have surplus water. It could sell some of that water to the Gaza Strip as well as to Israel and—by a process of substitution—to the West Bank as well. That idea resurrects the offer by the late President Anwar Sadat, who termed it "sacred water to the sacred city" (meaning, figuratively, Nile water to Jerusalem). However, not since President Sadat's tragic death in October 1981 has any Egyptian official indicated willingness to consider the transfer of Nile water to Israel or to the occupied territories, and several have expressed decided opposition to that possibility.

In the meanwhile, Egypt has taken the initiative to build an eastward canal to convey water from a distributary of the Nile Delta toward Sinai. That canal already has a length of 87 kilometers and reaches across the Suez Canal. The Egyptians hope to enlist the financial assistance of the Saudis and Kuwaitis to complete the canal into northern Sinai. They emphasize that the function of this canal will be confined to irrigating land in Sinai, though the same canal may well have the technical capacity to convey water to Gaza as well. (Just 5 percent of its designed capacity of 3 billion cubic meters per year would do very nicely for Gaza.)

Litani to Jordan Water Transfer

Compared with its neighbors, Lebanon has plentiful water resources, which could be traded to other nations when a peace agreement restores stability to the region. Lebanon's numerous rivers and underground reservoirs are reliably recharged from ample precipitation falling on its mountains. As a supplier of water, Lebanon could benefit and restore itself to the status of an important trader in the economy of the Middle East. For the time being, though, Lebanon's water resources are not fully developed and the country is suffering from water shortages in its major cities, from seawater intrusion into its overpumped coastal aquifer, and from farmland neglect due to lack of irrigation—all made worse by the havoc of internal and external wars.

The idea of purchasing Litani water from Lebanon and transferring it southward to Israel, to the West Bank, and possibly to Jordan, has been raised repeatedly over the years. The Lebanese themselves suggested this possibility informally during Eric Johnston's negotiations with them in 1955.

At present, the Litani River is being utilized mainly for power generation and only marginally for irrigation. The greater volume of its flow is now diverted to the Awali River and its hydroelectric facility, and then allowed to flow to the sea. When peace with Israel is achieved, the Lebanese will undoubtedly wish to develop the poverty-stricken southern district of their country, now partly under Israeli control. The Lebanese will then need a significant volume of the Litani water to irrigate land in that district. Estimates vary, but there is a possibility that even so there will remain a sufficient surplus to allow exporting some water southward. Here, again, the commitment need not be open-ended. An agreement spanning 30 years or so would be sufficient to tide Israel and the West Bank over the interim period until alternative sources (e.g., desalinization) become feasible.

Some 80 percent of the Litani's flow occurs during the six winter months when irrigation is not required. To utilize this water, it would need to be stored in a reservoir. One possibility is to divert some of the winter water (perhaps 100 MCM/Y) at or above the bend of the Litani into the Bagharit or Hasbani streams, which flow naturally into the Jordan Valley. There, the extra water could be stored in the Huleh basin or in the Sea of Galilee, from whence it could be pumped and conveyed to underground storage (aquifer recharge) in Israel and the Gaza Strip. Israel could then forgo an equitable volume of the Jordan or Yarmouk water it now uses in favor of the Kingdom of Jordan, or a share of the mountain aquifer it now uses in favor of the West Bank. Alternatively, the Litani water could be stored in the Syrian-Jordanian reservoir behind their envisaged Unity Dam, which could be enlarged to accommodate the additional water. The originally planned West Ghor Canal, intended to convey water to the West Bank of the Jordan River, might then be revived.

The disadvantage for the Lebanese of any such diversion of Litani water into the Jordan basin would be the loss of that water's energy potential. An alternative, therefore, would be to catch the water excess downstream, near the river's outlet to the sea and after its energy has been utilized. That water could then be piped southward along the coast of Israel and used either for direct irrigation in the spring season, or to recharge the coastal aquifer for utilization during the summer. In this case, too, an agreement might be reached whereby the amount of water received by Israel would require it to free up an appropriate volume of water from other sources for the use of the Palestinians and Jordanians.

The feasibility of any such project depends, in the first instance, upon the agreement of the Lebanese to transfer their surplus water, for which they must be adequately compensated. The feasibility also depends on engineering and environmental considerations, and must, of course, be worthwhile for both the providers and the recipients. Finally, it depends on the restoration of political stability within Lebanon, and between Lebanon and Israel.

Importing Water by Sea

As Israel searches farther afield for water, ideas have been advanced for schemes to import water by sea from such countries as Turkey, Bulgaria, or even the former Yugoslavia. Transporting water by sea may seem highly fanciful, but the idea was given serious consideration following the drought that afflicted Israel and some of its neighbors in the late 1980s. Conveying water in oil tankers might be prohibitive, but less expensive means may be devised. A Canadian firm designed inflatable plastic-walled barges, which it proposed to call "medusa bags" (presumably after the stinging jellyfish that is prevalent in the Mediterranean Sea). Such bags, large enough to contain many thousands of cubic meters, are to be towed by sea from the outlets of rivers, where the barges can be filled, to specially constructed terminals along the coasts of Israel and Gaza. The contents of the barges can be emptied through pipes into operating reservoirs for direct use, or allocated to groundwater recharge. The estimated costs of water conveyance from Turkey: $0.25 per cubic meter to the coast of Israel, and an additional $0.15/CM for transport from there to Amman.

The apparent advantage of conveyance by sea rather than overland is that it does not necessitate so heavy an initial investment as would a pipeline or canal, nor does it require the agreement of countries through whose territories the water must be piped. Sea transport should therefore be more flexible than overland conveyance, as it can make possible direct supply from source to recipient (provided both have access to the sea) and ad hoc arrangements between buyers and sellers rather than permanently binding commitments. However, these theoretical advantages may not work out in practice, since the choice of water suppliers for the Middle East is rather limited.

A case in point is what happened when the Israelis attempted in 1990 to "shop around" for potential sources of water in other coastal countries around the Mediterranean. One promising possibility seemed to be the Dalmatian coast of the Adriatic Sea, where numerous brimming streams discharge copious amounts of runoff, largely unused, from the highlands of former Yugoslavia. The Croatians, who control that coast, quoted a rather high price for the water, but that was not the main problem. To justify the expectable cost of building the necessary facilities, Israel needed to be assured of dependable supplies over a period of at least 20 years. No one in Croatia could give such an assurance, especially as the headwaters of those streams are located in unstable Bosnia and unfriendly Serbia, and subject to extraction or contamination in areas beyond Croatia's control.

Many observers believe that the ultimate long-term answer to the water shortage—in the Middle East as in other arid areas—will be the desalinization of seawater. It is, interestingly, a very old method. Aristotle taught that "salt water, when it turns into steam, becomes sweet, and the steam does not form salt water when it condenses." Julius Caesar used stills to convert seawater to drinking water for his thirsty legions during the siege of Alexandria.

The appeal of desalinization is easy to understand: it promises unlimited supplies. But that blessing carries the burden of high cost. Although it is entirely feasible technically, desalinization consumes prodigious amounts of energy. Hence it is practiced primarily in the few countries where fuel is abundant and cheap, while its large-scale application in other countries has been held up by the cost of energy. In the 1950s and 1960s, the age of desalinization seemed near: the world price of petroleum was low and was predicted to fall further, and there was great optimism regarding the potential availability of cheap nuclear energy. (Early in the development of nuclear power plants, enthusiasts predicted that nuclear power would be so cheap that it would not even need to be metered.) Since that time, however, the price of petroleum has fluctuated dramatically, as has public awareness of the environmental consequences of operating large fossil fuel–burning power plants. Moreover, the prospects for cheap nuclear power have dimmed, and support for the construction of new reactors has diminished, owing to escalating costs and to problems associated with safety and the disposal of radioactive waste.[14] So desalinization is not yet the panacea to water shortages that the public was once led to expect.

Still, the idea of desalinization continues to draw attention. Technical innovations hold the promise of reducing the cost of desalted water gradually, while the costs of obtaining additional water supplies from conventional or alternative sources continue to rise. Eventually, some nations will no longer be able to satisfy rising requirements from existing water resources that are, in any case, being depleted or contaminated. So the time is nearing, though it cannot be assigned an exact year or decade, when the imperative of supplying water to thirsty populations and developing economies will compel countries to undertake large-scale desalinization.

Certain countries suffering extreme water shortages but fortunate enough to have abundant energy resources have already turned to desalinization to supplement their limited natural sources of fresh water. Globally, there are some 8,000 desalinization plants that produce more than 100 cubic meters per day. Nearly half of the world's present desalinization capacity is in the countries of the Arabian Peninsula (mostly along the Persian Gulf), and 12 percent of it is in the United States. Saudi Arabia is the world leader in desalinization. It alone produces nearly 30 percent of the global output, which was estimated to have totaled some 14 million cubic meters per day (over 5 billion cubic meters per year) in 1990. About two-thirds of those plants convert seawater, while the others are used to purify waters of varying quality but generally lesser salinity, such as drainage effluent or river water. Kuwait and the other Gulf States depend almost totally on desalting plants for their supplies of fresh water. During the 1991 war with Iraq, the Saudis worried greatly over the vulnerability of their desalinization plants to sabotage, as well as to air or missile attack. The several hundred thousand American troops sent to this desert region also depended on desalinization of seawater and were similarly vulnerable.

In most countries, however, desalinization is still three to four times more expensive than conventional sources of fresh water. Hence it is still considered prohibitive for all but urban and specialized industrial uses in isolated areas otherwise devoid of local water resources. As yet, desalinized seawater is far too costly to be used for the irrigation of field crops.

To the cost of the desalinization process one must add the cost of conveying the water to the consumers, and those costs can be quite high for inland areas of high elevation. In the case of a practically landlocked country like Jordan, and a city so distant from the sea as Amman, the total costs are likely to be great indeed. However, as in a cascade, the supply of water at one spot helps to deliver water to another spot, in succession, so that the water shortage along the entire system may be alleviated. In concrete terms, if Israel produces desalinized water along its coast, it will be able to forgo some of the water it now takes from the mountain aquifer and more of the Jordan basin water for use by the Palestinians and the Jordanians, respectively. Water, like money, is fungible—that is to say, negotiable in kind by substitution—though only in the context of peace.

For coastal sites within proximity to prospective desalinization plants, the cost of desalting seawater may already be approaching the range of feasibility, at least for domestic and industrial uses (though not yet for agriculture). The caveat here is that the future cost of desalinization will depend largely on the price of fuel, and that is notoriously unpredictable.

Currently, the cost of desalinized seawater is of the order of $1 to $1.5 per cubic meter. Innovators in Israel and elsewhere claim to be able to lower the cost to about $0.70 per cubic meter.[15] In Israel, the real cost of present water supplies is of the order of about $0.45 per cubic meter. The price charged to farmers is of the order of $0.15–0.25, while the price charged to industrial and domestic users can be more than twice as great. For the latter users, therefore, the price discrepancy between desalinized seawater and conventional water supplies is no longer very great. The cost estimates cited are gross averages. In actual practice, the cost of desalinization depends on the method employed.

The classical and still most common method of desalinization is by distillation: evaporating the water, then recondensing the steam to obtain pure water. Since the vaporization of water requires much energy (nearly 600,000 kilocalories per cubic meter), the problem is how to recover as much as possible of the latent heat released during the recondensation of the vapor. There are several variants to the evaporation-condensation method of water desalinization, including flash distillation, multistage distillation, and vapor compression. The distillation process is generally most feasible where it is operated in conjunction with power-generating plants, which generally produce excess heat that, instead of being wasted, can be used to warm the water to be evaporated.

One alternative to evaporation-condensation is to freeze the water out of the saline solution, then wash and rethaw the ice to obtain more or less salt-free water. Since the latent heat of freezing is, on average, only about one-seventh the latent heat of vaporization, the energy requirement can be lower for the freezing method; on the other hand, the product is generally less pure. In either case, the residual brine is disposed of by returning it to the sea if the operation is carried out on the coast; or by evaporating it in open pans to dry out the salt and either finding some use for it or trucking it to the sea if the operation is carried out inland.

The modern alternatives to those methods of desalinization are electrodialysis and reverse osmosis. Both these methods employ selective membranes that can filter out the water from a saline solution, leaving the salts behind. Filtering water out of the salt solution through a membrane requires force. In the case of electrodialysis, an electromotive force is employed. In reverse osmosis, a mechanical force, or pressure, is applied to "squeeze" the water through the membrane.[16]

The cost of desalinization depends on the quality of the saline water source as well as on the method of desalinization. A distinction must be made between seawater, having a salt concentration of 3.5 percent to 4.5 percent (35,000 to 45,000 parts per million, or ppm) and therefore considered distinctly saline; and brackish waters, with a slight or moderate salinity ranging from 1,000 to 5,000 ppm (0.1 percent to 0.5 percent).

With the distillation method, there is relatively little difference in cost between saline and brackish water conversion, since in either case the water must be vaporized. In the membrane filtration methods, however, the energy expended increases with the salt content of the solution from which the purified water is to be extracted. On average, the desalinization of brackish water by reverse osmosis should cost roughly (depending, of course, on the degree of salinity) half as much as the desalinization of seawater. For each of the methods and for all salinities, the cost of desalting also depends on the required quality of the desalinized product, that is, on the purpose for which the water is to be used. The greater the degree of purity required, the more expensive the process. Total desalinization is much more expensive than partial desalinization, and the question of balancing water quality against cost is an exercise in optimization. Of course, the minimal quality standards must be tailored to the intended use. For drinking water quality, the salinity must generally be under 500 ppm of total soluble salts, and should preferably be under 300 ppm. For irrigation water quality, a somewhat higher salinity can be tolerated by some crops (such as dates, sugar beets, cotton, tomatoes, alfalfa), but not by sensitive crops (such as most tropical fruits). Finally, for certain (but not all) industrial processes an even higher level of salinity might be tolerable.

Early in the development of electrodialysis and reverse osmosis, the problem was how to produce membranes with the desired properties of selectivity (allowing the molecules of water to pass through while excluding those of the salts) and durability. In recent decades, improved membranes especially useful for reverse osmosis have been developed in the United States, Japan, and Israel. Such membranes are capable of transmitting the pressurized water while excluding the salts over extended periods of operation.[17]

As mentioned, desalinization is still far too expensive for the irrigation of most crops. If, however, the cost of desalinization is sufficiently reduced (through technical innovations, the economy of scale, and the more efficient use of energy—especially solar energy) while the prices of agricultural products and their yields per unit amount of water continue to rise, desalted water may eventually be used in the production of high-value crops. The time may also come when desalinization will be used not only to create new fresh water but also to improve the quality of the water already available. The measure of water quality will then no longer be based on the presence and concentration of dissolved salts per se but on the cost of removing them through desalinization.

Much depends on the future cost of energy. If the industrial world continues to depend on limited and dwindling supplies of fossil fuels obtained from unstable countries, then the prospects are not bright. If, on the other hand, cheaper sources of energy are developed (such as more efficient collection and utilization of solar radiation, or even nuclear fusion), then the cost of desalinization may be lowered to, say, $0.40 per cubic meter. That cost would approach the threshold of feasibility for agricultural use. All of the current desalinization plants use conventional energy sources: electrical or thermal energy obtained from the combustion of fuel, either directly or in conjunction with electricity-generating power plants.

There is, of course, growing interest in the potential use of alternative energy sources, such as solar or hydropower energy. Arid regions suffer the disadvantage of low rainfall, yet their very aridity offers the advantage of abundant sunshine. Sunshine can be converted to usable energy for generating electricity and for desalinization of brackish or saline water. Current developments in the science and art of collecting and utilizing solar energy offer exciting possibilities.

In recent years, attention has been given to the possible use of solar ponds. Such ponds consist of water layers with different degrees of salinity and density. Radiant energy from the sun is trapped in the deeper, denser layers, which can thus be heated to temperatures as high as 90 degrees Celsius. The temperature difference between the hot water at the bottom and the cool water at the top can permit the production of electricity. So far, only small-scale solar ponds have been tried, but researchers believe that they can be scaled up. Planners estimate that solar ponds of 100 hectares or more can be built such that will power low-pressure steam turbines capable of operating multistage distillation or reverse osmosis systems at a cost that will be competitive with conventional energy units. Another possibility now being explored is to utilize the evaporative power of desert winds for desalinization.

One of the oft-mentioned prospects for alleviating the water shortage of Israel, Jordan, the West Bank, and Gaza is the joint undertaking of water desalinization to treat brackish groundwater as well as seawater. Here the attractive feature of desalinization, notwithstanding its costs, is that it would enable the Israelis, Jordanians, and Palestinians, acting jointly, to obtain fresh water without having to depend on distant sources and delivery systems from across international borders. Hence desalinization seems more attractive, politically at least, than the alternative of importation.

The crucial problem is that of enlisting the very considerable financial resources necessary for initial installation and subsequent operation. Will other nations in the region, as well as international financing agencies, be willing to help make the necessary investment in peace? (Certainly, the required investment would be less than the cost of even a small war, and it would constitute a powerful incentive to avoid future wars in that part of the region.)

Since a desalinization plant, to achieve maximal efficiency, must be operated continuously throughout the year, whereas water demand fluctuates seasonally, some system of storage will be necessary. In principle, underground storage is preferable to surface storage, since it entails smaller losses due to evaporation and seepage, though—on the other hand—it may involve some loss of quality and it necessitates a larger investment of energy for subsequent pumping. A desalinization plant located on the Mediterranean coast could provide water to recharge the coastal and mountain aquifers during winter months, with a preagreed formula for sharing and scheduling of utilization by Israel and the Palestinians during the summer period of peak demand.

The option of seawater desalinization is more realistic for Israel, given its long Mediterranean coast and with the majority of its people living near the coast, than it is for the Kingdom of Jordan, whose only seacoast is the Red Sea some 300 kilometers of rugged terrain away from that country's population center. Jordan stands to gain access to the Mediterranean Sea when it enters into a peace agreement with Israel. Transferring seawater to the Jordan Valley and building desalinization facilities there will then become possible.

Perhaps the best way for Israel and the Kingdom of Jordan to begin cooperation, once a peace agreement is finalized, is for them to undertake joint desalinization on the border between Eilat and Aqabah. Both of those coastal towns on the Red Sea lack sufficient water to allow them to develop to their full potential. Such a project could be a prelude to the development of joint harbor and tourist facilities, to great mutual benefit. At present, both countries are constrained by the lack of space for separate development, with Israel's stretch of the Red Sea shore only about 10 kilometers long and Jordan's not very much longer. Additional Israeli-Jordanian desalinization plants might be located along the projected Red Sea–Dead Sea canal (which could produce its own power), if and when that visionary plan is realized.

Israel has already acquired considerable experience in desalinization. The country now has some 35 operating units. These facilities are for supplying water to remote locations and for research purposes. Eight of those units employ evaporation-condensation methods, 23 use reverse osmosis, and the remaining are devoted to trials on alternative methods. All of the operating facilities convert brackish water rather than seawater. An early system built at Eilat to desalinize seawater, using the freezing-thawing process, has been discontinued in view of the high cost relative to that of converting locally available brackish water using reverse osmosis. The total capacity of all desalinization plants in Israel today is only about 50,000 cubic meters per day, amounting to no more than 1 percent of the country's total supply. However, Israeli water managers anticipate that the time is nearing when the experience gained from these small-scale efforts will be applied on a much larger scale to help alleviate the growing water shortage of Israel and its immediate neighbors.

Since Israel and its immediate neighbors must import most of their fuel from abroad, the costs of desalinization using fossil fuels are still very high. However, energy can be generated by taking advantage of an important natural feature of the local landscape: the difference in elevation between the open sea and the bottom of the rift valley wherein lies the Dead Sea. Linking the Dead Sea with either the Mediterranean or the Red

Sea can provide considerable benefits, including the preservation of the Dead Sea itself as well as the generation of electric power and the desalinization of seawater. This is an intriguing project that can, of course, only be carried out in full cooperation between the Jordanians and the Israelis, preferably including the Palestinians.

Geologically, the territory of Israel and the West Bank may be viewed as a natural dam that holds back the Mediterranean Sea and the Red Sea (and through them, the world's ocean waters) from rushing down into the deepest chasm on the earth's continental surface. The dramatic 400-meter elevation difference between the seas and the chasm is a topographical asset that can be the energy equivalent of several large oil fields. That possibility was conceived long ago by visionaries like Walter Clay Lowdermilk. To realize that potential, engineers will need to repeat on a grand scale what Moses did with his staff: to smite the rock and make water gush forth, with the aid of modern technology. A passageway must be cut through the intervening hump of bedrock for seawater to flow to where it can be dropped vertically, in a series of cascades, to the Dead Sea below. It has been estimated that a flow of 1 billion to 2 billion cubic meters annually would enable turbines to generate enough electricity to desalinize several hundred million cubic meters of water.

The scheme is neither simple nor cheap. For it to work, the Dead Sea must evaporate the full amount of water to be added to it. Prior to the diversion of water from the upper Jordan River by Israel and from the Yarmouk River and Zerqa Rivers by the Kingdom of Jordan, the surface area of the Dead Sea was about 1,000 square kilometers (though it fluctuated considerably over the preceding decades and centuries). Assuming an average evaporation rate of 1.6 meters per year, a surface area of 1,000 square kilometers (= 1 billion square meters) would evaporate about 1.6 billion cubic meters per year. Since the diversion of the northern streams that had fed it (the upper Jordan River by Israel and the Yarmouk River by the Syrians and Jordanians), the level of the Dead Sea has fallen over the last three decades by some 19 meters (from 390 to 409 meters below sea level). Consequently its area has shrunk to little over 700 square kilometers. The main body of the Dead Sea has receded from the two chemical plants established separately by the Israelis and the Jordanians on what was the southern shore of the lake. Now the brine must be pumped and conveyed to these stranded facilities from a distance of many kilometers, at considerable cost. What remains of the southern basin of the Dead Sea is now being maintained artificially by the Israeli Dead Sea Works, which uses the old lake bed for its evaporation pans to concentrate and extract valuable minerals. The Dead Sea will continue to shrink as long as the amount of water evaporating from it exceeds the amount flowing into it, albeit at a slower rate since the northern basin of the lake is much deeper than the southern basin that has already dried up.

As a first approximation, the amount of seawater to be added to compensate for the diverted streams might be about 800 million cubic meters per year. That amount would prevent any further shrinkage of the lake. A larger amount would raise the Dead Sea gradually toward its original level and size, but thereafter could not be continued at the same rate without endangering existing facilities (industrial, recreational, transportational). The estimated 800 MCM/Y does not account for evaporation from the artificial lakes that might be established along the Jordan Valley or the Arava Valley. A somewhat larger transfer of water will be required if those lakes are established, depending on their surface areas. The total amount of seawater to be transferred must also include the volume of water to be desalinized. In total, the program might involve a transfer of some 1,600 MCM/Y of seawater from the Mediterranean or the Red Sea.

Israel and the Kingdom of Jordan share the Dead Sea. Both now use the lake as an industrial resource, for the extraction of potash (in the form of potassium chloride, a valuable fertilizer) and other chemicals (e.g., bromides). Their facilities on both sides of the lake would be affected by the change in the level as well as the quality of the lake water that would result from the importation of seawater. Since the industrial plants have long since adapted to the lower level of the lake, restoration of the original level would require new and expensive readjustments.[18] In the long run, however, the benefits of such a change may well outweigh the drawbacks.

There are serious questions regarding the most favorable route for the

Figure 11.9 The Arava Valley, with a small oasis in the foreground.

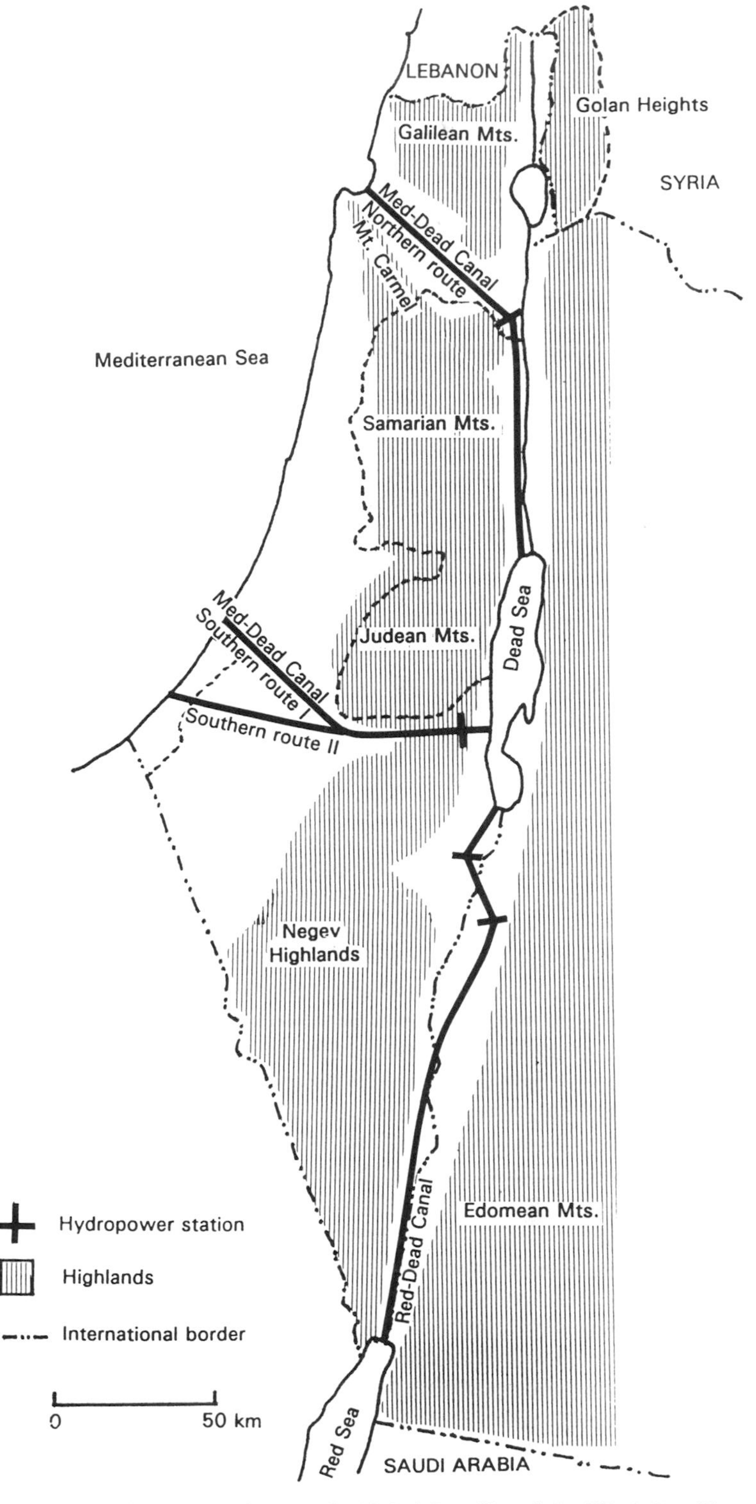

Figure 11.10 Alternative schemes: the "Med-Dead" and the "Red-Dead" canals.

seawater conduit. The "Red-Dead" route proposes a canal from the Red Sea (between Aqaba and Eilat) along the floor of the Arava Valley to the Dead Sea. The length of the canal would be somewhat under 200 kilometers. En route, the water must be lifted over a topographical hump, as the floor of the Arava Valley rises gradually to a height of over 200 meters above sea level, about 80 kilometers from the Red Sea. North of that rise, the water can be dropped through a series of cascades to generate either electric power or hydrostatic pressure for the direct desalinization of water by reverse osmosis, or both.

A seawater canal through the Arava Valley could also supply water to a series of lakes and ponds for recreation and aquaculture (fish, shrimp, and algae production), and perhaps for generating additional electricity by means of solar ponds. The potentially negative aspects of the project would be the drastic alteration of a unique desert environment and the possible seepage of saline water into the small but locally significant aquifers of fresh groundwater.

The alternative to the Red-Dead project is the proposed "Med-Dead" aqueduct, to convey water from the Mediterranean Sea to the Dead Sea via one of two proposed routes: (1) from the southern coast of Israel (near Gaza) through the district of Arad to the southern basin of the Dead Sea near Sodom; or (2) from Haifa Bay through the Jezreel and Beit Shean valleys to the Jordan Valley, and thence—parallel to the lower Jordan River—to the Dead Sea. The latter route follows the plan originally conceived by Walter Clay Lowdermilk over 50 years ago.

The first of these routes will involve overcoming a higher barrier (over 500 meters above sea level) by means of a tunnel, which is certain to render the scheme more expensive. However, the advantage for Israel is that the route will not traverse the West Bank, nor densely settled areas within Israel, so it will involve fewer political and logistical complications. The final drop to the Dead Sea from the Arad area will be very steep, a fact that could facilitate the efficient generation of electricity or of hydrostatic pressure for direct desalinization by reverse osmosis.

The second route, though longer, requires less tunneling and much less lifting of the water, since the area to be traversed is a series of valleys with a maximum elevation below 100 meters. However, the Haifa Bay Valley and the Jezreel Valley are densely settled, so the construction will involve greater disruption of population and infrastructure. And the facilities required in the lower Jordan Valley (a section of the West Bank) may be politically problematic.

Proponents of the Beit Shean route[19] have proposed desalinizing as much as 800 million cubic meters of water per year in the first stage of the project.[20] This quantity (which seems rather optimistic) would suffice to satisfy the foreseeable requirements of Israel, Jordan, and the West Bank for additional supplies of fresh water; and it would allow irrigating extensive areas of the Jordan Valley. As proven by the few existing settlements, the lower Jordan Valley, when irrigated, can be as productive as the Imperial Valley of southern California. Its hothouse climate permits the production of excellent out-of-season (elsewhere) vegetables, fruits, flowers, and seeds; as well as herbs, spices, and aromatic and medicinal plants. Such high-value products can be exported both to Europe and to the lucrative markets of the Arabian Peninsula. All the valley lacks is water, expert management, and extra care to prevent salinization. At present, only a small fraction of the potentially arable land is actually irrigated, while the larger area of the valley lies barren.

In summer, when water requirements (for domestic, industrial, and agricultural purposes) are highest, the desalinized water could be used directly. In winter, when water demands are reduced (especially for agriculture), the desalinized water must be stored, preferably by recharging underground aquifers. Alternatively, the project in winter could be operated so as to produce less desalinization and more electricity, thus saving the cost of the imported oil or coal otherwise used to generate the same amount of electricity.

The Jezreel-Beit Shean canal proposal also envisages establishing four dams along the descending course of the Jordan River. These dams could be used to generate additional electricity. The lakes formed by the dams would occupy an area of about 80 square kilometers (about half the area of Lake Kinneret). The proposed lakes would have approximately 150 kilometers of shoreline (three times the shoreline of Lake Kinneret), and could contribute to the life of the valley as centers of recreation and mariculture, including fishing.

Whichever the route chosen, the entire seawater canal project could be one of the great engineering enterprises of modern times. Several years would be required to develop the plans, assess the environmental impacts, obtain the approval of all concerned, arrange the financing, assemble the specialists and equipment, complete the excavations, build the conduits and generators, and produce the power for desalinization. Both the Red-Dead and the Med-Dead projects require close cooperation between Israel and the Kingdom of Jordan. Now both countries are suffering an acute water shortage, both have imposed water rationing, and both are facing the painful need to restrict their agricultural sectors. Neither country has remaining water resources that can be developed independently. Therefore, these countries must begin to cooperate with each other, as well as with the Palestinians, in developing shared resources.

The realization of one or another of the elaborate plans proposed is also conditional on the willingness of the international development agencies to help finance the venture, appropriately enlisting the participation of the oil-rich states within the region. Even if such a project will not justify its costs[21] on narrowly construed economic grounds, it may well be worth supporting for the sake of promoting peace and securing the great economic benefits that are sure to result from peace.

WATER IN THE MIDDLE EAST: POTENTIAL AND ACTUAL SUPPLIES (Gross Estimates)

	Bahrain	Cyprus	Egypt	Iraq	Israel	Jordan	Kuwait	Lebanon	Libya
Total land area (km²)	675	12,500	995,500	434,300	20,500	89,200	17,700	10,400	1,743,000
Cropland	20	2,000	25,800	54,000	4,330	3,960	40	2,980	21,500
Pasture	38	1,500	-	39,650	1,450	7,840	1,330	1,000	132,000
Woodland	-	1,600	300	17,750	1,100	700	-	800	6,800
Wilderness	617	7,400	969,400	320,000	13,620	76,700	16,330	5,520	1,582,700
Population (millions)	0.5	0.9	58.0	19.0	5.0	4.2	2.1	3.0	4.8
Growth rate (% per year)	3.2	1.5	2.6	3.3	1.9	3.4	4.0	2.5	3.5
Urban (%)	90	70	50	65	90	75	95	60	80
Gross nat'l product ($M/Y)	3,700	6,200	68,000	35,000	59,000	6,800	20,000	3,800	26,000
Per capita ($/Y)	7,100	8,600	1,170	1,840	11,800	1,620	9,500	1,270	5,400
Renewable wtr.res. (MCM/Y)	180	1,100	75,500	54,000	1,800	950	130	6,000	1,000
Surface	100	150	55,500	42,000	600	500	50	4,000	600
Groundwater	50	650	5,000	8,000	900	350	50	1,000	250
Recycling	30	150	15,000	4,000	300	150	30	1,000	150
Per capita (CM/Y)	360	1,220	1,300	2,840	360	225	60	2,000	210
Actual water use (MCM/Y)	420	500	68,000	39,200	1,900	850	420	900	2,900
Surface	-	100	58,000	33,700	600	400	-	700	400
Groundwater	20	300	2,000	5,000	1,100	400	20	200	2,300
Recycling	-	50	8,000	500	200	100	-	-	100
Desalination	400	-	-	-	20	-	400	-	100
Ag/ind/munic&domes (%)	4/36/60	91/2/7	88/5/7	92/5/3	76/6/18	65/6/29	4/32/64	85/4/11	75/9/16
Exotic (%)	-	-	98	70	40	25	-	5	-
Projected demand (MCM/Y)	600	600	72,000	45,000	2,200	1,200	500	1,200	3,500
Surplus/deficit (yr 2000)	-180	-150	+3,500	+9,000	-400	-250	-370	+4,800	-2,500
Co-riparians: Sharing streams	-	-	Sudan Ethiopia Uganda	Turkey Syria Iran?	Syria Jordan Lebanon	Syria Israel	-	Syria Israel	-
Sharing aquifers	-	-	Libya Sudan?	-	West Bank	S.Arabia	S.Arabia	-	Egypt Chad?

WATER IN THE MIDDLE EAST: POTENTIAL AND ACTUAL SUPPLIES (Gross Estimates)

	Oman	S.Arabia	Sudan	Syria	Turkey	U.A.E.	W.Bank&G.	Yemen
Total land area (km^2)	212,500	2,130,000	2,483,000	183,000	771,500	82,100	6,500	537,000
Cropland	1,600	24,000	127,800	55,800	276,560	400	2,500	16,000
Pasture	10,900	844,000	1,090,000	78,000	84,230	1,980	1,600	160,000
Woodland		12,000	444,000	7,200	200,150		200	40,000
Wilderness	200,000	1,250,000	820,000	42,000	210,560	79,720	2,200	321,000
Population (millions)	1.7	15.4	28.0	14.0	60.0	2.4	1.8	13.0
Growth rate (% per year)	3.8	3.7	3.0	3.5	2.4	3.5	3.5	3.7
Urban (%)	30	60	40	40	50	60	70	40
Gross nat'l product ($M/Y)	9,500	105,200	10,600	22,000	190,000	28,000	3,400	7,000
Per capita ($/Y)	5,600	6,900	380	1,570	3,170	11,670	1,900	540
Renewable wtr.res. (MCM/Y)	1,800	4,500	30,000	20,000	180,000	550	440	4,500
Surface	1,000	2,000	20,000	15,000	130,000	300	60	2,500
Groundwater	600	2,000	6,000	3,000	40,000	200	300	1,500
Recycling	200	500	4,000	2,000	10,000	50	80	500
Per capita (CM/Y)	1,060	290	1,070	1,430	3,000	230	240	390
Actual water use (MCM/Y)	750	10,000	13,000	12,000	20,000	820	300	2,550
Surface	400	500	12,500	10,000	16,000	200	10	2,000
Groundwater	250	6,000	400	2,000	3,000	200	280	500
Recycling	-	500	100	-	1,000	20	10	50
Desalination	100	3,000	-	-	-	400	-	-
Ag/ind/munic&domes (%)	90/5/5	47/8/45	98/1/1	83/10/7	75/5/20	80/9/11	70/5/25	90/4/6
Exotic (%)	20	-	80	75	2	20	10	20
Projected demand (MCM/Y)	850	12,000	22,000	15,000	30,000	900	600	3,000
Surplus/deficit (yr 2000)	+950	-7,500	+8,000	+5,000	+150,000	-350	-160	+1,500
Co-riparians: Sharing streams	-	-	Egypt Ethiopia Uganda	Turkey Iraq Jordan Israel	Syria Iraq	-	Israel	-
Sharing aquifers	S.Arabia U.A.E.	Jordan Oman U.A.E.	Egypt?	-	-	S.Arabia	Israel	-

12

Criteria for Sharing International Waters

Those who believe and work righteousness will be admitted to gardens beneath which rivers flow. . . . Their greeting therein shall be: "Peace!"
Koran XIV:23

With the end of the Cold War, previously ignored environmental and resource issues have risen to prominence in the international arena. The obsessive preoccupation of nations with competitive military security is giving way, tentatively yet none too soon, to a common realization of the need for cooperative ecological security. Some of the issues now coming to the fore are global, including depletion of the ozone layer and greenhouse warming; others are regional, including acid rain and pollution of lakes and seas.

Especially difficult to redress are inequities among nations over shared resources such as water, which is often scarce and subject to both depletion and degradation. Many rivers and aquifers serving as sources of water are common to two or more nations. Among these are the Nile, Jordan, Orontes, Yarmouk, Euphrates, and Tigris in the Middle East; the Indus, Ganges, Brahmaputra, and Mekong in southern and southeastern Asia; the Parana in South America; and the Colorado and Rio Grande in North America.

Societal needs for water are growing everywhere. In many parts of the world, fresh water resources have become foci of rivalry, especially among countries that already use water at a rate faster than natural processes can replenish it. Such rivalries seem certain to intensify as human populations grow, as rising standards of living (and possibly a change in climate) increase demands for water, and as human activity depletes water supplies. Apart from the disputes regarding the quantities of water, new disputes are arising over the contamination of water resources by some users and its effect on the quality of the water available to others. Further complications arise regarding uses of rivers for energy generation and for navigation. Finally, there are potential disputes over the management of watersheds by countries insofar as they affect downstream sedimentation, flooding, land drainage, and biological resources in estuaries and other wetlands.

Unlike petroleum, water has no substitutes and cannot be purchased in a world market that has many alternative suppliers. This situation makes rivalries over water even more fateful than over oil. Hence ways must be found to mitigate such rivalries before they flare into armed conflict.

One shudders at the thought of what consequences may follow a failure to resolve the worsening water shortage in the Middle East: hunger, privation, disruption of international oil supplies, war (with the next war in the Middle East all too likely to involve means of mass destruction, perhaps even nuclear). And if those threats are insufficient to draw the attention of the world at large, consider another possibility: that millions of desperate refugees from the Middle East and North Africa, dispossessed by the collapse of their countries' economies, will seek refuge in Europe and America early in the twenty-first century. The recent (and current) agony of the people of Ethiopia, Somalia, and Sudan—as well as the plight of Haitians trying to escape the poverty of their degraded homeland by smuggling themselves into the United States—may be but harbingers of the mass migrations expectable from the arid countries of the Middle East and elsewhere unless their problems are redressed.

The international community should be aware of the threats to global welfare posed by current and impending water shortages and environmental degradation and do its part to help remedy those circumstances where they are most likely to occur. Unfortunately, international institutions do not yet have adequate mechanisms to anticipate resource problems, mediate disputes before they degenerate into violent confrontations, initiate and finance effective programs, and help organize the management of multistate resources. Therefore, the initiative resides in the affected (and often conflicted) countries themselves and depends on their ability to cooperate and to elicit the technical and financial assistance of international agencies.

History, both ancient and modern, is replete with examples of how water supply systems became targets of war, and were sometimes used as the means by which to wage war. The Book of Genesis describes the struggles of the patriarchs Abraham and Isaac with the Philistines over wells of water in the arid Negev. Herodotus, in writing about the Greek wars against Persia, described the way towns were subdued by filling in their wells or plugging their water supply tunnels. Judean King Hezekiah was able to withstand Assyrian King Sennacherib's siege of Jerusalem by diverting Siloam Spring into the city and denying it to the enemy. The Muslim leader Saladin defeated the Crusaders decisively at the Horns of Hattin in the summer of 1187 by denying them access to water until they and their horses were utterly enervated by thirst.[1]

Water has played a role in modern wars as well. In 1938, Chiang Kai-shek ordered the destruction of the dikes confining the Yellow River to thwart the advance of the Japanese invaders of China, thus using the river as a weapon of mass destruction.[2] Reservoirs and hydroelectric dams were routinely bombed during World War II and the Korean War. Irrigation systems in North Vietnam were bombed by the American forces in the late 1960s. When Syria tried to divert the headwaters of the Jordan River in the mid 1960s, Israel launched air strikes and ground attacks against the diversion works, and that conflict contributed to the outbreak of the 1967 Six Day War. In the subsequent 1973 Arab-Israeli War, the Egyptians used water cannons spraying powerful jets of water to breach the Israeli-built embankment on the eastern side of the Suez Canal, and they worried greatly lest the Israelis retaliate by sabotaging the Aswan High Dam to unleash a devastating flood over the lower Nile Valley.

During the Persian Gulf War of 1991, both sides targeted waterworks such as dams, desalinization plants, and water conveyance systems. Most of Kuwait's desalinization capacity was destroyed by the retreating Iraqis. In turn, the Iraqis suffered severe destruction of Baghdad's water supply and sanitation systems. At the beginning of that war, consideration was given to using Turkish dams to deprive Iraq of a portion of its water supply in order to pressure it to withdraw from Kuwait. Although no such action was taken at the time, the threat of the Turkish "water weapon" remains hanging over both Iraq and Syria.

Since the major rivers and many aquifers in the Middle East cross international borders, the riparian disputes in the region contribute significantly to its instability.

The Nile River is an important testing issue. The nine riparian nations of the Nile are Zaire, Tanzania, Uganda, Burundi, Rwanda, Kenya, Ethiopia, Sudan, and Egypt. The river flows through one of the most arid parts of northern Africa and is absolutely vital for agricultural production in Egypt and Sudan. Nearly all of Egypt's water supply comes from the Nile, and nearly all of it originates outside Egypt. A treaty signed in 1959 resolves a number of important issues, but it was negotiated and signed by only two nations, Egypt and Sudan. Additional water development in the other upstream nations, particularly Ethiopia, could greatly increase tensions over water in this arid region.

Following the conclusion of his peace agreement with Israel, President Anwar Sadat declared in 1979: "The only matter that could take Egypt to war again is water." A short while later, none other than Boutros Boutros-Ghali, then Egypt's foreign minister and now secretary general of the United Nations, made the same statement with even stronger emphasis: "The next war in our region will be over the waters of the Nile." The two of them were referring to Ethiopia's purported intent to utilize some of the water of the Blue Nile within its territory. Their words were echoed in the warning sounded by Jordanian King Hussein to Israel in May 1990: "The only issue over which Jordan might go to war is the issue of water."

Another incipient issue is the growing tension among Turkey, Syria, and Iraq over the Euphrates River. In 1974 Iraq threatened to bomb the al-Thawra dam in Syria and massed troops along the border, alleging that the flow of water to Iraq had been reduced by the dam. In 1990 Turkey completed construction of the giant Ataturk Dam, the largest of the 21 dams proposed as part of its Southeast Anatolia Project. That year, Turkish President Ozal threatened to restrict water flow to Syria if it did not withdraw its support for the Kurdish rebels raiding southeast Turkey and disrupting the water development project there. The threat was a plausible one, as Turkey had in fact interrupted the flow of the Euphrates for a month to partly fill the reservoir. Despite advance notification by Turkey of the temporariness of the cutoff, Syria and Iraq both protested the closure and threatened to retaliate. Active hostilities were averted at the time, yet the problem continues to fester and may erupt once again at any time.

Antecedents to rules regulating water allocation can be found in the Middle East itself. Bedouin tribes maintain a strict order of priority in the use of a well or spring, with larger families allowed first rights. Islamic law had evolved a sophisticated set of principles to regulate water management in order to minimize conflict. Accordingly, wells are protected by specifying that the digging of additional wells must be distanced apart sufficiently to avoid mutual interference. Furthermore, an owner of a well must allow others living in the same area a share of the well's water for their domestic needs.

Medieval Islamic literature contains numerous examples of *fatwas* (religious edicts) concerning disputes over water rights. They usually take the form of specific and detailed questions posed to religious sages, followed by their reasoned judgments.

In one set of edicts[3] issued in North Africa in the fifteenth century, the following question (freely translated) was asked of Abu Musa Bin Minas: A group of water users built a succession of dams, one above the other, along a certain stream, for the purpose of allowing each user access to the water. In a particular year the water dwindled so that none was available to the users downstream, and the upstream users refused to release any of the water impounded in their dams. Do the downstream users have the right to breach the upstream dams so as to obtain the water thus impounded? The edict was that the upstream dams may indeed be breached, but only if the upstream users have had their fill and the downstream users can make effective use of the water.

A different edict in the same set is attributed to Abu Muhammad Bin Mahsud. Here the question pertained to the rights of downstream versus upstream users along a seasonal stream that provided abundant water in winter but only meager water in summer (the main growing season). The sage's decision in this case clearly favored the upstream users as having first rights to the water.

A third judgment in the same set was given by Abu Muhammad Abu Zeid. The issue in contention was whether people whose water channel had lost its function could by right build a new channel through the land belonging to a neighbor in return for the payment of compensation to that neighbor. The unequivocal answer was that they may not do so without the neighbor's explicit permission.

In still another edict, Abu Said Bin Lub established that in case of a dispute over water, the contender who can prove prior rights (ownership) to the water shall prevail. If no one has proven ownership then the upstream settlers should be given priority in usage over the downstream settlers. However, if the various settlers had a prior agreement among themselves regarding the allocation of the water, then the terms of that agreement must be observed.

A different set of *fatwas*[4] contains a significant judgment by Sheikh Isa Bin Ali al-Alami. He was asked what are the rights of a person who had been a resident and water user along a lower section of a stream, when subsequently new settlers took possession of an upstream section and began to use the water to the detriment of the prior user. The answer was that the earlier user had the right to continue using water, and that the new settlers could avail themselves of upstream water only to the extent that they did not jeopardize the preexisting downstream plantation.

Edicts issued by medieval Jewish sages were, in general, similar to those of their Islamic contemporaries. (This should not be surprising, in view of the close communication that existed between Muslims and Jews in Spain and throughout the Middle East during the period of their common cultural "golden age.") Two edicts issued in the eleventh century by Rabbis Sharira, Hai Gaon, and Yitzhak Alfasi[5] dealt with water rights under rather complicated circumstances. In the first of these cases, a man hypothetically named Reuven owned a tree orchard irrigated by a channel running through a field belonging to Shimon.[6] Reuven wished to uproot his trees and plant annual crops in their stead. Shimon refused to let him do so, contending that the orchard was irrigated only once a month, whereas field crops require more frequent irrigation. The ruling was that if the neighbors did not have an explicit prior agreement limiting the quantity (or frequency) of water delivered in the channel then Reuven has the full right to replant his own land and irrigate his crops as he sees fit. But if their original agreement had specified the quantity of water then Reuven did not have the right to exceed it.

The second deliberation concerned Reuven's wish to clean out his channel to ensure unobstructed flow of water to his field. Reuven was not permitted by Shimon to traverse his field freely, lest the latter's crops be damaged. So Reuven could only wade within his own channel. He contended, however, that the branches of Shimon's plantings obstructed his passage there, so he wished to remove those overhanging branches and—furthermore—to dredge out his channel once a year by digging out the silt and laying it on the banks. The rabbis ruled in favor of Reuven for the reason that he had no choice but to clean out his channel periodically.

The various traditional rulings are impelled by a sense of communal responsibility regarding water rights. In a modern interpretation of these traditions, nearly all the states of the Middle East have established the principle that water is a national resource to be regulated by collective interests rather than given to unbridled private ownership. Its proper management and use are in most cases the responsibility of the state, which can grant rights of use but not permanent ownership to individuals and groups. However, the region's traditions have not yet been extended to international rivalries over water rights.

An international river is one whose water course traverses, or whose catchment basin lies in, the territories of two or more sovereign states. The United Nations recognizes at least 100 such rivers. An examination of the map shows that most of the world's major rivers are in fact international rivers.

The catchment basin of a river constitutes a natural hydrological entity. Ideally, such a basin should be managed as a unit. The division of a catchment among different states (called "riparians," from the Latin *ripa,* meaning river bank) hinders the integrated management and efficient utilization of water and land resources. The use of water by any of the riparian states may degrade the river by reducing the quantity of its flow or the quality of its waters.

In considering the issue of degrading use, a distinction must be made between upstream and downstream states. Obviously, degrading use by a state controlling the headwaters of a river can affect the usability of the river by states located along the lower course of the river. This includes the effects of extracting water from the river for irrigation as well as the disposal into the river of wastes containing pathogenic or toxic residues.

Upstream riparians have a natural advantage over their downstream neighbors. In some cases, the latter can overcome their disadvantage by exercising, or threatening (whether explicitly or implicitly) to exercise, military or economic pressure. Such is the case of Egypt, for instance, vis-à-vis Sudan and Ethiopia. In other cases, an upstream nation enjoys economic or military superiority (as in the case of the United States versus Mexico, or India versus Bangladesh), and is therefore in a position to maximize its use of the water resource even at the expense of the downstream riparian. The latter nation must then depend on voluntary cooperation to secure its fair portion of the water.

Ideally, riparian states should enter into an agreement for coordinated or joint management of a shared river so as to protect the needs and interests of all the riparians, as well as the larger environment—including watersheds, woods, grasslands, banks, estuaries, floodplains, wetlands, lakes, and shores. Obvious as this principle may seem, it is rather difficult to apply in practice. Neighboring states are apt to have conflicting interests, not only with regard to water but also with regard to a host of other issues affecting the territory at issue. Such issues may include boundaries, mineral resources, the fate of ethnic minorities, and the protection of sites having historical, religious, or military significance to one side or the other.

An upstream state is tempted to claim an absolute right to use the water resources within its territory in whatever way suits its purposes, regardless of the consequences to downstream riparians. This idea was enunciated by the United States in the infamous Harmon Doctrine of the 1890s, which claimed that this country could do what it pleased with the Rio Grande, irrespective of downstream consequences to neighboring Mexico. That doctrine was universally criticized and subsequently was repudiated by the United States itself in a dispute with Canada over the Columbia River, in which the United States found itself in the downstream position.

A downstream state typically takes the position that the natural course and flow of the river must be respected and preserved. Furthermore, a downstream state is also apt to claim that its prior use of the water along the lower course of a river endows it with historical rights that must be respected by the upstream state.

Such diametrically opposing positions are likely to result in conflict in the absence of recognized criteria and mechanisms for adjudicating them. Any effort to devise the necessary mechanisms must consider and resolve the contradictions among following concepts: (1) *absolute sovereignty*—each state has an unrestrained right to use the water resources within its own territory; (2) *riparian rights*—every state along the course of a river has an inherent right to the water of that river, which is not to be diminished or degraded without that state's concurrence; (3) *river integrity*—all riparians are required to preserve the natural course of the river and utilize it within the natural watershed rather than divert waters out of the basin; (4) *historical rights*—a state that had used the resource consistently in the past has the right to continue using it; (5) *optimal development*—each river basin should be developed optimally as an integrated hydrological unit.

When such concepts or doctrines become tangled or mutually negating—as when state "A" lays claim to water on the basis of one doctrine and state "B" claims the same water on the basis of another doctrine—they obviously cannot be applied simultaneously. A negotiated or mediated compromise is the only plausible solution.

International peace and amity, as well as economic progress, depend on having a set of just and realistic principles and mechanisms to help in resolving disputes over water whenever the nations directly involved cannot agree among themselves. In such cases, international technical and financial institutions can contribute data, expertise, and resources to promote cooperation rather than strife. In a civilized world, there should be universally accepted norms or criteria by which to resolve international disputes peaceably. These norms should be based on universally recognized principles of justice and they should be codified formally in a set of laws and procedures. Ideally, there should also be institutions capable of applying international law to adjudicate disputes over environmental assets or resources such as water.

The Environmental Modification Convention of 1977, negotiated under the auspices of the United Nations, specifies in Article I.1: "Each state party to this convention undertakes not to engage in military or any other hostile use of environmental modification techniques having widespread, long-lasting or severe effects as the means of destruction, damage or injury to any other state." The 1977 Bern Geneva Convention on the Protection of Victims of International Armed Conflicts (additional to the Geneva Conventions of 1949) states in Article LV.1: "Care shall be taken in warfare to protect natural environments from widespread long-term and severe damage. This protection includes a prohibition of the use of methods or means of warfare which are intended or may be expected to cause such damage to the natural environment and thereby to prejudice the health or survival of the population." Such declarations, however well intentioned, carry little weight in the international arena when situations of conflict arise, since there is as yet no mechanism for enforcing them. A prime example of a blatant environmental crime was the action of the Iraqi forces in setting fire to oil wells and causing oil spills into the sea during the 1991 Persian Gulf War.

Similarly, no satisfactory water law has been developed that is binding on all nations. The first step in such development must be the promulgation of a set of principles on which to base a legal code. For international river basins, the question is which nation should be given priority: whether the one upstream or its neighbor downstream; the nation whose territory contributes most to the resource or the one that occupies the greater proportion of the watershed; the nation with the greater need (if, indeed, need can assessed objectively) or the one with the greater population; the nation that has used the resource longest or the one with the fewer alternative sources of water. In the absence of an agreed set of principles, it is typically the nation with the greater power or the stronger international alliances whose interests predominate.

The charter of the United Nations requires nations to resolve all disputes, ipso facto including those over water resources, without resorting to force. This principle, unfortunately, has all too often been honored in the breach.

In recent decades, international organizations have attempted to devise more specific principles and concepts governing shared fresh water resources. The International Law Association, an agency established under U.N. auspices that formulated the Helsinki Rules of 1966, established the hydrological principle that a water course must be considered in terms of a drainage basin, or catchment. It then attempted to formulate general rules for the usage of internationally shared water courses.

Article IV of the Helsinki Rules specifies the guiding rule: "Each basin state is entitled, within its territory, to a *reasonable* and *equitable* share in the beneficial uses of the waters of an international basin." However well intentioned and high-minded, that statement begs a number of issues. What is reasonable? What is equitable? What is beneficial use?

In a further effort to provide criteria by which those ambiguous terms can be applied in practice, Article V of the Helsinki Rules lists several factors to be considered. Loosely defined, these include consideration of each state's (1) proportion of the catchment area; (2) proportionate contribution to the annual flow; (3) prevailing climate; (4) prior and present patterns of water use; (5) economic and social needs; (6) costs of alternative means for satisfying those economic and social needs; (7) size of the population dependent on the water resource; (8) availability of alternative water resources; (9) actions to avoid waste of water; (10) possibility of providing compensation to redress conflicts over water; (11) possibility of satisfying water needs without harm to another state.

The Helsinki convention stops short of specifying an order of priority among these various criteria for establishing "equitable," "reasonable," and "beneficial" use of the water of a shared basin, some of which overlap while others may conflict. Nevertheless, these rules—albeit vague and ambivalent—are important in that they tend to shift international water disputes from contests of power to considerations of fair rights and mutual obligations. They imply more than they specify. Inherent in the rules is the responsibility of each state to use water efficiently and to avoid depriving or damaging a coriparian state.

The Helsinki Rules were a promising start, but only a start, toward creation of a comprehensive global code of law to govern the management of internationally shared water resources. Wishing to go beyond general principles toward developing a comprehensive legal structure for settling international disputes over water resources, the United Nations General Assembly in 1970 directed the International Law Commission to consider "codification of the law on water courses for purposes other than navigation." That task has evidently turned out to be more difficult and intricate than anticipated. Despite the renewed pleas for such codification made at the U.N.-sponsored International Water Conference at Mar del Plata (Argentina) in 1977, the task is still incomplete. In any case, the work of the International Law Commission is only preliminary, and even after its code is complete it will not assume the status of law until debated and ratified by the General Assembly.

In its work to date, the International Law Commission has attempted to clarify and sharpen the concepts first formulated by the Helsinki Rules. Its corollaries to those rules confirm that states with a common water resource are expected to share in its use, cooperate in its development, and protect the resource from depletion or deterioration. Moreover, states are expected to limit their sovereignty whenever it conflicts with those ends, and to bear responsibility for actions in their territories that might cause harm to other states.

In 1991 the International Law Commission completed the drafting and provisional adoption of 32 articles on the Law of the Non-Navigational Uses of International Watercourses. Article 5 provides that all states sharing an international water course have an obligation to use it in an "equitable and reasonable manner," and to do so, according to Article 7, at "optimal utilization" on the basis of mutual benefit and ecological protection. This formulation reflects the international community's growing (and long overdue) sensitivity to the issues of resource conservation and environmental protection. (Funding agencies, likewise, are now concerned not only with the expectable economic return from proposed projects but also with their social and environmental consequences.) Such rules formalize already existing international legal obligations. As part of "customary international law" they are, in principle, not dependent on treaties. Nevertheless, questions still remain regarding the actual implementation and enforcement of the stated principles whenever conflicting interpretations arise out of opposing national interests.

The most important guiding principle of the evolving international law on disputed water resources is that of *equitable utilization.* The principle so designated is that the interests and needs of all riparians be taken into account when allocating or regulating international water resources. Each of the riparian states having access to the common water resource is entitled to a "reasonable and equitable" share in the beneficial use of the water. As such, this principle directly negates the old Harmon Doctrine, which held that each sovereign state is entirely free to use water within its borders without restriction, even if that use might injure a neighbor. While some upstream nations still cite the Harmon Doctrine, more than 100 extant river treaties restrict the freedom of action of upstream nations.

The term "equitable" does not necessarily imply equal division between the nations involved. Rather, its application to specific cases must be flexible, taking into account a wide range of additional factors (including prior use, alternative sources of water, efficiency of use, and population needs). So, in the final analysis, the allocation of water remains a subject of negotiation and—in cases of disagreement—of arbitration and judgment by an objective intermediary.

Apart from its explicit requirements, the doctrine carries several logical implications. Fundamental among these is the obligation of each state to *prevent significant harm to other states.* This principle is embodied in the maxim *sic utere tuo ut alienum non laedas*—use your property in a way not to injure others. A corollary to this principle is that if a state has engaged in actions harmful to other states, it is under obligation to compensate for and to mitigate the effects of its actions.[7]

The duty to curb adverse effects is particularly relevant to water pollution, downstream sedimentation, and flooding. The doctrine has been adopted in practice by the World Bank, which now requires an assessment of potential "appreciable" harm before approving funding for projects involving international water courses. This formulation, of course, begs the question of defining what level of harm is "appreciable," a matter that must be evaluated by a knowledgeable third party in disputed cases. A serious difficulty here is to assess or quantify the space-removed and time-removed environmental and economic impacts that may result from upstream activities.

International cooperation requires that each state provide *notification and information* to its coriparians regarding any activities likely to affect them. Such notification, if given in advance, should allow an affected state to negotiate or to take timely action to prevent or minimize the harmful effects that might otherwise be caused to it.

A corollary to the last principle is the *sharing of data*. Each nation is expected to provide its neighbors with relevant information regarding shared resources and their utilization, including sectoral water supplies and demands. Moreover, there must be interchange and cooperation on the collection and interpretation of such data. This, in turn, requires communication not only on present usage but also on planned future usage, conducted in a spirit of mutual trust, good faith, and cooperation.

Obviously, unless coriparian states share hydrological data, no satisfactory agreement on the quantities available and their equitable division (particularly during periods of shortage) can be achieved. At present, many countries regard water resources data as classified and therefore to be withheld from neighboring nations. The collection and dissemination of information can be greatly facilitated by the involvement of international scientific and technical institutions, including agencies of the United Nations and the World Bank.

The obligation of all riparian states to share in the management of international rivers implies actual participation in the development, use, and protection of common water resources. The ideal approach to the efficient utilization of international water courses while ensuring equity and preventing harm would be to organize integrated and joint development schemes rather than separate and competitive ones. In the real world, however, joint development is too often thwarted by disputes over sovereignty, ownership of waterworks, jurisdiction, and financing. Competitive nationalistic concern for each country's exclusive rights can be a major obstacle to achieving integrated development. Even where states are willing to cooperate for mutual benefit, each generally desires to preserve as much control as possible over territories and resources within its own boundaries. Hence joint organizations or institutions are usually given coordination roles rather than authority to manage the resource independently. A common mechanism for such coordination is a joint river basin commission[8] charged with the negotiation of disputes, the allocation, of resources and the administrative management of the river for navigation, flood control, power generation, water supply, irrigation, and drainage.

As originally conceived, the principles governing the management and utilization of international water resources pertained primarily to surface water bodies—rivers and lakes. Only lately have international bodies begun to address the issue of shared groundwater resources.

By their nature, surface water resources are easier to define, regulate, and oversee than underground waters. Since rivers and their tributaries are visible on the ground and are drawn on maps, and their flow rates and directions are readily measurable, the notion of a river basin is easy to grasp. Aquifers, however, being invisible, are difficult to conceive, let alone to quantify with respect to water volume, recharge rate, flow direction, and water quality. Hence it is not surprising that recognition of international aquifers and their inclusion in the context of international water law were late in coming. The Helsinki Rules mentioned groundwater only cursorily.

Many groundwater basins or aquifers, like rivers, underlie two or more countries and are therefore international resources. Like surface water, groundwater typically ignores political boundaries. Withdrawals by one country can drain water from the shared aquifer and thereby deprive a neighboring country. Consequently, a conflict may erupt. In many areas, shortages or poor quality of surface waters have caused farmers, municipalities, and industries to expand the use of groundwater, especially in arid and semiarid regions. The frequent result has been depletion of the aquifers, with consequent deterioration of water quality.

Among international aquifers in the Middle East are the Saharan aquifer shared by Libya, Egypt, Chad, and Sudan; Arabian Peninsula aquifers that the Saudis share with Bahrain, Qatar, the United Arab Emirates, and Jordan; and the mountain aquifer shared by Israel and the West Bank.[9] There is a paucity of aquifer-wide data on most such groundwaters. Many of them are already subject to overuse, and are in zones where intensified development is likely to take place in the future.

Treaty provisions and international agencies with jurisdiction over groundwater are few and limited in scope, as international law and treaty practices regarding groundwater lag far behind the progress of international law regarding surface waters.

With the goal of advancing the cause of international law in this area, a multidisciplinary group of specialists, working over an eight-year period, has drafted an international groundwater convention.[10] The draft, entitled the Bellagio Draft Treaty, is based on the proposition that rights to international aquifers should be determined by mutual agreement rather than be the subject of uncontrolled, unilateral taking; and that rational conservation and protection actions require joint resource management machinery. It suggested mechanisms for international aquifers in critical areas to be managed by mutual agreement so as to avoid competitive extraction.

The issues covered by the draft include contamination, depletion, drought, and transboundary transfers, as well as withdrawal and recharge. The fundamental goal is to achieve joint, optimum utilization of shared aquifers, and to avoid or resolve disputes over such aquifers in a time of ever-increasing pressures on this precious resource.

The ideal principles described, pertaining to both surface and underground water resources, though widely acknowledged, are yet to be embodied in binding and enforceable laws. Nevertheless, they are useful inasmuch as they serve to guide the formulation of bilateral or regional water treaties covering specific water basins. Many such international treaties are now in existence, having been achieved by direct negotiation among the parties concerned, without the intervention of any international agencies. Sometimes, however, longstanding treaties begin to lose relevance as changing circumstances alter the water needs of nations and regions. When conflicts threaten to erupt even among erstwhile cooperating nations, a need arises for renegotiating preexisting treaties.

International law has generally developed slowly through cumulative experience, both good and bad. The law reflects the acquired wisdom of nations on the most fair and effective approaches toward regulating their reciprocal relations so as to reduce conflict. It also embodies the commonly felt need for a more stable, dependable, and reasonable modus in the relations among nations. As such, international law is a still evolving system of treaties, customs, precedents, and principles accepted by civilized nations, recognizing the rights and obligations of all toward one another and the common responsibility to promote peace and welfare on the basis of equity and universal justice.

The system of norms constituting international law is necessarily couched in ambiguous language, leaving room for interpretation and counterinterpretation. General principles are difficult to apply to particular cases involving water resources, since the hydrology of every river and aquifer is site-specific and often poorly defined. Moreover, the application of international law is constrained by the absence of enforcement procedures to compel the compliance of reluctant disputants. Therefore, despite the growing accord on basic substantive principles, the procedural means for their effective application to specific situations is generally subject to ad hoc negotiations and is not free of political, economic, and military pressures.

The International Court of Justice (also called the World Court), located in the Hague, is considered the final arbiter of international legal disputes. However, it does not consider cases unless all sides to a dispute accept its jurisdiction and agree to abide by its decisions. The World Court takes no initiative to intervene in disputes and takes no submissions from other than the parties directly involved. As its hearing of a case depends entirely on the voluntary appeal of the disputants, action by the World

Court presupposes a measure of conciliation among them, or at least a willingness to settle their dispute peaceably. Moreover, there is no international authority able to enforce a decision of the World Court if one or another of the parties refuses to comply with it.

The aspects of common water resource management that are most often contentious are water use, allocation, quality, and joint development. In the absence of binding rules, each riparian generally strives to take as much water as possible, to limit the rights of competitor nations, and to dispose of negative "externalities" such as pollution. Such behaviors are inimical to international harmony and peace, and are therefore impossible to justify or ultimately to perpetuate. The only reasonable way is to replace the untenable claims to "absolute rights" with more flexible approaches, based on the principles of limited sovereignty, community of interests, and cooperation among riparians.

Self-limitation to sovereignty is made acceptable by the perceived benefits to each country from international cooperation. Such benefits must be seen to be greater than what each country could achieve on its own. In many cases, such as a dam across a river that forms the boundary between countries, the only possible option for development is cooperation, its sole alternative being no development at all. The ultimate degree of cooperation is acceptance of common ownership of the resource and formation of an autonomous river basin authority to plan, implement, and administer the joint development project.

The development of waterworks is generally costly, and the integrated development of entire river basins is likely to require a greater investment than any one of the riparian countries can enlist on its own. The participation of international institutions is therefore necessary, and such participation (technical and financial) should be predicated on honoring the principles of equitable and unharmful use. The policies of the international community can thus promote and encourage the inflow of capital into the development of joint waterworks.

The most important financial institution for funding international water programs is the International Bank for Reconstruction and Development (IBRD, commonly called The World Bank). It has two affiliated institutions, the International Finance Corporation (IFC), and the International Development Association (IDA). These affiliates aim to address the needs of the least affluent nations by providing more concessional conditions for loans than does the World Bank in its regular operations. The bank itself adheres to more stringent policies since it must obtain most of its funds through borrowing in the world's financial markets, and must also consider the policies of its sister institution, the International Monetary Fund. The World Bank has been involved with a considerable number of international water programs, the most well known being the Indus River (India-Pakistan). In recent years, however, the bank has been subjected to increasing criticism over its apparent preference for large dam projects, so it has become wary of programs that do not enjoy the consensus of all concerned parties.

The formulation of an explicit, detailed, and binding international water law is an exceedingly complex and difficult task. The discussions leading to a law of the seas have lasted more than a decade, and are still unfinished. We cannot expect that a complete water law will become available immediately. Progress toward such a law is gradual. In the meanwhile, the most practical approach to the resolution of water problems is likely to be regional rather than global. And for such an approach to stand any chance of success it should be motivated by a spirit of cooperation. A belligerent attitude is counterproductive. That is why attempts to resolve the water issue in the Middle East have not been very successful in the past. But times and attitudes are changing. Necessarily so, because the problem of water is becoming too acute to be ignored much longer. Resolving it constructively is a matter of survival for all concerned.

13

Water for Peace

Spread upon us thy tabernacle of peace.
Ancient Hebrew Prayer

If thy enemy incline toward peace,
do thou also incline toward peace.
Koran VIII:61

Arabic lore features a mythical character named Juha, a clever prankster and folk hero who, pretending to be a simpleton, delights in exposing the hypocrisy of the rich and powerful. Juha is said to have been present in many places and times. In one of his innumerable incarnations, Juha lived in the Ottoman Empire early in the twentieth century. One day, he was exasperated by the tardiness of the train service, so he petitioned the Sultan, claiming to have an idea to make all the trains in the Ottoman Empire run on time. To be sure, not many trains existed in that Empire, and those that did exist never ran on time and some days didn't run at all. So, when the Sultan received the petition, he immediately summoned Juha to present his idea before the Imperial Board of Transportation.

The dignified board members sat around the great oval table as the Grand Sultan invited Juha to present his idea. Juha then rose and declaimed: "To Your Imperial Highness, King of Kings, Protector of the Faith, Benefactor of All the People, and to your honorable administrators and engineers, I—your humble servant—hereby present my idea for the greater glory of the Empire, to make all the trains in the realm run on time." So saying, Juha sat down at the right-hand side of the Sultan and waited. There was a tense silence as the assembled dignitaries fidgeted, until one of them blurted nervously, "Well, what is your idea?" So Juha rose again and repeated: "As I said, my idea is to make all the trains run on time." "But how?" asked the baffled members of the Imperial Board of Transportation. "Why do you ask me?" cried Juha in righteous indignation, "I'm not an engineer, I only conceived the idea. You are the experts, so you should be able to figure out how to carry out *my idea!*"

Those who pretend to solve the problems of the Middle East, especially the water shortage in the context of the Arab-Israeli dispute, by suggesting grandiose schemes while being too hasty or superficial to examine their schemes in detail, remind me sometimes of Juha's prank. I have tried to avoid the temptation to present a pet scheme of my own. All seemingly plausible regional schemes must be considered rigorously by multidisciplinary and multinational teams able to analyze the engineering, economic, environmental, and political feasibilities.

In the foregoing chapters I have endeavored to show that promising options do exist for resolving the water problems of the region, and to explain some of those options in realistic yet positive terms. My aim has been to present not merely a compendium of facts but also an interpretive explanation, not only to inform but also to develop understanding and insight. Only when fully aware of the opportunities to undertake constructive action, and of the perils of failing to do so, might people in the Middle East and in the international community at large join forces to initiate an effective program of development capable of retransforming this troubled region into the Eden of culture and progress that it once was and can become again.

Why did the Ottoman trains not run on time? Because nothing else ran on time. And why not? Because the trains did not run on time. A similarly circular cause-and-effect, chicken-and-egg dilemma has seemed to thwart progress in the Middle East. The water problem and the peace process are inextricably intertwined. The one cannot be resolved without the other. For too long, the peoples of the region, especially the Arabs and Israelis, have been locked in an agonizing embrace, paralyzed with fear, unable to disengage or to cooperate.

The situation of the Middle East is a seemingly endless series of problems within problems—like Russian matrushka dolls, each nested inside another. But unlike those dolls, which diminish progressively as they are revealed one by one, the predicaments of the Middle East appear to grow larger, like the fat cows that were contained in the thin cows in Pharaoh's portentous dream. We need a *Zaphnath-paaneah*[1] to break the vicious cycle and help extricate the region from the quandary of its self-encapsulation.

The troubles besetting the nations of the Middle East are many and varied. Among them are wide disparities of wealth between an affluent few and an impoverished multitude, thwarted national aspirations, disputed borders, religious fanaticism, indiscriminate terrorism, ethnic civil wars, and megalomanic dictators. These afflictions and more combine to make the Middle East one of the world's most unstable regions, where an intensifying arms race poses the real danger that the next war will be fought with weapons of mass destruction (chemical, biological, and nuclear). Such a catastrophe must be averted at all costs.

Pessimism[2] regarding the Middle East has long masqueraded as realism, claiming that eternal strife is inherent in the region and can never be resolved. A particularly harsh view of the region's peoples and their mentality was articulated by T. E. Lawrence, who lived and fought in the Middle East during the first decades of the twentieth century. In his book, *Seven Pillars of Wisdom,* he wrote of the Semites (presumed to include both Arabs and Jews): "They were a people of primary colours, or rather of black and white . . . a dogmatic people, despising doubt. . . . They knew only truth and untruth, belief and unbelief. . . . Their thoughts were at ease only in extremes. . . . Sometimes inconsistencies seemed to possess them at once . . . but they never compromised: they pursued the logic of several incompatible opinions to absurd ends, without perceiving the incongruity. . . . They were a people of spasms, of upheavals. . . . Their . . . revealed religions . . . were assertions, not arguments."

An extremist perception has indeed gained currency in recent years among the disillusioned people of the Middle East. It seems to many that the secular ideologies of nationalism and socialism, which dominated the struggle for independence during most of the twentieth century, have failed to improve the lot of the people. While the nationalistic struggles continue, anti-Western slogans and the appeals to ethnic pride are now couched increasingly in the language of resurgent religion. The notions of progress and democracy are besieged by the bitter opposition of the religious establishment on the one hand, and by the corrosive dissillusionment among the impoverished classes on the other hand. Many in the Middle East have turned back to their ancestral faith for a sense of solace and security. Among the Arabs, there is a powerful nostalgia for the Islamic civilization that reached its apex in the ninth and tenth centuries—a mobilizing, conquering faith, combining spiritual fervor, social virtue, cultural achievement, and military prowess.

The rise of political Islam is fueled in part by a rage against the West, which introduced secularism and supported corrupt regimes throughout the region, provided they were "pro-Western." It is also impelled by a widespread belief that modern Western political and economic structures, as well as social mores, are alien to Arab culture, of which Islam is the most authentic expression. As reflected in films and personified by tourists, the image of an affluent, smug, and libertine ("decadent") West provokes both jealousy and revulsion. The ancient traditions of a strongly patriarchal society are especially threatened by what appear to be lax morals and excessive personal freedoms.

With the end of the Cold War, the overriding issue is no longer seen to be a global struggle between socialism and capitalism. The internal conflict besetting the region threatens to revert to its most atavistic form: exclusive and xenophobic particularism, sanctified by religious zealotry, a tendency to blame and demonize "others," all fed by pervasive poverty and a sense of deprivation.

Some may dismiss this as a passing fad that can be contained, accommodated, or repressed; a trend that in any case will self-moderate in time. In fact, it is far from having lost its momentum. Fed by wellsprings of deep frustration and resentment, it may not pass soon unless a more hopeful situation develops in the region, especially for young people. Simple repression is bound to fail in the absence of a genuine alternative to extremists or religious fundamentalists, in the form of a convincing program of development and a tangible move to improve the lot of the people. Otherwise, the only remaining choice will be between a weak "democratic" regime of politicians unable to generate reform and an authoritarian regime (a police state, with or without purported religious justification) that is equally inept at the task of true development, indeed a choice between the bad and the worse.

Peace has seemed elusive because the basic cause of the suffering remained unrecognized and unaddressed. That root cause, I submit, is the destruction of the region's traditional way of life, brought about by the twin onslaughts of ill-fitting modernization and environmental degradation, exacerbated by the population explosion. Multitudes of people, displaced from the land, migrate into overcrowded cities lacking infrastructure or even elementary services. They are unemployed and practically unemployable, and therefore easy prey to the lures of extremist demagoguery. The most prone to violence are young people, embittered by poverty and deprived of prospects for a better life. They need education, training, and opportunities for productive employment.

To strengthen the progressive moderates in their struggle against the backward-looking extremists, the Middle East requires a constructive and comprehensive program of economic development and environmental rehabilitation, offering progress and the hope of a better life for *all.* We stress the need for a program that will benefit *all* the people, because the culture of the Middle East differs from that of the West, particularly of the United States, in one important respect. American society stresses the role and identity of the individual, endowed with rights, opportunities, and the challenge to compete within a totally secular market economy. Middle Eastern society, in contrast, affirms the primacy of a communal identity, based on family, clan, tribe, religion, or nation. Hence a program of development here cannot be based on purely capitalistic notions of a competitive market economy. It must be more inclusive.

To be credible, the peace process must begin to produce tangible results, not merely in the political arena (where symbolism often prevails over reality) but also—and more imortantly—in the concrete economic realm of living standards, employment, and infrastructure development.

An essential ingredient of any development program is the provision of fresh water to nations otherwise growing thirstier by the year. For each of the nations involved in one or another of the region's numerous water conflicts, independent control of water resources is equated with control over the national destiny. However, hydrographic realities have deprived most of the countries of the region of exclusive water resources sufficient to their needs. Where nature conspired to provide common resources, there can be no ultimate independence, only mutual dependence.

As the rivalry over common waters between neighboring states and territories intensifies, the Middle East is poised literally on a precarious watershed divide, between war and peace. For just as water in nature can be either a bearer of life or an agent of death, so in a desperately thirsty region the issue of water can either bring the parties together or set them apart. It will take a strong measure of wisdom and goodwill to choose the one course over the other, to tame the wild cascade of destructive contention and turn it into the still waters of fruitful cooperation.

Among numerous Arabic aphorisms pertaining to water, one seems particularly apt: "*Halli yado bilmai mish zai halli yado bilnar,*" meaning "One whose hand is in water is not like one whose hand is in fire." Although different interpretations are possible, I understand that saying to suggest that those engaged in water management are not likely to engage in war. Far better to have one's hands in water than in fire. The Arabs also say, *"Ilmai ni'ima wa naqma"*—water is a blessing and a curse. Indeed, it can be either, depending on how it is managed or mismanaged, shared or divided, used judiciously or carelessly.

The hydrological imperative thus presents a challenge and an opportunity. Water can catalyze and lubricate the peace process, smooth the rough edges, and soften the transition to regional cooperation. The thirst for water may be more persuasive than the impulse toward conflict. For the people of the Middle East, who have lost faith in empty promises, peace will be more attractive if it offers tangible prospects for progress and prosperity, first and foremost in the development and equitable sharing of water supplies.

The envisaged program must enlist international assistance and challenge the oil-rich countries within the region itself to allocate a part of their wealth to alleviate the plight of the region's poor. That might well prove to be their best investment yet, and it will strengthen the intraregional cement of trust and generosity that has been so sadly lacking. The best guaranty of regional security is prosperity on both sides of every boundary.

Increasing supplies and, although thus far to a lesser degree, reducing demand are tasks now on the national agendas of most states in the region. Among the methods employed or considered are the tapping of fossil groundwater, damming of seasonal streams, using brackish water for irrigation, collecting and reusing wastewaters, rainfall inducement, runoff water-harvesting, desalinization of brackish or saline waters, and importation of water from humid regions. Measures to save water include introducing efficient systems for water delivery (such as lining of canals and using pressurized pipelines), monitoring and repairing of leakages, modernizing irrigation methods, and—most importantly—educating the public to regard water as a precious and expensive commodity to be used sparingly. All such measures can best be carried out conjointly.

No measures to develop and conserve water resources, however well conceived, can suffice in the long run if the present rate of population growth continues in the region. The increased demands of ever-growing populations on limited waters will certainly plunge the region into incurable misery. All economic development must therefore be accompanied by a consistent decrease in the rate of population growth.

The conflicts of the Middle East, in particular that between Israel and its Arab neighbors (being a tangled web of grievances and suspicions), may seem too complex to settle in totality all at once. A wise course might be to define its components and seek ways to resolve them incrementally.

Henry Kissinger has long been a proponent and exemplar of the step-by-step approach to thorny international problems. That approach begins with small initial steps to address the most readily solvable issues, thereby acting to gain trust and goodwill through "confidence-building measures." Once the atmosphere improves, more difficult issues can gradually be considered, and perhaps resolved in turn. Critics of this approach claim that it merely "nibbles at the edges" of the most important political issues, which will tend to fester unless addressed directly.

The United States has tried the gradual approach repeatedly during the last few decades. In May 1967, on the very eve of the Six Day War, the United States government sponsored the first International Conference on Water for Peace in Washington. It enlisted the participation of thousands of representatives from nearly 100 countries, significantly including Israel as well as the Arab states. Away from the glare of publicity, technical experts from both sides made one another's acquaintance and initiated a dialogue that has continued and grown ever since.

After the Six Day War, former President Dwight Eisenhower, together with Lewis Strauss (head of the U.S. Atomic Energy Commission) and Alvin Weinberg (director of the Oak Ridge National Laboratory), proposed a "Water for Peace" program for the Middle East. It envisaged the construction of several nuclear reactors to provide power and desalinize seawater for the development of "agro-industrial" centers. By such development, the sponsors had hoped to provide employment to refugees, alleviate water scarcity, and defuse political tensions in the region. In explaining the rationale for the agro-industrial scheme, Strauss stated: "Water is an eloquent advocate for reason."

The plan was endorsed by Senate Resolution 155, which projected three sites for development, one each in Egypt, Israel, and Jordan. Notwithstanding the best intentions of its sponsors, however, the plan has not been implemented, either because the intended beneficiaries were unready to cooperate or because, on reconsideration, the United States hesitated to promote the introduction of nuclear technology into the region.

Nonetheless, might that visionary program have been a harbinger of what could now (albeit in a modified form not necessarily based on nuclear technology) become possible at long last? Several changes have taken place since those early efforts. Israel, Jordan, and Syria went ahead with individual programs to develop the Jordan basin. Lacking any agreed framework for coordinating the distribution of the water, the various competing projects have spawned additional flash points.

Other significant trends have affected prospects for agreement. Agricultural water use is no longer so decisive a criterion as it once was. The share of agriculture in the national economy of each of the riparian states has diminished, while that of the urban (residential, service, and industrial) sector has grown. Moreover, the crop water requirements per unit of land and per unit of yield have also diminished, thanks to the development of better varieties and water conservation techniques. At the same time, however, the population of the countries and territories surrounding the Jordan basin has more than doubled. Now that all the natural supplies of fresh water obtainable separately within each country are being used or overused, and with population continuing to grow, uncoordinated sectarian water use can no longer be sustained.

Although past attempts to redress the water shortage have not been completely successful, times and attitudes have changed. The availability or the lack of water is now seen to be a major determinant of peace or war. The cause of Middle Eastern peace is too important to be left to the generals or to the politicians. Now the technical experts—the hydrologists, agronomists, and engineers dedicated to problem solving on a practical level—must be given a role. As scientists sharing common professional concerns and concepts, and accustomed to dealing with facts and figures without emotion or intrigue, they might well find it easier to reach a modus vivendi or at least a modus operandi. They might thereby set an example. Working together can be habit forming. For just as violence leads to more violence, so may a small measure of peace lead to greater peace.

Those who wish to resolve the water issue must go beyond the bickering over the competitive allocation of existing supplies for traditional uses and concentrate on longer-term programs. Such programs should be aimed at augmenting supplies, avoiding waste, improving efficiency of utilization, and capturing wastewater for treatment and reuse.

The least costly option for alleviating shortages is by means of conservation and rational management of demand, along with the limitation of population growth. Immediate and substantial water conservation can be achieved by rehabilitating conveyance and storage systems to avoid leakage and evaporation losses. (At present, the leakage loss alone may amount to half the water supplied to several of the major cities in the region.) Water can further be saved by increasing the efficiency of use in the domestic and industrial sectors, in ornamental gardens, and—above all—in the agricultural sector. Increased efficiency involves charging realistic prices for water, installing water-saving devices (including water meters and automatic shutoff valves, designed to apply the precise optimum of water for each purpose), and educating the public regarding the principle of water conservation as an economic imperative and a national duty.

Efficiency requires that water be allocated preferentially to sectors able to produce higher returns per unit volume of water—even though this principle may contravene the vested interests of traditional water users. Conservation also demands that wastewaters be recovered and treated to allow reuse, especially in agriculture and industry. Experts believe that as much as 75 percent of urban water supplies can be recycled for use in irrigation, and that future agriculture in the countries of the Jordan basin will be based primarily on recycled water.

In the area of water conservation, Israel has made pioneering innovations, which could be applied to great benefit in other countries throughout the region. However, because of the differing circumstances in each country, no techniques should be simply transferred from one country to another without careful adaptation. The further development of water conservation technology demands close cooperation in research and commercial application.

Apart from conservation, the augmentation of water supplies can also be achieved by a variety of other means. The latter include weather modification by cloud seeding, capturing and storing seasonal floods, inducing and harvesting runoff from designated slopes (especially in desert areas), managing aquifers (by promoting natural and artificial recharge, judicious extraction, and protection of water quality), and possibly importing water from various sources within and outside the region.

Prospects for multinational water transfers seemed to dim when the offer of a Peace Pipeline to alleviate the thirst of the arid countries of the Middle East, first made by Turkey's late President Turgut Ozal, was repudiated by his longtime rival and successor as president, Suleiman Demirel. Yet the idea is not dead: Turkish Prime Minister Tansu Ciller appeared to revive it publicly during her visit to Washington in October 1993. However, such schemes cannot be decided upon unilaterally. To be viable, they must enlist the concurrence and assure the benefit of all parties involved. These goals have been difficult to achieve heretofore, but times and attitudes are changing.

The ultimate option for augmenting supplies is desalinization. Here, too, there are advantages to be gained from international cooperation, as—for example—between the Kingdom of Jordan, which is practically landlocked, and the State of Israel, which can provide Jordan with access to the Mediterranean Sea. Paradoxically, the very presence of a deep rift between them (the Dead Sea and Arava Valley) offers an opportunity for generating the energy needed to accomplish desalinization at relatively low cost. Experts on both sides look to the possibility of jointly building large-scale desalinization projects along the border between them, with international assistance that cannot be procured by either country alone.

Investments in desalinization and other modes of water resource development, as well as in water conservation technology, however expensive, are certain to be much less costly than perpetuation of rivalry and tension, possibly leading to yet another war.[3] Water control systems are not all that expensive when compared to the costs of tanks, warplanes, missiles, and—God forbid—nuclear bombs. War is surely the worst of all possible alternatives, a waste of human hopes and lives as well as of financial resources.

In a peace arrangement, various compromises and trade-offs are possible. Israel could forgo its claim to a portion of the Yarmouk in return for Jordan's acceptance of responsibility for supplying a commensurate amount of water to the West Bank. Such an arrangement would obviate Israel's fears over Palestinian exploitation of the mountain aquifer, and remove its objections to the construction of the Jordanian-Syrian Unity Dam. The remaining obstacle to that dam would be the increased Syrian abstractions of water from the upper tributaries of the Yarmouk. Syrian action reduces the amount of water that would otherwise flow into the projected reservoir and hence casts doubt upon its hydraulic and economic feasibility. To relinquish the water that they are now impounding, the Syrians would need some inducement, perhaps in the form of a greater assured supply of water from Turkey. Though this series of compensatory water transfers might seem like a game of musical chairs, it could indeed become practical in the context of a truly comprehensive regional accommodation.

Not the least of the benefits of peace will be the interchange of information and the initiation of joint research. Heretofore, the Arab countries' boycott of Israeli research has prevented the farmers of most Arab countries from availing themselves of Israeli varieties and technical innovations, and prevented Israelis from utilizing the experience and the genetic resources (the species and varieties of crops and livestock) of Arab agriculture. Even international research centers (such as the International Center for Agricultural Research in the Dry Areas—ICARDA—located in Aleppo, Syria) have had to comply with that self-thwarting boycott. Other arid-zone countries around the world could also benefit from such regional cooperation.

A dispute over water resources need not lead to violent conflict. Rather than fester as *causus belli,* it may well serve as *casus pacis,* an inducement to negotiate a non-violent resolution through compromise and collaboration. "The nations of the region will act rationally," observed Abba Eban, "once they've run out of all other possibilities." Their options for alleviating the water shortage separately have already run out.

The water issue has been excessively politicized throughout the Middle East. In some of the countries, data on water availability and usage are considered state secrets. Even in Israel, which takes pride in its open democracy and freedom of information, there has been an attempt to stifle public consideration of the state's water policy vis-à-vis its Arab neighbors. Ultimately, that attempt failed, but the incident and the discussions it engendered are instructive.

In November 1991, the Center for Strategic Studies, an academic institute attached to the University of Tel Aviv, was preparing to release a comprehensive study entitled "The Water Problem in the Context of Arrangements Between Israel and the Arabs." The aim of the study, conducted in cooperation with Israel's Water Planning Authority (Tahal), was to examine options for resolving disputes over water, consider the possibilities for cooperative water development, and assess the costs of such development relative to the costs that would accrue otherwise. Included in the report were two maps, specifying how far Israel might withdraw from occupied territories without jeopardizing its crucial water interests.

At the last moment, Israel's then minister of agriculture, Raphael Eytan (who had earlier published full-page newspaper advertisements opposing *any* withdrawal from those territories on grounds that it would indeed jeopardize Israel's water supplies) intervened to prevent dissemination of the report. He claimed the authority to do so because Tahal was funded by his ministry to carry out purely technical planning and therefore should not make any political pronouncements regarding the possibility of Israeli withdrawal from the territories.

In 1992, however, Mr. Eytan resigned his position, and subsequently a national election resulted in the accession of a new government more amenable to compromise with the Arabs. Yet the ban over publication of the report continued. This time the rationale for the suppression was that, in view of the ongoing negotiations with the Arabs, revealing Israel's options might constrain its bargaining position. Not until October 1993 was the substance of the report (including the controversial maps) made public.

The report provided an interesting summary of the issues in contention. Though written from an Israeli point of view, it suggested that Israel can go quite far to accommodate its neighbors, but only with reciprocal assurances. The following paragraphs summarize the published version of the report.

The issue that has long been considered paramount is that of land, which in turn is related to security. Narrowly construed, the notion of security refers to the avoidance of military threats. In a larger sense, however, a state's security depends on its economic viability, which in turn rests on an assured supply of fresh water.

Israel is reluctant to yield tangible water assets unless it is given clear political and economic incentives for doing so, along with solid guarantees of equivalent water supplies either by importation from abroad or by internationally financed desalinization plants. Another equally essential condition for Israeli withdrawal is effective international supervision of the rivers and their drainage basins in the territories to be evacuated, so as to avoid pollution and ensure water quality.

The report predicts that if Israel returns the entire Golan to Syria unconditionally, it stands to lose at least 40 million cubic meters per year (MCM/Y) from its present supplies. Moreover, there is danger that the Syrians might try once again to divert the upper sources of the Jordan away from Israel, as they had tried to do prior to 1967. If the Syrians resettle and industrialize the Golan plateau after Israel's evacuation, the area might become a source of pollution endangering the water quality of the Sea of Galilee.

Under the circumstances, the report suggests that Israel set two alternative boundaries in the Golan: (1) a "maximal boundary," designed to protect all of Israel's current water supplies and allowing only minor territorial adjustment of Israel's present position on the Golan; and (2) a "minimal boundary" that concedes more territory and leaves Israel in control of just the water courses and their banks, barely enough to ensure that future Syrian activity in the area will not have a *critical* effect on Israel's water supplies. Only later, if and when full cooperation between Syria and Israel is achieved, might Israel consider relinquishing its direct control over the latter part of the territory in favor of international supervision of the shared water resources.

Regarding the West Bank, the report suggests that Israel's withdrawal should be predicated upon safeguards to prevent significantly enhanced withdrawals of water from the mountain aquifers. And as for the acute water shortage in Gaza, the report insists that it be relieved by desalinization or by transfer of water from the Nile, but not from Israeli sources.

Hydrologically, Israel and the West Bank are now mutually dependent. Prior to the Israeli occupation in 1967, the residents of the West Bank utilized only some 5 percent of the potential yield of the mountain aquifers. Today they utilize between 10 percent and 15 percent (estimates vary), while the remainder—amounting to some 300–400 MCM/Y—is utilized by Israel. The Israelis fear that, should they withdraw unconditionally, the Palestinians are likely to increase their extraction of water from the upper parts of the shared aquifers and thereby reduce the supply to Israel's main population centers, obtained from wells located in Israel proper. Israelis are also concerned that water quality will be affected if unrestricted urban, industrial, and agricultural development takes place in the West Bank.

The report recommends therefore that Israel should forgo direct hydrological control over the shared aquifers of the West Bank only if full cooperation is instituted to ensure the quality and quantity of Israeli groundwater supplies. As in the case of the Golan, the report suggests a critical line, located some kilometers east of, and parallel to, the so-called green line (the pre-1967 boundary), beyond which Israel should not withdraw at this stage. On the positive side, the report stresses that when true peace is achieved, all sides can benefit from joint action to promote groundwater recharge, judicious extraction, and protection of water quality.

Such is the intensity of the public debate now taking place in Israel over the issue of water, that soon after publication of the described report, a team of economists from the very same Tel Aviv University strongly criticized its territorialist approach. They consider the issue of water in purely economic terms, as a resource over which different sectors and users (regardless of political or administrative affiliation) should be allowed to compete freely on the basis of cost and efficiency of use.

The economists contend that the best way to allocate water efficiently is to use market mechanisms. The price set for water should consist of two components: a uniform basic price for the commodity at its source, and a transfer price to reflect the varying cost of conveying water to different districts. Accordingly, the total price of fresh water would be least near its sources, such as the Sea of Galilee or the Yarkon Springs, and greatest in areas that are distant from the sources and that are at a high elevation.

To illustrate the principle, the same economists divided Israel, the West Bank, and the Gaza Strip into 19 districts. For each of these districts, they estimated the potential water demands by the agricultural, domestic, and industrial sectors, as well as the local price for water. Thus, the hypothetical price per cubic meter might be about $0.23 in Tiberias, $0.32 in Haifa, $0.35 in the central coastal plain of Israel, $0.43 in Tel Aviv, and $0.54 in Gaza and Beersheba. Employing the same criteria, they calculated that the price of water in the West Bank should be $0.37 in the north, $0.43 in the center, and $0.48 in the south (Hebron).

For the price mechanism to work effectively and impartially, water must be treated as a freely traded commodity. According to these economists, the most effective way to accomplish this is to establish a joint Israeli-Palestinian water authority for the mountain aquifer. Such an authority will set the basic price for water, as well as determine the realistic price for conveying it to users in different areas; and it will invest its earnings in developing, maintaining, and protecting the water resource. The same authority may also supplement its own supplies by importing fresh water or by desalinizing brackish water. Israel, on its part, will undertake to sell water to West Bank users from its own sources at a price to be set by the same criteria that apply to its own users. In other words, Israel and the West Bank will enter into a free trade zone for water (as well, perhaps, as for other goods).

With such an arrangement in place, Israel will no longer need to exercise exclusive control over water resources and the territories from which they originate. Thus, potential conflicts over water and land may be prevented. Water will no longer be considered a strategic asset, nor a potentially explosive obstacle to peace. On the contrary, the entire mystique of land and water ownership, with its inevitable burden of nationalistic emotion, will be transmuted into a dispassionate issue of economics involving costs and returns.

Finally, the Israeli economists who promulgate this approach set into context the financial dimension of the water issue. The aquifer in contention between Israel and the Palestinians, they wrote, yields an average of, say, 300 million cubic meters per year. Given the present price of desalinization (about $0.80 per cubic meter), the cost of that quantity of water is of the order of $240 million per year. That is a trivial sum, when considered in terms of the price to be paid for war, or even for continuation of the present state of tension between Israel and its neighbors. Surely, the great potential benefits of peace—economic and otherwise—should not be jeopardized by such a sum, especially when an option exists that seems so simple, elegant, and feasible. Thus spake the economists.

Yet the argument continues. In early November 1993, a group calling itself "The Movement to Save the Waters of Israel" issued a public appeal. It contends that as much as one-third of Israel's fresh water supply is obtained from the mountain aquifer, and this amount will be in jeopardy if Israel cedes that water resource to the Palestinians. In strongly emotional terms, the appeal warns that the loss of that water will be a death knell to Israeli farming and will result in soil salinization and water pollution in Israel, in the drying of public parks, and in the rationing of water for domestic purposes. Therefore, the appeal demands that access to the mountain aquifer (consisting of some 20 percent of the area of the West Bank) be retained within the jurisdiction of Israel even after the granting of Palestinian autonomy.

Diametrically opposite views regarding the fate of water underlying the West Bank have been expressed by Palestinians experts. Fortunately, the differences, though serious, are not irreconcilable. During the last couple of years, Palestinian and Israeli hydrologists have been meeting regularly, both formally and informally (mostly away from the glare of publicity), to share information and to search actively for compromise and mutual accommodation.[4] Their constructive effort is progressing.

Although it is far from being the only, or even the most acute, conflict among the nations of the Middle East, the Arab-Israeli dispute has seemed the most prominent. It has raged for the greater part of this century and is only lately beginning to abate. Most Arabs and Israelis have grown weary of the endless accumulation and recitation of reciprocal grievances and recriminations. They, and the entire world, yearn for an end to it.

To some, the hostility between the Arabs and Israel seemed chronic. A favorite cliche of commentators, tending to disregard the far bloodier quarrels elsewhere (such as those that have raged between Iraq and Iran, within Iraq, in Syria, in eastern Turkey, in Lebanon, in southern Sudan, in Kuwait, and in Yemen), has been that this one conflict is the "root cause" of the trouble in the entire region. In fact, it has received such disproportionate attention as to seem larger than it really is. What has been the true nature of that conflict? Is it an age-old existential rivalry bequeathed by Abraham's two sons, Isaac and Ishmael, to their presumed descendants—the Jews and Arabs? Is it a continuation of the ancient wars between the Israelites and the Canaanites, or between the Israelites and the Philistines? Or of Muhammad's quarrel with the Jews of Medina? Too many "expert" observers have opined that the conflict is so deeply rooted in history as to be unresolvable. And legions of propagandists and polemicists on both sides have inflamed and polarized the emotive issue to the point of making it seem intractable indeed.

The Arab-Israeli dispute is in fact a recent one. Arabs and Jews had lived together for centuries in relative peace, and there were periods of close association and harmony. Certainly, the Arabs—who have been the majority in the Middle East for over a millennium—were never as virulently prejudiced and as cruelly oppressive toward the Jews as were many European Christians.

To be sure, the Palestinian-Israeli conflict is real. Just as it should not be exaggerated, it must not be minimized. It is made all the harder to resolve because it is a conflict between two rights. Both nations are attached to the same sliver of land with historical justification and with an emotional intensity that can only be described as desperate. After the Holocaust in Europe and the progressive assimilation of their Diaspora elsewhere, the Israelis consider their Zionist enterprise to be the last chance for Jewish survival and rebirth. To the Palestinians, the trauma of displacement from the parts of Palestine taken over by Israel seems like an equivalent holocaust. They cannot be persuaded to forgo their grievance on the contention that they should have accepted the U.N. partition plan of 1947 and taken half the country, instead of fighting for all and losing most of it. What happened happened and cannot be reversed, yet the pain and distrust linger.

As the common cliche goes, time heals all wounds. But how long will it take to heal the Palestinian-Israeli rift? Nearly half a century has elapsed since the 1948–49 war that displaced hundreds of thousands of Palestinians and turned them into a nation of refugees. It seems that time alone works too slowly. It must be helped, and the way to do so is by a proactive policy of accommodation. The Palestinians deserve and must be granted their inherent right to nationhood, and to autonomy and eventual sovereignty over the West Bank and the Gaza Strip. It is in the interest of Israel, as well, to stop subjugating this nation and to befriend it by positive acts of generosity.

On the issue of rights and wrongs, there is an apt Hasidic tale. Two disputants once sought the judgment of a venerable rabbi. He listened patiently to the litany of complaints recited by each, then pronounced his verdict: "You are both right." When the perplexed litigants left, the rabbi's wife reproached him: "You did not make a real judgment!" To which the weary old man replied: "You are also right."

When I first heard this story, many years ago, I thought it quite absurd. With age, however, come interesting realizations. The rabbi may have been wise to leave the antagonists to work out their differences directly and pragmatically, unencumbered by external opinions. In general, it seems, few disputes are between right and wrong, most being between partial or sectarian rights. Hence the effort to define absolute justice is so often a sterile exercise. Casting a dispute in terms of rights and wrongs ignores the opportunities. Left to their own devices, disputants may well find it to their mutual advantage to forgo exclusive claims, cool the ardor of their indignation, and find ways to accommodate each other by compromise. Only thus can life go on and the benefits of coexistence be realized. Eventually the contradictory assertions that fueled old disputes tend to fade into irrelevance as new situations arise.

The Arabs and the Israelis are, in a profound sense, twins, having common roots and a common fate. They share similar subliminal perceptions, memories, and aspirations. Both are blessed or afflicted (as one chooses to see it) by a sense of destiny arising out of the conviction that they are heirs to a great faith and culture. And that conviction is couched for both in a Semitic language of powerful imagery and allusion. The imagery captivates, summons, inspires, and may lead astray, away from present reality. The metaphors can deceive, the parables can delude. The classical language is surrealistic, full of vivid hyperbole and historical evocations of heroic deeds and divine interventions. Its basic vocabulary still conveys subliminal mythological associations. It possesses the magical power of poetry to clothe harsh ideas with the beauty of rhetoric. Hence the narrative of the region's woes, as told by the partisans of one or another of the warring factions, is usually cast in febrile and melodramatic prose that smolders with righteous indignation.

The character of the people of the region is shaped by a checkered past of alternating ascendancy and subjugation, glory and humiliation. Torn between conflicting perceptions of superiority and inferiority, the Jews and Arabs are both driven to recapture past glories and to erase recent humiliations. To both, the ancient notion of honor is especially important. It is a vague yet deeply passionate notion, for which its adherents willingly—even eagerly—lay down their lives and the lives of their children, and for which some leaders are prepared to risk all.

The Jewish and Arabic peoples have both been victimized. The main oppressors of each were Europeans, who persecuted the Jews and exploited the Arabs. The tragedy is that instead of realizing their common predicament and joining forces to rebuild their region and themselves, they tend to vent their rage at each other. It seems doubly tragic when one considers that the Arabs and Jews, over the centuries, had gotten along with each other better by far than either had with the Europeans. Their pent-up anger is in part misdirected, projected at the wrong targets, at each other, while the formerly imperialist but now smug and sanctimonious Europeans sit in moralistic judgment.

An essential leitmotif in Middle Eastern culture is the notion of pride or manhood, both personal and national. All too often that notion is associated with the mystique of heroism and the glorification of war. Pride often seems more important than economic benefit. To some Arabs, perhaps the greatest insult is that their armies were defeated in the battlefield not by the European powers themselves but by their seemingly weakest surrogates, the remnants of those who were the defenseless victims and the most downtrodden of Europe for countless generations. This view ignores the fact that the Jews of Europe were weak physically (though they excelled culturally) only because they were a small minority there, and that the very searing memory of their weakness drove them to strength in pursuit of national renewal. The same may be said of the Palestinians at the present time.

The fundamental questions are often avoided. Why is the Middle East so conflicted? The simplistic answers are: because of Israeli intransigence, or Arab belligerence, or foreign machinations, or religious extremists, or tribal divisions, or the polarization between rich and poor. All these are true, in part. In another sense, the area is a battleground because of the past. It is still entrapped in its own memories. History, both brilliant and tragic, is etched in the face of the land and seared into the souls of the people. Unfortunately, nations are still fighting at present and risking the future over a past that cannot be changed. (The obsession with the past was captured by a wag who commented that in the Middle East, too many people see the light at the end of the tunnel only by looking backward.) In the end one is impelled to shout, as did Prime Minister Rabin in September 1993 at the signing of the historic accord with the Palestinians: "Enough!" Enough with the past, on to the future!

Underlying the basic urge to war—to any war—is the rejection of compromise. The side instigating the war believes it cannot compromise its "rights" when it can get "all" by force. But the act of war is a risky gamble of all or nothing. The losing side stands not merely to get nothing further but, indeed, to lose what it already has. The Arab extremists (a minority, but vociferous) still believe that Israel has no rights at all, being a foreign intruder. Their diametric opposites in Israel (also a minority, but similarly vociferous) use the Bible and the fact that the Arabs had started the 1948–49 war against Israel to justify giving the Palestinians nothing.

The two sides have misperceived each other. The early Zionists did not recognize the Palestinians as a distinct nation, actual or even potential, seeing them instead as a mere fragment of the larger Pan-Arab nation that had won independence over the greater part of the Middle East. Why then should the great Arab nation begrudge the Jews tiny Palestine? In so thinking, they failed to realize that the Palestinian Arabs were and shall always be strongly and specifically attached to the country in which they had lived for well over a millennium, and that they are a talented, tenacious, and resilient people whose nationhood and aspiration for independence cannot be denied any longer.

The Palestinians, on their part, considered the Jews to be foreign intruders, as ephemeral as the European Crusaders. The Palestinians thus ignored the inextricable connection between the Jewish people and the place that was their one and only homeland, in which their nation and culture were born. The Jews are not alien to the region, but an integral part of it. Although many Jews had lived in Europe and even farther, their culture was still rooted in the place of their origin. Moreover, large Jewish communities continued to live throughout the region, including in Palestine, and to adhere to their culture there throughout the course of history. Deeply rooted in the region, elements of Jewish culture are ingrained in Islamic tradition itself, and are part and parcel of the collective Semitic culture of the Middle East as a whole. Yet Jewish culture is not merely a historic curiosity, but indeed a vibrant force able to reconstitute a nation even after the greatest of all traumas, the Holocaust.

Polemics aside, the incontrovertible fact is that neither side can eliminate the other, nor deny the other's continuing existence and right to progress. Hence, as fatigue sets in and the utter futility of strife becomes evident, the majority on each side has come to the point of realization: better to negotiate a peaceful compromise. Rather than continue to engage in a futile quixotic effort to redress past grievances by reversing the course of history, begin to look toward and work for a better future.

Now that the initial steps have been taken to break the psychic barrier, the two nations can begin, tentatively at first but with increasing commitment and hope, to move beyond a grudging detente toward a willing reconciliation. The Israelis and the Palestinians will come to recognize their common humanity and their spiritual kinship, based not only on their related languages and lore but also on the fact that they have been forged by parallel experiences. Apart from their attachment to the same land and its ways, there are the searing memories of exile and dispersion, and the yearnings for a redeeming future. The two nations will come to understand each other's cultural and spiritual connection to the land. The self-centered idea that either can have that land all for itself is now recognized to have been a dangerous and ultimately self-defeating delusion. The futures of the two nations are interdependent. Reciprocal recognition and affirmation of this interrelatedness will allow the growth of a new dynamism. Only thus can democracy, progress, and environmental rehabilitation redress the abuses of war, oppression, land degradation, and—not the least—water depletion.

The provisional understanding reached between the Palestinians and the Israelis is only a beginning. Any political arrangement is merely a framework, needing to be filled with substantive content. As the negotiations proceed and intentions are tested, the agendas of the parties may conflict over such issues as boundaries, statehood, the status of Jerusalem, return of refugees, compensation, et cetera. All these are important and highly emotive. Yet more immediately vital than any of these is the increasingly acute problem of water. Fortunately, as we have endeavored to show, this problem can be resolved equitably to mutual benefit, and its successful resolution can help create an atmosphere of trust that will facilitate the resolution of the other issues as well.

Regarding trust and mistrust, the Bedouin tell this tale: One bright day a daring young tribesman named Hassan rode his camel across a steep mountain range. The trail was narrow and treacherous but brave Hassan was unafraid. Exuberantly, he drove his camel at a gallop, oblivious to all danger. Suddenly, as he skirted the edge of a sheer cliff, he caught sight of another rider galloping toward him, teetering erratically on the brink of the precipice. Hassan recognized the other rider as reckless Abdul of the rival tribe, so he braced himself for an imminent collision from which only one of them could possibly emerge alive. At the last moment, however, Abdul steadied his camel and passed Hassan without mishap. As they crossed paths, at the very moment of their near-disaster, Abdul glanced at Hassan excitedly and cried: "Mule!" Incensed at what he considered a great insult, Hassan raised himself high on his saddle, and—looking back at the receding Abdul—shouted his reply: "Ass!" Thus absorbed in pungent vituperation, Hassan failed to notice the real mule that stood astride the path just beyond the next turn. The resulting impact hurled Hassan into the chasm, a victim of his own distrust of the man who had tried to warn him.

In the Middle East, said Israel's David Ben-Gurion, anyone who does not believe in miracles is not a realist. But what is a miracle? In an age of innocent faith, people believed that divine intervention can induce supernatural events. Today, skepticism prevails, and many people have ceased believing not only in the supernatural, but even in what is entirely natural; namely, that erstwhile enemies can—out of self-interest if not pure magnanimity—become cooperative neighbors. Miracles do not happen spontaneously any more, but they can be created. The essential ingredients are mutual respect and trust in the possibility of a better future, achievable through peaceful cooperation. Surely, love will not replace hate immediately. The best for which we can hope at this stage is pragmatic coexistence. True reconciliation will come later.[5]

In the Hebrew Bible, no less a figure than Moses, whose very name meant "Drawn out of Water," once erred fatefully through lack of faith over the issue of water. As related in chapter 20 of the Book of Numbers, the people of Israel strove with Moses after he had led them into the waterless Wilderness of Zin: "Why have ye brought us into this wilderness to die here?" Seeing their plight, the Lord then instructed Moses: "Speak ye to the rock before their eyes, that it give forth its water." But instead of speaking to the rock gently, Moses, probably made impatient by extreme thirst, lifted his hand and smote the rock with his staff. Water did come forth, momentarily, but at a terrible price. For so transgressing, Moses and his entire generation were condemned to die in the desert rather than enter the Promised Land. The passage concludes with the words: "These are the Waters of Strife, where Israel strove with the Lord." Now, no less than in the time of Moses, more water can be had by speaking than by striking.

At certain pregnant moments in history, peace may break out suddenly, as if by miracle, just when skeptics are most convinced it will never come. At the right time, the Cold War between the United States and the former Soviet Union ended suddenly, and so it was between Israel and Egypt. Now is the time for Israel to make peace with its remaining Arab neighbors—the Palestinians, Jordanians, Syrians, and Lebanese. The seeds of peace have already been planted in the parched ground. With a little water, steadily applied drop by drop, they will sprout and bear fruit. And it may yet prove to be an exemplary peace, to be followed by the other contenders over the region's great and small rivers and aquifers.

In the Semitic languages, the derivations and associations of words can be discerned from their shared three-consonant roots. Accordingly, the words *shalom* in Hebrew and *salaam* in Arabic are associated with words implying wholeness, perfection, well-being, and acceptance (the latter being at the very root of the term *Islam*). So *shalom* and *salaam* mean much more than the mere absence of war; their rich meaning conveys the full promise of recovery and reconstruction. Beyond military security, beyond politics, peace is transcendent. With peace, and only with peace, many problems can be resolved, including the water shortage.

The Book of Job states: "As the river is drained dry, so man lieth down and riseth not." Ultimately, the choice is between that bleak prospect and the rousing vision of Isaiah: "Behold, I will extend peace to her like a river, and the wealth of the nations like a mighty stream. . . . And when ye see this, your heart shall rejoice."

Because they say unto thee: Thou land art a devourer of people, and hast been a bereaver of thy nations; therefore thou shalt devour people no more, neither bereave thy nations any more, saith the Lord God.

And I will sprinkle clean water upon thee, and ye shall be clean. . . . A new heart also will I give thee, and a new spirit. . . . And I will call for the corn, and will increase it, and lay no famine upon thee. And I will multiply the fruit of the tree, and the increase of the field. . . . I will cause the cities to be inhabited, and the waste places shall be builded. . . . And they shall say: This land that was desolate is become like the Garden of Eden. Ezekiel 36

Notes

1 An Overview

1. The entire scene gave literal meaning to the haunting lines from T. S. Eliot's *The Waste Land:* "Where the dead tree gives no shelter . . . and the dry stone no sound of water . . . I will show you fear in a handful of dust."

2. Of course, the diversion of water from the Sudd will have negative effects, too, in the destruction of parts of that wetland. (See Chapter 6.)

2. Waters of Life

1. Hillel, D. 1982. *Negev: Land, Water and Life in a Desert Environment.* New York: Praeger.

2. Hillel, D. 1987. *The Efficient Use of Water in Irrigation: Principles and Practices for Improving Irrigation in Arid and Semiarid Regions.* Washington, D.C.: World Bank.

3. The evident need for coordination in large riverine irrigation systems led anthropologist Karl Wittfogel to postulate that irrigation-based societies, which he called "hydraulic civilizations," necessarily tended to centralized control and even to despotic rule. Although simplistic and lately much criticized, the concept contains a grain of plausibility. (See Wittfogel, 1956, in the bibliography.)

4. The name Rachel itself means "ewe."

5. The concept of rain as an expression of mercy is echoed in Shakespeare's *The Merchant of Venice,* in which Portia pleads: "The quality of mercy is not strained,—it droppeth as the gentle rain from heaven upon the place beneath: it is twice blessed,—it blesseth him that gives, and him that takes."

6. Plato held that all surface bodies of water were connected to, and fed spontaneously via subterranean channels by a vast, bottomless reservoir of primordial waters called Tartarus.

7. The spring 1993 issue of *Research and Exploration* (a scholarly publication of the National Geographic Society) provides a comprehensive and well-balanced account of the prospects for global warming.

8. Hillel, D., and C. Rosenzweig. 1989. *The Greenhouse Effect and Its Implications Regarding Global Agriculture.* Amherst: College of Food and Natural Resources, University of Massachusetts.

9. Lonergan, S., and B. Kavanagh. "Climate Change, Water Resources and Security in the Middle East." *Global Environmental Change,* Sept. 1991, pp. 272–290.

10. Strzepek, K. M., and J. Smith, eds. 1993. *As Climate Changes: International Impacts and Implications.* Cambridge, England: Cambridge University Press.

3. Ancient Civilizations

1. The backbreaking work that was required is described in the Bible and later Hebrew writings. Especially poignant is the allegorical outcry in Chapter 5 of Isaiah: "Let me sing of my beloved, a song concerning his vineyard. My beloved had a vineyard on a fruitful hill. And he digged it, cleared it of stones, planted it with the choicest vines, built a tower in the midst of it, and hewed out a vat therein. And he hoped that it would yield good grapes, but it yielded sour grapes. . . . Now come, I will tell you what I will do to my vineyard: I will take away the hedge thereof, and it shall be eaten up; I will break down the fence thereof, and it shall lay waste. It shall not be pruned nor hoed, but there shall come up briers and thorns, and I will command the clouds that they rain no rain upon it."

2. Leviticus 25: "And the Lord spoke unto Moses in Mount Sinai, saying . . . when ye come into the land which I give you . . . six years thou shalt sow thy field . . . but on the seventh year . . . thou shalt not sow thy field. It shall be a year of solemn rest for the land."

3. There is a contrary view regarding the Middle Eastern goat, which may have been vilified and condemned too harshly in the court of expert opinion as an incorrigible overgrazer, a denuder of vegetation, a promoter of soil erosion, a cause of desertification, altogether a menace to civilization. Perhaps that is too much guilt for one small scapegoat to bear. The truth of the matter is that overgrazing is caused by the people who put too many animals for too long on an area of rangeland incapable of sustaining them. With proper grazing management, the goat is no special menace, and no worse than its cousin the sheep. However, with excessive grazing, and given no choice, the goat will subsist on anything it can find, including thorny and dry brush, and it will survive long after the sheep has died of starvation. The goat is thus a remarkably hardy animal, a spirited and whimsical individualist who fully justifies the application of its name, *capra,* to the adjective "capricious." Through no fault of its own, however, this lovable animal became an agent of great destruction throughout the Middle East.

4. The Gilgamesh epic was first discovered in the extensive remains of the Assyrian king, Ashurbanipal, who reigned in Nineveh from 668 to 626 B.C.E. The epic, evidently borrowed from the Babylonian or Sumerian original, was inscribed in cuneiform on clay tablets. These tablets were scorched and shattered when the

city was sacked, and were subsequently disrupted by careless excavators and further damaged by the intrusion and evaporation of salt-bearing groundwater. Nevertheless, George Smith of the British Museum was able to piece together (circa 1870) the multitude of fragments to recompose the twelve tablets of the epic of Gilgamesh, which contains the story of the deluge. Several older versions of the epic were later discovered in southern Mesopotamia.

The story of the flood as related in the Gilgamesh epic apparently predates the biblical story of Noah's ark. There can be no doubt that the story of the flood was indigenous to southern Mesopotamia, which is ever susceptible to catastrophic inundations. (One particularly suggestive item is the smearing of the ark with pitch or bitumen, a substance that was abundant in Mesopotamia but rare elsewhere.) In the Mesopotamian story, Gilgamesh, the king of Erech, became despondent following the death of his friend Enkidu, and sought the advice of his immortal ancestor, Ut-napishtim (often referred to as the "Babylonian Noah"). Ut-napishtim then tells how he survived the great deluge (unleashed by the gods to destroy humankind) by building a great ship and by gathering into it all his family and all the beasts of the field. When the tempest began, all light was turned into darkness, water overwhelmed the land, and all humanity was returned into clay. . . . When the flood subsided, the ship settled on dry land and Ut-napishtim, with his household, began life anew.

5. In the ancient Middle East, grain multiplication was considered the criterion of fertility and agricultural success. (Witness the account in Genesis 26:12: "And Isaac sowed in that land [Gerar, located in the northern Negev], and found in the same year a hundredfold; and the Lord blessed him.") Sowing technique was apparently based on the use of a funnel-plow, which placed the seeds in widely spaced rows that promoted the tillering of grain crops such as wheat and barley. In our times, yield per unit of land area (e.g., kilograms per hectare or bushels per acre) is the more common criterion. In water-scarce irrigated areas, however, the yield per unit of water volume applied may prove to be the more appropriate criterion.

6. Legendary Queen Semiramis, wife of Ninus and ruler after him, was, incidentally, noted for her great beauty, wisdom, and power. She was reputed to have built many cities, conquered Egypt and much of southwestern Asia, and unsuccessfully attacked India. Other Assyrian kings, incidentally, took similar pride in their waterworks. One example is Ashurnasirpal, who boasted: "I dug a canal from the Upper Zab River; I cut straight through the mountain; I called it Patti hegalli ('Channel of Abundance'); I provided lowlands along the Tigris with irrigation; I planted orchards at its [the city's] outskirts, with all sorts of fruit trees. . . . In the gardens, the irrigation weirs [distributed the water evenly]; pomegranates glow in the pleasure garden like the stars in the sky, they are interwoven like grapes on the vine . . . in the garden of happiness they flourish like cedar trees." Such Assyrian texts echo earlier Sumerian writings, such as the hymn to Ninurta as god of vegetation: "In the river there flowed fresh water, in the field grew rich grain. . . . The watered gardens were filled with honey and wine, in the palace grew long life." (See Pritchard, 1958, in the bibliography.)

7. The carrying capacity of water for suspended silt depends greatly on the speed of flow. Steeply descending mountain streams are swift and capable of carrying great loads of silt. When the speed is reduced, however, the silt settles. A dramatic example is the Yellow River of China. (See the article by the author in *Natural History Magazine,* 1991, listed in the bibliography.)

8. Artzy, M., and D. Hillel. 1988. A Defense of the Theory of Progressive Soil Salinization in Ancient Southern Mesopotamia. *Geoarchaeology,* Vol 3, pp. 235–238.

9. See Pritchard, J. B., ed. 1958. *The Ancient Near East: An Anthology of Texts and Pictures.* Princeton: Princeton University Press.

10. In the vassal treaties of Assyrian King Esarhaddon, there are telling curses laid upon those who would dare to be disloyal. Typical ones are: "May Adad, the canal inspector of heaven and earth, put an end to [vegetation] in your land. . . . May he hit your land with a severe destructive downpour. . . . May the great gods deprive your spirit of water. . . . May a flood, an irresistible deluge, rise from the bowels of the earth and devastate you. . . . May Ea, lord of the springs, give you deadly water to drink. . . . Just as rain does not fall from a copper sky, so may there come neither rain nor dew upon your fields and meadows. . . . Just as you can blow water out of a tube, so may they blow away you, your women, your sons, your daughters, may they make your rivers, your springs, and their wells flow backward."

11. Adams, R. McC. 1981. *Heartland of Cities: Surveys of Ancient Settlement and Land Use on the Central Floodplain of the Euphrates.* Chicago: University of Chicago Press.

12. Cisterns, of course, were not at all unique to the Negev, but were in fact prevalent throughout the Middle East. The city of Jerusalem, for instance, was sustained for many centuries mainly by cisterns hewn in the bedrock and fed by runoff from roofs, courtyards, and streets.

13. An example of the importance of water supplies in the Negev is given in Joshua 15:18, 19. After Caleb gave his daughter Achsah to Othniel for a wife, she "persuaded him to ask of her father a field; and she alighted from her ass, and Caleb said unto her: 'what wouldst thou?' And she said: 'Give me a blessing; for that thou hast set me in the Negev, give me therefore springs of water.' So he gave her the upper springs and the lower springs."

14. Hillel, D. 1982. *Negev: Land, Water, and Life in a Desert Environment.* New York: Praeger.

15. Evidence of the association of agricultural land ownership with water rights to the adjacent tracts of sloping land is present in the papyri discovered at Nessana. The documents show that the inhabitants of this remote desert outpost in the Negev were sustained by a highly developed agricultural economy during the sixth and seventh centuries C.E. The documents reflect the explicit concern of the Nessanites with seed-land, vineyards, gardens; as well as crops of wheat, barley, aracus, olives, and dates; and—most significantly—with water rights, water channels, and water reservoirs. See Mayerson, 1960.

16. See Tadmor, Shanan, Evenari, and Hillel, 1958; *Ktavim* (Israel) 8:127–151; as well as Hillel, 1982 (listed in the bibliography).

17. Later on, some Muslims would claim that a member of Muhammad's family should have been chosen. As Muhammad had no surviving sons, they held that his successor should have been his cousin and son-in-law, Ali, the son of Abu-Talib. Those who believe that Ali was the rightful heir are called Shiites. Those who adhere to the line of succession beginning with Abu-Bakr are called Sunnis.

18. Amazingly, the original Arabian armies were small, usually under a thousand men, certainly fewer in numbers and less well equipped than their Byzantine or Persian foes. But they fought with great zeal, and their horses and camels gave them speed and endurance. A common Arab tactic was to draw enemy forces into a wadi (valley) and then use the terrain to trap them. One of the Arabs' decisive

victories, the Battle of the Yarmouk River in 636, resulted from a dust storm that concealed the desert fighters from the "civilized" Byzantines. This victory gave the Arabs control over Syria. Another dust storm helped the Arabs to defeat the Persians in 637 at Qadisiyah and, hence, to overrun Mesopotamia.

19. Ironically, that process was accelerated by the Crusades: the European Crusaders, who purportedly came in the name of Christianity, were in actuality so cruel to the local inhabitants—even those who had clung to the Byzantine Christian tradition—that Christianity itself became hateful to many of their victims.

4. Modern States

1. Although Britain had relatively few Jewish subjects of its own, it wished in 1917 to enlist the support of American Jews (among others) for America's active entry into the war. Supporting Zionism was one way to do it. The leading Zionist spokesman in England at the time was Dr. Chaim Weizmann, a chemist who had contributed greatly to Britain's war effort by synthesizing acetone (a chemical hitherto imported from Germany that was needed for making explosives). Weizmann's discoveries made him known to leading journalists and eventually to cabinet ministers. Prime Minister David Lloyd George was sympathetic to the Zionist cause from having studied the Bible. So was the foreign minister, Lord Arthur Balfour. It was he who communicated to Lord Rothschild, then head of Britain's Zionist organization, the cabinet's decision to support Zionist aspirations in a letter that came to be known as the Balfour Declaration. The letter stated: "His Majesty's Government view with favor the establishment in Palestine of a national home for the Jewish people, and will use their best endeavors to facilitate the achievement of this object, it being clearly understood that nothing shall be done which may prejudice the civil and religious rights of the existing non-Jewish communities in Palestine, or the rights and political status enjoyed by Jews in any other country." That declaration was seconded in 1922 by a unanimous resolution of the United States sixty-seventh Congress and later approved by the League of Nations.

2. The name Palaistine had been given to the country (previously called Canaan, Israel, or Judaea) by the Romans (following Greek precedent) after their suppression of the Jewish rebellions in the first and second centuries C.E. Their aim was to obscure and deny the Jewish character of the country. To do so, they resurrected the memory and name of the long-extinct Philistines who had briefly occupied the southern coastal plain of the country during the end of the second millennium B.C.E.

3. The notion of a specific Palestinian nation distinct from the larger Arab nations (or from the notion of a "Greater Syria") did not come into being until after the Balfour Declaration and the efforts of the Zionist movement to establish the country as a Jewish National Home. At first, in fact, the nationalist Arabs in Palestine opposed the political separation of the country from Syria. The Ottomans, who had ruled over the entire Middle East during the four preceding centuries, never defined the country as an administrative unit, preferring to govern part of it out of Damascus and part out of Sidon.

4. By the end of the British mandate following the Second World War, the Arab population had grown to 1.2 million and the Jewish population to 600,000. As of 1993, the population of Israel was near 5 million (84 percent Jewish, 15 percent Arab, and 1 percent other) and that of the Palestinians in the West Bank and Gaza about 1.8 million.

5. The mandate over Palestine granted to Britain by the League of Nations made specific reference to the Balfour Declaration and to the British commitment to facilitate the establishment of a Jewish National Home there, provided it did not prejudice the civil and religious rights of existing non-Jewish communities in Palestine, or the rights and political status of Jews in any other country. On the other hand, the terms of the mandate did not specify any concrete procedure or timetable for the transition to independence. The British administration of Palestine, in any case, did not adhere to the spirit of the Balfour Declaration. By playing an ambivalent game between the Jews and the Arabs, alternately seeming to favor one side and then the other, the British may have hoped to "divide and rule," but eventually succeeded only in losing credibility with both sides.

6. On March 1, 1919 Prince Feisal wrote to Felix Frankfurter (a prominent jurist and long active in American Jewish affairs, then serving as an advisor to President Wilson at the Paris Peace Conference and later to be appointed by President Roosevelt to the U.S. Supreme Court), confirming his agreement with Weizmann. In his letter, Feisal noted that the Arabs and Jews share a common origin, that both had suffered oppression, and that they can now achieve their national aspirations together. He furthermore expressed sympathy for the Jewish desire to return home, noting that the Arab and Jewish nations may live side by side in peace in a region that can accommodate both.

7. The population of Jordan is now over 4 million—a 20-fold increase in some 70 years.

8. The decision to anoint Abdallah as Emir of Transjordan and his younger brother Feisal as King of Iraq was made by none other than Winston Churchill, who became colonial secretary in January 1921. In March of that year, he convened a conference in the Middle East, in which he and his advisors (including T. E. Lawrence) completed the task of carving up the region in accordance with the interests of the British. Lawrence would later brag that he, Churchill, and a few others had designed the modern Middle East over dinner. Seventy years later, in the tense confrontations now besetting the region, the question is whether the people of the region can continue living with that design.

9. A particularly spectacular example of how water was used as a weapon of mass destruction was described in an article published in the August 1991 issue of *Natural History Magazine* (Hillel, 1991). In June 1938, the Chinese forces under the command of Chiang Kai-shek breached the dikes of the Yellow River in a desperate effort to stop the Japanese advance. From a military standpoint, the maneuver was a success: part of the invading army was destroyed and its heavy equipment mired in thick mud. For the farmers living in the floodplain, however, the flood thus unleashed greatly magnified the misery already caused by the war. With no high ground and with all escape routes cut off by the rising water, nearly a million Chinese perished. The crops in this fertile region were also destroyed, and the land was smothered beneath a thick blanket of silt. The change in the river's course disrupted irrigation over a vast region, so that countless others suffered from famine during the following years.

5. *The Twin Rivers*

1. The conventional designations "left bank" and "right bank" for rivers assume that the observer is facing downstream.

2. As in other great rivers—e.g., the Huang He of China and the Mississippi

of the United States—efforts to contain the capricious flow within a narrow course may eventually be self-defeating. The sediment contained in the confined river tends to raise the river bed, thus increasing the risk that a large flood, certain to occur sooner or later, will be all the more devastating.

3. Among the traditional water-lifting devices still in use, particularly on the middle Euphrates, are tall waterwheels, driven by the force of the current, which raise the water in earthenware jars attached to their rims.

4. Bombings during the course of the 1991 Gulf War caused great damage to Iraq's hydraulic works, among other facilities. The destruction of dams and pumping installations, water purification plants and power stations has deprived the cities of clean drinking water, and deprived farming areas of water supplies as well as of fertilizers, pesticides, and spare parts for machinery and pumps. As a result, agricultural production has diminished markedly.

5. Turkey's official view is that the Tigris and Euphrates are its sovereign resources, to be exploited as Turkey sees fit. In the words of Suleiman Demirel: "Water is an upstream resource and downstream users cannot tell us how to use our resource. By the same token oil is an upstream resource in many Arab countries and we do not tell them how to use it." More reasonably, the Turks argue that there is sufficient water in the Tigris-Euphrates basin for all countries if it is used properly, but Turkey contends that Syria and Iraq have been wasting their water for many years.

6. GAP's stated aims are quite ambitious: "to mobilize regional resources, to eradicate regional disparities, to enhance productivity, to create employment opportunities, to raise income levels, to develop urban centers and to ensure economic growth and social stability in the region." In fact, Turkey's water projects are motivated by the need for energy as much as the need for water as such. Turkey imports half its annual energy requirements, and about 25 percent of its electricity production depends on imported fuel (for which it paid $3.5 billion in 1990). GAP is expected to supply as much as 25 percent of the country's electricity.

7. There are significant Kurdish minorities in Turkey, Syria, Iraq, and Iran: Kurds account for some 23 percent of the population in Iraq and 19 percent in Turkey, where they are in fact the majority in the Southeastern Anatolia province.

8. The separatist Kurdish Workers Party (PKK) has repeatedly declared its opposition to GAP, regarding it as Turkish theft of Kurdish waters, and has attempted to sabotage its works.

9. The name "Lake Assad," granted to the impoundment behind the Tabqa Dam on the Euphrates, certifies the importance of that reservoir, as do "Lake Nasser" in Egypt and "Ataturk Dam" in Turkey. Even the Iraqis have gotten into the act by naming a dam on the Tigris near Mosul after Saddam Hussein.

10. Syria's current efforts to dam the headwaters of the Yarmouk River are causing great concern in the Kingdom of Jordan, for which the Yarmouk is the principal source of water.

11. Since it claims sovereignty over the Hatay district (which the French had ceded to Turkey in the late 1930s), Syria does not consider the Asi (Orontes) to be an international river. Hence Syria has appropriated nearly all the water of that river, leaving very little for the Turkish farmers of the Hatay district. Turkey has attempted to link negotiations over the Euphrates with the issue of the Asi, but Syria has rejected such linkage since it implies de facto recognition of Turkish sovereignty over Hatay. Like many of the other territorial-hydraulic disputes in the

region, this issue cannot be resolved unless the sides are willing to compromise and cooperate.

12. Syria's position regarding water rights is self-contradictory. In exploiting the Asi (Orontes) River and the headwaters of the Yarmouk (as well as in attempting to divert the headwaters of the Jordan River away from Israel in the 1960s), Syria has assumed for itself the upstream right of sovereignty over water that originates in its territory. That is precisely the right that it denies to Turkey over the Euphrates and Tigris.

6. The Mighty Nile

1. Strabo's observation regarding the Nile is especially noteworthy: "Its rising and its mouths . . . are amongst the most wonderful and most worthy of recording of all the peculiarities of Egypt."

2. This Diogenes is not to be confused with the Athenian philosopher and eccentric who roamed the streets of his city holding a lit candle, searching in vain for an honest man.

3. What interseasonal variation does occur in the headwaters of the White Nile is further buffered by the Sudd swamps, which absorb variable flows and release a nearly steady stream.

4. Alternatively, the name Nile may have evolved from *nil,* a word of uncertain etymology meaning "blue" or "indigo" (as in the word *anil,* derived from the Arabic *an-nil,* related to the Sanskrit *nila* = "dark blue").

5. As in the case of Mesopotamia, the Arabic name Gezira, meaning "island" or "peninsula," is used in Sudan to describe a land between two rivers—the Blue Nile and the White Nile.

6. Since the late 1950s the streams have been choked even more by the uncontrolled spread of an accidentally imported plant: the South American water hyacinth.

7. The average annual discharge of the White Nile at Mogren (south of Khartoum) is 26 BCM/Y, that of the Blue Nile and its tributaries (Rahad and Dinder) is 51 BCM/Y, and that of the Atbara is 12 BCM/Y. The Blue Nile and Atbara flows are markedly seasonal, occurring mainly from August to December.

8. The instability of Nile flow is illustrated in the following statistics. Between 1961 and 1964, above-average rainfall over Lake Victoria raised the level of that lake by two meters, thereby increasing While Nile flows by 32 percent above their pre-1961 average. Meanwhile, rainfall in Ethiopia steadily declined, and this reduced the flow of the Blue Nile by 16 percent. When the White Nile flows returned to their average levels in the late 1980s after having previously compensated for the low Blue Nile flows, Egypt and Sudan were hit by water shortages that were alleviated only by unexpectedly heavy rains occurring over Ethiopia and eastern Sudan in August 1988.

9. The Greek letter delta itself was derived from the fourth letter of the Phoenician (and Hebrew) alphabet. The letter was originally a pictograph of a *delet,* meaning "door."

10. Muhammad Ali, a tobacco merchant of Turkish origin, born in Albania in 1769, with no education but endowed with ability amounting to genius, became a soldier of fortune in the Ottoman army. By 1805, he had become pasha, or governor, and soon secured virtual independence in Egypt. He then founded a dynasty that reigned in Egypt until 1952.

11. The soil of the Gezira plain (a former lake bed) has striking properties. It is a highly fertile alluvial clay that absorbs and retains much water when irrigated, but shrinks markedly and cracks deeply if it is allowed to desiccate.

12. According to the visionary plan conceived by the British engineer Hurst, comprehensive drainage of the Sudd could add as much as 13 billion cubic meters to the annual flow of the White Nile. (See Waterbury, 1979, in the bibliography.)

13. If the lake were maintained at its largest possible size, the evaporation might exceed 15 BCM/Y rather than the average loss of 10 BCM/Y, which is still very high.

14. The water deficit of summer 1988 was unprecedented since the construction of the High Dam. In July 1988, the volume of live storage in Lake Nasser had fallen to 41 billion cubic meters. The level of the lake then was only three meters above its "red line," at which the hydroelectric turbines must be shut off and irrigation water must be curtailed. The torrential rains of August 1988 restored the live storage to 75.3 billion cubic meters, to great relief.

15. The importance of subsoil drainage is expressed in an old Egyptian adage, translated into English as: "Sand on clay is money thrown away; clay on sand is money on the hand." Modern soil physicists can confirm that a sandy subsoil is more permeable, and hence more readily drainable.

16. A distinct shift of international opinion has taken place in recent years in relation to the development of water resources, particularly with respect to large dams, large-scale irrigation projects, and the exploitation of wetlands. An awareness has grown of the previously unacknowledged costs, including the displacement and disruption of indigenous populations, the damage to wildlife and the endangerment of species, the submergence of archaeological sites and other cultural and scenic locations, and the downstream effects on river valleys and estuaries. Current opinion favors small-scale, low-technology projects that fit the social and environmental fabric, that conserve resources and that are sustainable.

17. In addition to its annual quota of 55.5 billion cubic meters of water from the Nile, Egypt has been receiving about 6 BCM/Y of Sudan's quota, since Sudan has not been able to use its full allocation. Having become accustomed to this additional water, Egypt will find have difficulty giving it up.

18. World Bank Tables, 1990–91.

19. In its agricultural modernization efforts, Egypt is aided by technical cooperation with the United States, as well as with neighboring Israel (whose record of innovation in the area of land and water management is highly relevant to the development of agriculture in other countries of the Middle East). The latter cooperation, largely unpublicized, is but one example of the important benefits to be gained from regional peace and technological interchange.

20. This assumes an average water use in reclamation projects of about 17,000 cubic meters per hectare. Actual irrigation water use ranges from 19,000 to 25,000 cubic meters per hectare in Egypt. More efficient irrigation methods and strict water conservation may reduce water requirements considerably for carefully selected cropping sequences.

21. The area under cultivation in Egypt totals 3.1 million hectares, covering only 3 percent of the total land area. Of this, 2.4 m ha are in the Nile valley and the Delta—the so-called Old Lands—and the rest is in the reclaimed areas known as the "New Lands." The latter are located mainly along the western fringes of the Delta, near the Suez Canal, in northern Sinai, and in several oases in the Western Desert. Reclamation efforts carried out during the 1960s met with only partial

success: of the 383,000 ha reclaimed, at least 25 percent later became unproductive owing to waterlogging and salinization (caused by mismanagement, over-irrigation, and poor drainage), and only 30 percent could be considered economically productive. Nonetheless, efforts at land reclamation were renewed in 1978, and have continued ever since. Egypt's long-term goal is the reclamation of an additional 1.25 million hectares, of which 385,000 are included in the third five-year plan that began in 1992. Plans also call for the transfer of 3 BCM/Y of Nile water eastward via al-Salaam Canal across the Suez Canal to northern Sinai, with the aim of reclaiming some 400,000 ha there. Western Desert reclamation is also proceeding. Plans call for the reclamation of 33,700 ha in the remote Oweinat area, to be irrigated with water drawn from the Nubian sandstone aquifer.

22. On the face of it, Ethiopia seems to be in a fortunate position hydrologically. Its total surface water supply exceeds 110 BCM/Y. There are 14 major river basins, 11 of which convey approximately 100 BCM/Y across Ethiopia's borders. The groundwater potential is not known precisely, but it undoubtedly amounts to several additional BCM/Y. However, Ethiopia's water resources are unevenly distributed, being mostly in the less populated and topographically rugged western provinces. Over a third of the country is semi-arid or arid, while other parts are vulnerable to flooding. Nonetheless, there is a considerable area of land (estimated at 3.7 million hectares) suitable for irrigation, which is practiced only to a limited extent at present.

23. The evaporation rate in the Ethiopian highlands is of the order of 1 meter per year, whereas that at Aswan is over twice as great. Given the topography of those highlands, storage reservoirs there will have a greater mean depth than at Aswan, hence a smaller surface will be exposed for unit volume of stored water, a factor that is likely to reduce the quantity of evaporation even further.

24. Whittington and McClelland (1992).

25. Unfortunately, the close cooperation that prevailed between Egypt and Sudan throughout the 1970s and early 1980s has turned to hostility during the late 1980s and early 1990s. The Egyptian government has accused the radical Islamic government of Sudan (with its links to Libya and Iran) of training and abetting violent opposition groups in Egypt, while the Sudanese have made the counterclaim that the enemies of their own regime (including the rebels in southern Sudan) are being aided by Egypt. The two governments supported opposite sides in the Gulf War and are now engaged in a territorial dispute over the Halaib area on the Red Sea coast. Nevertheless, the two governments are mindful of the need to preserve an essential measure of cooperation over the waters of the Nile.

7. The River Jordan

1. Deuteronomy 8:7–9.

2. That epic sea battle, fought in the estuaries of Egypt's Delta, is commemorated in the great mortuary temple of Rameses III at Medinet Habu.

3. The First Book of Samuel, Chapter 13, tells the story: "Now there was no smith throughout all the Land of Israel, for the Philistines said: 'Lest the Hebrews make swords or spears.' So every Israelite had to go down to the Philistines to sharpen his plowshare, his spade, his axe, and his mattock. . . . So it came to pass in the day of battle that there was neither sword nor spear found in the hand of any of the people that were with [King] Saul."

4. One of the interesting aspects of the Babylonian exile was its confusing

effect on the Jewish calendar, which was ecologically determined. In rainfed Canaan, the onset of the rainy season in autumn initiates the growing season, and hence naturally represents the beginning of the new year. In irrigated Mesopotamia, by contrast, it is the flooding of the twin rivers in springtime that constitutes the beginning of the agricultural season. This can explain the apparent contradiction between the prevalent tradition of considering Tishrei (September-October) to be the first month of the Jewish calendar with Rosh Hashanah as its first day, and the biblical passages representing that same month as the seventh month (e.g., Numbers 29:1, 7).

5. Thereafter, the bulk of the Jewish people lived in the Diaspora, scattered through most of the countries of southwestern Asia, northern Africa, southern Europe, and—increasingly—western, central, and eastern Europe. Lacking permanent security in any of the countries of their refuge, and generally prohibited from owning land, the Jews, who had been primarily herders and farmers in their own homeland, perforce became city dwellers and merchants. Throughout the many centuries and lands of their Diaspora, they retained their collective memory of the ancestral land and yearned to return there and reconstitute their national life.

6. The sand along the shore of Israel apparently originated in Nubia and was brought to the Mediterranean coast of Egypt by the Nile. From there it was carried by offshore currents and deposited along the coast of southern Israel. The accretion of sand there has diminished since the construction of the Aswan High Dam, which blocks most of the sediment formerly carried by the Nile.

7. More amusing is Twain's account of his dip in the heavy brine of the Dead Sea: "It was a funny bath. We could not sink. . . . You can sit with your knees drawn up to your chin and your arms clasped around them, but you are bound to turn over presently because you are top-heavy in that position. You can stand up straight in water that is over your head, and from the middle of your breast upward you will not be wet. . . . Then [we] came out coated with salt till we shone like icicles."

8. Most of the land for Jewish settlement was purchased and developed by the Jewish National Fund and became a collective property of the nation, to be leased on a long-term basis to deserving users who would be responsible for its sustainable and productive use. That concept was later applied to water: soon after the State of Israel came into being, a law was passed declaring water to be the property of the nation.

9. To the early Zionists, agriculture (particularly irrigated agriculture) was not merely a means of subsistence, but indeed an ideology. Farming was considered the noblest and most honest and natural occupation, a way of life that literally reattaches the Jewish people, who had long ago been forcibly deracinated, to the land of their ancestry and destiny. By transforming the sons and daughters of merchants in the Diaspora into a nation of tillers of the soil, the pioneer Zionists hoped to "re-productivize" the Jewish people. In their view, only those who redeem the land by reclaiming and tilling it, and by living on it and of it, can truly and deservedly possess it. They considered farming to be a return to the land in the fullest and most literal sense, and working the land to be a spiritual fulfillment as well as a physical task.

10. The Banias, providing about 120 MCM/Y, has, since 1967, been in the possession of Israel, which has in fact declared its annexation of that territory (included with the Golan). According to the Israeli view, therefore, Israel's con-

tribution to the Jordan basin amounts to about 41 percent, whereas that of Syria is only about 22 percent. This hydro-territorial dispute will presumably be negotiated and resolved in the context of the Syrian-Israeli peace talks.

11. In the area of the old British Mandate (including Israel, Jordan, the West Bank, and Gaza), which the British "experts" once claimed could not support more than 2 to 2.5 million people, there now live nearly 12 million.

12. A similar possibility was envisioned half a century earlier by Dr. Theodor Herzl, the founder of modern Zionism.

13. Because so much of the country was arid, the need for irrigation water development became an imperative equal to that of land development. In the words of Moshe Sharett, Israel's second prime minister: "Water for Israel is life itself. It is bread for the nation, and not only bread. Without large irrigation works we will not reach high production levels . . . to achieve economic independence. . . . And without agriculture . . . we will not be a nation rooted in its land, sure of its survival, stable in character . . . with material and spiritual resources."

14. The term *eutrophication* refers to a condition in which nutrient-enriched waters induce the proliferation of algae and consequent depletion of dissolved oxygen, much to the detriment of fish and other aquatic fauna.

15. By 1990, following several years of drought, the Sea of Galilee had fallen four meters to its lowest level in memory: −212.48 meters below sea level (just above the "red line" of −213 meters). The National Water Carrier was shut down in January 1990, and the country faced a deficit of some 400 MCM/Y. In summer 1991, water allocations to farmers were cut by as much as 50 percent, leading to widespread reductions of the area of irrigated cotton and orchards. The two relatively rainy seasons that followed (especially the heavy rains of January–March 1992) restored the water level and even caused widespread flooding because of the lack of sufficient reservoir capacity. However, those seasons allowed the pumping of about 1.5 MCM per day for groundwater recharge.

16. The Dead Sea consists of two basin: the southern, which is very shallow, is now largely dry; and the northern, which is much deeper. Hence the shrinkage of the Dead Sea is not likely to continue at the same pace as during the last few decades.

17. The following is a summary of Israel's annual water supplies: upper Jordan, 580 MCM; Yarmouk, 70–100 MCM; flash floods, 15 MCM; groundwater, 1,000–1,200 MCM; and reusable wastewater, 30–110 MCM; minus unrecovered conveyance and production losses, 60 MCM. The total is between 1,650 and 1,950 MCM.

18. Until 1967, Syria used less than 80 MCM/Y of the Yarmouk's waters. Since 1975, however, Syria has been drawing more and more water out of the Yarmouk for purposes of irrigation. As of 1991, total Yarmouk water use by Syria had grown to some 153 MCM/Y, and if all 25 planned projects are implemented, that amount will increase to 200 MCM/Y. Obviously, such developments are very worrisome to the kingdom of Jordan.

19. The lack of adequate storage capacity is one of the major problems besetting agriculture in Jordan. The country's total storage capacity is 115–120 million cubic meters, but actual storage is generally much less. Jordan's target is to be able to store as much as 388 million cubic meters, but that figure is predicated on the implementation of the Wahda ("Unity") dam, which may or may not be built in cooperation with Syria and with Israel's concurrence.

20. The Disi fossil-water aquifer, which Jordan shares with Saudi Arabia, is capable of producing 100 MCM/Y of good-quality drinking water, but that water

will be expensive to convey to Amman. Saudi Arabia itself, incidentally, has been withdrawing increasing amounts of water (reportedly, about 250 MCM/Y) from that aquifer at Tabuk, only 50 kilometers from the Jordanian border. If both countries draw more and more water from the same aquifer, its reserves may not last more than two or three decades.

21. Recycled waste water now yields much less than 100 MCM/Y. Jordan plans to exceed that figure by the year 2000 by establishing tertiary waste treatment plants to remove mineral and organic contaminants so as to make the effluent safe for use in irrigation and in industry.

22. Agriculture plays a diminishing role in the national economy of Jordan, where it now accounts for no more than 6.8 percent of the gross national product and employs less than 7.6 percent of the workforce.

8. *The Flowing Streams of Lebanon*

1. Hourani, 1988; Farsoun, 1988; Owen, 1988; Toksoz, 1986.

2. A feasibility study of transferring Litani water to the upper Jordan was conducted by Cotton, an American engineer, shortly after the establishment of Israel. The idea lingered for some years and was mentioned wistfully by Levi Eshkol, Israel's premier, in 1967. However, no action was ever taken by Israel to actualize that early interest in the Litani.

3. The notion that Israel could divert the waters of a major river through a rugged terrain and do so in secrecy without revealing its actions to aerial or satellite detection, or to local observers on the ground (including United Nations forces), would be ridiculous if it were not so often repeated. Conspiracy theories are part of the region's mentality.

9. *Fountains of the Deep*

1. Anyone who takes notice of the ironies of history cannot but savor the anomaly of the fiery Arab nationalist and zealous Muslim, Qaddafi, working with the pragmatic and cosmopolitan Jew, the late Armand Hammer, to draw water from the depths of the Sahara and make a section of the Libyan desert bloom.

2. Included among the minor aquifers are the following: (1) A western Galilee aquifer yielding some 110 MCM/Y; (2) An eastern Galilee aquifer, yielding about 20 MCM/Y; (3) The Carmel basin aquifers, yielding about 60 MCM/Y; (4) The Gilboa-Shean aquifers, providing some 100 MCM/Y; and (5) The Arava (desert) aquifers, which yield about 10 MCM/Y of more or less fresh water. Finally, there is the fossil water aquifer contained in the Nubian sandstone underlying parts of the southern Negev (as well as extensive areas in neighboring countries), which can be mined for a period of some decades to exploit perhaps 60 MCM/Y. These estimates pertain only to fairly good quality water. Additional quantities are available of brackish waters, totaling about 80 MCM/Y.

3. It is a curious fact of physics that a 1 meter fall of the freshwater level in a well results in a 9 meter rise of saline seawater up the bottom of the well, since the density ratio of fresh water to seawater is about 1/1.1.

4. Salinization of well water in the Gaza Strip has occurred at an average rate of 15–20 parts per million per year. Most of the local water in the Gaza Strip now exceeds the salinity level of 500 ppm considered the upper threshold for safe drinking water and in some areas the salinity has even reached 1,500 ppm. Consequently the situation regarding water there is nearly intolerable. Irrigation consumes about

100 MCM/Y. Domestic use, now amounting to some 20 MCM/Y, is forecast to increase to 35 MCM/Y by the year 2000 and to 55 MCM/Y by the year 2010. At the same time, industrial consumption, now negligible, is likely to increase to 5 MCM/Y or more. Inevitably, therefore, the amount of water used in irrigation (especially of citrus, which has long been the principal crop) must be reduced.

5. The names used to designate territories convey strong political connotations. The term "West Bank" was assigned to the territory in question by the Kingdom of Jordan when it took possession of it in 1948–49 and had thus created a state with an "East Bank" (formerly Transjordan) and a "West Bank." When—in 1989—Jordan's King Hussein renounced his claim to this territory, the term "West Bank" became, in effect, anachronistic. Still, it is used by the Palestinians living there to refer to the territory over which they hope to achieve sovereignty. The Israeli term "Judea and Samaria" has a historical and a geographical basis, but it also implies a national association with the ancient Israelites.

6. Limestone and dolomite rocks, unlike sandstone, are not generally very porous. However, they are subject to chemical dissolution by rainwater seeping into fissures, so these rock formations tend to form passageways and caverns through which water flows rapidly. Such formations are called *karstic* (after Karst, a region near Trieste, Italy, where they were first described).

7. In the winter of 1992/93, following the copious rains of that season and the preceding one, the Yarkon springs emerged to the surface for the first time in many years, resulting in the outflow of tens of millions of cubic meters to the sea. This, however, was an unusual occurrence.

8. In addition to the Yarkon-Taninim aquifer that flows westward, the Gilboa-Beit Shean aquifer that flows northeastward from the same mountain range yields about 140 MCM/Y. Still another aquifer that may be associated with the mountain aquifer complex was discovered in the 1970s in the piedmont of the western Negev, with a potential yield estimated at 40 MCM/Y. The water of this aquifer is brackish, but experiments conducted by the Ben Gurion University of the Negev have shown that it can be used to irrigate salt-tolerant crops of high quality.

9. Palestinians have accused the Israelis of lowering the water table in the West Bank by pumpage, to the point of drying up some local springs and wells. The Israelis contend that they pump only from the deep aquifer, and that the decreased yield of water in local springs and shallow wells was simply a consequence of the prolonged drought of the late 1980s.

10. The River of Waste

1. In some of the cities of the Middle East, the water distribution networks are so faulty that some 50 percent of the water is unaccounted for—lost in delivery through leaky channels or pipes.

2. In principle, the price of water should be based on demand and supply. However, the water market cannot be made entirely free, nor can the price be allowed to fluctuate unpredictably. A farmer who invests in the planting of an orchard expects it to become productive in, say, 5 years, and to return the investment in, say, 15 years. Such a farmer needs price stability, or at least predictability, to operate rationally.

3. The problem of maintaining water quality is acute in every country in the region. In Israel and Jordan, for instance, the per hectare applications of pesticides and fertilizers are among the highest in the world. Coupled with land disposal of

liquid and solid wastes, such practices inevitably result in the progressive pollution of water resources. One example is Israel's shallow coastal aquifer. Here, nitrate concentrations are rising toward levels at which local well water may no longer be fit to drink.

4. The average costs of installing alternative irrigation systems are as follows: $500–$700 per hectare for surface irrigation, $800–$1,000 for sprinkler irrigation, and $1,300–$1,500 for drip or microsprayer irrigation. This are rough estimates, however, as actual costs vary widely within each system, depending on local conditions and requirements.

5. Israeli methods are already being adopted in the Arab countries, albeit in an indirect and largely unacknowledged way. With peace, much closer relations and sharing of experience will be possible.

6. The significant development of irrigated agriculture in Israel has been made possible by large investments of scientific, human, and financial resources directed at improving irrigation technology. Over the years, innovations by scientists and engineers, and—above all—by farmers, have spawned the growth of several factories to produce and commercialize advanced drip, microsprayer, sprinkler, filtration, volume control, weather-monitoring, water-quality-monitoring, and fertilizer injection devices. These innovations have made Israel a world leader in irrigation technology, a status that not only brought benefits to the country's own agriculture but also led to the growth of an export industry that has spread the benefits of such technology throughout the world.

7. I first noticed this fact in the early 1960s in the data obtained by a graduate student then preparing his doctoral thesis under my supervision, and I encouraged him to publish his findings even though they seemed to contradict prevailing views of crop-soil-water relations. (E. Rawitz, 1969, *Soil Sci* 110:172–82.)

8. Remarkably, Israel uses only one-third more water today than it did in 1967, when its population was less than half of today's. Still, the country cannot rest on its laurels: the population continues to grow and so does the water shortage.

9. Even though agricultural water use efficiency in Israel is already among the highest in the world, there is still latitude for improvement. Proof of that was given during the summers of 1990 and 1991, when the allocations of water for irrigation were reduced by about 30 percent, yet the overall yields were reduced only about 10 percent.

10. See, for example, Falkenmark and Lindh, 1976.

11. In the United States, incidentally, the potential fresh water supplies amount to nearly 10,000 cubic meters per capita. Canada's potential supplies are even greater.

11. Augmenting Supplies

1. So important was the prayer for timely rain that the Jews continue to intone it daily even after two millennia of exile from their homeland. In the Bible, God's exhortation to the Israelites to be faithful is followed immediately by the threat that else He might "withhold the skies that there be no rain, so the people will perish quickly from the good land." And when David inveighed against Mount Gilboa for being the site of King Saul's defeat by the Philistines, he could find no harsher a curse than: "May no rain or dew fall over thee."

2. In fact, dry seasons tend to be more frequent than wet ones in arid regions, since a few seasons of unusually high rainfall skew the mean rainfall so that it typically exceeds the median.

3. Universally, the hydrological cycle must be balanced. That is to say, all water evaporated must return as precipitation sooner or later. This does not apply, however, to specific locations, where the amount of evaporation may exceed precipitation or vice versa. So, while the overall amount of rainfall cannot be augmented, the local amount can be enhanced in certain locations.

4. Some years later, and probably not by total happenstance, Bernard's brother, Kurt, published an apocalyptic novel, *Cat's Cradle,* about a "superseed" chemical called "ice-nine" that freezes the entire world.

5. Rosenfeld and Farbstein. 1992. Possible Influence of Desert Dust on Seedability of Clouds in Israel. *J. Appl. Meterol.* 31: pp. 22–731.

6. The accumulation of loess in Israel is related to the proximity to the band of deserts and their winds, which are the source of the eolian dust.

7. Hillel, D. 1982. *Negev: Land, Water and Life in a Desert Environment.* New York: Praeger Scientific.

8. Hillel, D. 1967. Runoff Inducement in Arid Lands: Report to the U.S. Department of Agriculture. Jerusalem: Hebrew University Special Publication. Also: Hillel, D. 1980. *Applications of Soil Physics.* New York: Academic Press.

9. This is more than 100 times the water resources available to Israel, whereas Turkey's population is only 8 times Israel's, so its per capita reserves are 12 times those of Israel.

10. The important question is whether or not the envisaged Peace Pipeline as originally conceived is feasible from an engineering and, particularly, from an economic point of view. And then, of course, there remains the question of the political feasibility of what would require a complex multinational agreement among long-term rivals in the tension-laden Middle East. One must conclude that under present circumstances the supply of Turkish water to Israel is improbable, both because of the difficulty of securing the agreement of all intermediate countries involved, and because of Israel's own reluctance to be dependent on a distant source of supply that could be cut off at any time by potentially hostile neighbors. However, even if the Turkish project only supplies the additional water needs of Syria, Jordan, and the Palestinians, it can make a major contribution to alleviating the water problems of the area.

11. An alternative to the Peace Pipeline as originally proposed for interbasin water transfer would be to change the course of some of the rivers that flow from Anatolia in Turkey northward into the Black Sea. These rivers have a total annual discharge of some 36 billion cubic meters, whereas the irrigable lands in their basins total only 1.3 million hectares. Because rainfall is relatively abundant, their irrigation needs probably do not exceed 10 billion cubic meters per year. The surplus water can be diverted southward into the Euphrates basin where there is already sufficient storage capacity to regulate the increased flow. The economic costs and environmental impacts have not been assessed, however.

12. Some Israelis regret that their representatives at the Camp David negotiations over the terms of that peace agreement ignored the water problem of the Gaza Strip. That area had been under Egyptian administration, and therefore Egypt's responsibility, prior to the war of 1967. The Egyptian-Israeli treaty was supposed to restore the territorial pre-1967 status quo ante. Had the Gaza Strip been returned to Egyptian control (along with the Sinai), the Egyptians would probably supply surplus water there (with international financing, to be sure) as a matter of course, and without incurring any political problem in doing so. Moreover, the association with Egypt might have provided the people of Gaza with better

economic and cultural opportunities than did their forced association with Israel. The Gazans themselves, however, now prefer to associate with their Palestinian brethren in the West Bank, in a hoped-for Palestinian state, rather than with Egypt.

13. A group of Israeli economists and hydrologists, under the aegis of the Armand Hammer Center at Tel Aviv University, issued a report in 1989, entitled "Economic Cooperation and Peace in the Middle East."

14. In addition to the "ordinary" ecological and economic problems associated with nuclear power, there are special political and security objections to building nuclear facilities in so sensitive a region as the Middle East. Hence there is little chance for international agreement and financing of such projects. Hydroelectric power, however, is a possibility, especially in Israel, thanks to its unique topographical feature, the Dead Sea valley, which is the lowest valley on the continental surface of the earth.

15. Personal communication from Prof. Dan Zaslavsky, former Water Commissioner of Israel.

16. The principle of this method is, literally, to reverse the process of osmosis. That process occurs whenever a saline solution is brought into hydraulic contact with pure water while the two are separated by a "selective" membrane. Such a membrane permits the passage of water molecules but inhibits the passage of solute molecules. When this happens, water in the purer solution diffuses spontaneously through the membrane and in so doing dilutes the more-concentrated solution. That process is not brought to completion if a counterpressure on the side of the concentrated solution prevents the further osmosis of water into it. A counterpressure just sufficient to prevent osmosis is called the osmotic pressure, and it is proportional to the effective concentration difference between the two bodies of water. If the counterpressure is made greater than the osmotic pressure, then water moves in the opposite direction; it is pushed out of the concentrated solution through the membrane into the pure water phase. By this means, saline water can be converted to sweet water, provided sufficient pressure can be applied and suitable membranes are available.

17. A desalinization unit using such improved membranes has been in operation since the early 1980s just north of the Israeli city of Eilat, and is supplying that city with 15,000 cubic meters of water per day. The energy requirements of that plant have actually decreased gradually, as the process has been made more efficient.

18. The present management of the Israeli Dead Sea Works is opposed to the entire idea of introducing seawater into the Dead Sea. In a conversation with me, Mr. Uri Ben-Nun (the current director) raised several concerns: (1) The tunnels and other facilities of the projected seawater canal will necessarily traverse geologically unstable fault-lines that are vulnerable to potentially disastrous earthquakes. (2) The introduced water, rather than float on top of the resident Dead Sea brine, might mix with it and dilute that mineral-rich fluid to a depth of 10 meters or more, thereby reducing the efficiency of the finely tuned mineral extraction process. Such an effect would force the industry to draw the brine from a greater depth and distance. (3) The introduced seawater might react chemically with the resident brine to produce undesirable precipitates that could clog the conveyance works. Such concerns call for thorough research before the seawater project is approved.

19. Gur, S. *Water and the Peace Process: A View from Israel.* Research Memorandum No. 20, The Washington Institute for Near East Policy, September 1992.

20. The elevation differential between the Mediterranean Sea and the floor of the Beit Shean Valley is about 275 meters. This differential can be used to create sufficient hydrostatic pressure to save about 60 percent of the energy required for desalinization by the reverse osmosis method.

21. How costly is the project likely to be, and how much will the desalinized water cost per unit volume? At this stage, the estimates are very uncertain. The costs must include the construction of the canals, tunnels, pumping stations, dams, and excavated lakes; as well as the costs of the desalinization facilities themselves. Preliminary estimates (Glueckstern, 1991) suggest that the cost of the Beit Shean Valley desalinization plants would be approximately $2.8 billion, and the cost per cubic meter of desalinized water would be $0.50 based on a 3.5 percent discount rate, or $0.58 based on a 6 percent rate.

12. Criteria for Sharing International Waters

1. The battle of Hattin was the dramatic climax in the contest between the Crusaders and the Muslim forces under the leadership of Salah-ed-Din Yusuf ibn Ayyub, known to Europeans as Saladin. The battle took place on the arid plateau of eastern Galilee during the hottest and driest season of the year, on the 3rd and 4th of July 1187. During the night that preceded the decisive battle, the Muslims encircled the Crusaders, blocking their access to nearby springs. The Muslims, in contrast, had ample supplies of water from the springs they had seized, as well as unhindered access to the Jordan River.

The following morning, the Muslims waited until the heat became oppressive, then drove the Crusaders up the slopes of an extinct volcano with dual peaks, called the Horns of Hattin. There the thirst of the Crusaders was aggravated by the tantalizing sight of the Sea of Galilee in the unreachable distance. Making the situation even worse, the Muslims ignited the dry brush that covered the ground. The wind-whipped flames and billowing smoke seared and choked the weary Europeans and their heavy-laden horses, while the Muslims on their light horses slashed and shot arrows at their enemies. Tormented beyond endurance, the Europeans were finally forced to surrender.

The Crusaders were thus defeated not so much by the sheer force of the Muslims as by the latter's superior grasp of the terrain, climate, and water requirements of the area in contention.

Saladin's own florid description of the battle is typical of the region's traditional penchant for glorifying war, repeatedly using hydraulic metaphors (Kedar 1992):

> May the enemies [of Allah] never cease from finding the cup of death poured out where swords are flowers . . . streaming in the blazing fire of noon. . . . The infantry and cavalry [of the enemy] took one of the waters [springs] but the devil seduced them . . . so they left the water . . . which was then seized [by our forces]. . . . Thy servant then kindled against them fire . . . a reminder of what Allah has prepared for them in the next world. He then met them in battle, when the fires of thirst had tormented them. . . . Rivers of swords sought out their livers, as though wanting to water what was diseased there. They drank the cup of fate when the swords came to water. . . . Showers of arrows . . . were sent down on them. . . . Their king was captured. . . . Glory to Allah. . . .

2. Hillel, D. Lash of the Dragon. *Natural History* 8/91, pp. 28–37.

3. Al Wansharisi, Ahmad Ibn Yahya (1403–1508): *Al Minyar al-Mughrib wal-Jami al-Mughrib 'an Fatawa Ahl Ifriqiyah wal-Andalus wal-Mughrib.* Issued in 13 volumes by the Islamic Tradition Commission, Rabat (Morocco), 1981.

4. Al-Alami, Sheikh Isa Bin Ali al-Hasani (11–12 century). *Kitab al-Nawazil,* Rabat (Morocco) 1986. 3 volumes.

5. Harkabi, A. E. 1966. *Zikhron Rishonim vegam Aharonim.* Berlin: Zvi Hershtkowski.

6. The names Reuven and Shimon (commonly rendered Reuben and Simeon in Latin transliteration) were used frequently in classical Hebrew writings to designate any two characters or litigants. The names were originally given to the first two sons of Jacob and Leah (Genesis 29:32, 33): "And Leah conceived, and bore a son, and she called his name Reuben (literally: "Behold, a son!"), for she said: 'Because the Lord hath looked upon my affliction; for now my husband will love me.' And she conceived again, and bore a son; and said: 'Because the Lord heard that I am despised, He hath therefore given me this son also.' And she called his name Simeon (literally: "Heard.")"

7. Solanes, M. 1992. Legal and Institutional Aspects of River Basin Development. *Water International* 17:pp. 116–123.

8. The classical example is the Rhine River Commission, which came into being in 1831. It incorporates six states, whose representatives consult and recommend cooperative action to manage the river for mutual benefit. Though it has not always worked in perfect harmony, the Rhine River Commission has served as the prototype for numerous other shared rivers in Europe, the Americas, Asia, and Africa. Such commissions have even been set up between otherwise hostile neighbors, such as India and Pakistan, and have succeeded in defusing threatening conflicts. Hostile parties must, however, enter into the necessary negotiations with a sufficient measure of a priori goodwill. The chances for successful negotiations leading to a lasting arrangement can often be enhanced by the participation of a third party able not merely to mediate but also to offer positive inducements in the form of financial or technical assistance.

9. In the argument over the waters of the mountain aquifer, the Palestinians claim territorial rights to the waters recharging the aquifer in their domain, and cite the Fourth Geneva Convention, which prohibits an occupying power from exporting the resources of the occupied territory. On their part, the Israelis claim that they are not exporting the water, which flows naturally to their territory, and that they have historical rights to the use of the aquifer (which they had begun to tap in their own domain long before 1967 and their occupation of the West Bank).

10. The draft, entitled Transboundary Groundwaters, was first published as an article in *Natural Resources Journal* (1989) 29: pp.663–722, and was issued as a booklet by the International Transboundary Resources Center under the aegis of the University of New Mexico School of Law.

13. Water for Peace

1. Zaphnath-Paaneah ("Decipherer of the Hidden Meaning") was the name assigned by Pharaoh to Joseph the Wise (Genesis 41:45) in gratitude for his interpretation of the monarch's dream. Pharaoh then appointed Joseph to rule over Egypt and prepare it for the impending drought.

2. A particularly cynical joke describes the arrival in heaven of a deceased American diplomat. Before taking his honored place there, the diplomat turns to the angel with an urgent plea: "I cannot rest until I know: will there be *real* peace in the Middle East?" The angel goes off to seek the advice of a higher authority, then returns with the definitive reply: "God says yes, there will definitely be *real* peace in the Middle East, though not necessarily in His lifetime." A slightly less cynical joke defines a liar as one who claims either to be immortal or to understand the Middle East.

3. Two years after the end of the Persian Gulf War, the Arab Monetary Fund released an estimate of the cost of that conflict in 1990 and 1991. That estimate was printed in the *New York Times* of April 25, 1993. Not counting the vast damage to the environment and a continuing suppression of the rate of economic growth in the region, the cost totaled an incredible $676 billion. We cannot but wonder at so vast an expenditure devoted to a war that—far from improving the lot of the region's people—only made matters worse. Nor can we ignore the costs of prior wars in the region and the present cost to the nations there of preparing for the eventuality of yet another war, all in the interest of seeking "security." How much good could that money bring if it were devoted to water resources development, land reclamation, environmental rehabilitation, infrastructure, industry, education, health, and the promotion of peaceful cooperation!

4. Assaf, K., al Khatib, N., Kally, E., and Shuval, H. (1993). *A Proposal for the Development of a Regional Water Master Plan*. Israel/Palestine Center for Research and Information, Jerusalem.

5. An old Arab aphorism seems particularly apt: "Ilh̄ub baàd làdaweh ah̄la min 'lh̄alaweh"—the love that follows hostility [can be] sweeter than the sweetest Middle Eastern confection.

Bibliography

Abdel Haleem, M. (1989). Water in the Qur'an. *The Islamic Quarterly* 33:34–50.

Abu Zeid, M., and Biswas, A. (1990). Impacts of irrigation on water quality. *Water International* 15.

Abu Taleb, M. F., Deason, J. P., and Salameh, E. (1991). *Water Resources Planning and Development in Jordan.* Report presented to the International Workshop on Water Resources Policy, the World Bank, Washington.

Adams, R. McC. (1981). *Heartland of Cities: Surveys of Ancient Settlements and Land Use on the Central Floodplain of the Euphrates.* University of Chicago Press, Chicago.

Ajami, F. (1981). *The Arab Predicament: Arab Political Thought and Practice Since 1967.* Cambridge University Press, New York.

Ali, M. K. (1977). *The Projects for the Increase of the Nile Yield, with Special Reference to the Jonglei Project.* Report to U.N. Water Conference, Mar del Plata. Argentina.

Allam, M. N. (1987). A cost allocation approach for irrigation water in upper Egypt. *Water Resources Management* 1.

Allen, R. (1974). *Imperialism and Nationalism in the Fertile Crescent: Sources and Prospects of the Arab-Israeli Conflict.* Oxford University Press, New York.

Al-Weshah, R. A. (1992). Jordan's water resources: technical perspectives. *Water International* 17:124–132.

Amit, D., Hirshfeld, Y., and Patrich, J., eds. (1989). *The Ancient Aqueducts in Israel.* Yad Ben Zvi, Jerusalem.

Antonius, G. (1938). *The Arab Awakening: The Story of the Arab National Movement.* Hamilton, New York.

Arlosoroff, S. (1977). *Israel—A Model of Efficient Utilization of a Country's Water Resources.* Report to U.N. Water Conference, Mar del Plata, Argentina.

Arnon, I., and Raviv, M. (1980). *From Fellah to Farmer.* Agricultural Research Organization, Bet Dagan, Israel.

Artzy, M., and Hillel, D. (1988). A defense of the theory of progressive soil salinization in ancient southern Mesopotamia. *Geoarchaeology* 3:235–238.

Assaf, K., alKhatib, N., Kally, E., and Shuval, H. (1993). *A Proposal for the Development of a Regional Water Master Plan.* Israel/Palestine Center for Research and Information, Jerusalem.

Avineri, S. (1981). *The Making of Modern Zionism: The Intellectual Origins of the Jewish State.* Basic Books, New York.

Baasiri, M., and Ryan, J. (1986). *Irrigation in Lebanon: Research, Practices, and Potential.* National Council for Scientific Research, Beirut, Lebanon.

Baly, D. (1957). *The Geography of the Bible: A Study in Historical Geography.* Harper, New York.

Bates, D., and Rassam, A. (1983). *Peoples and Cultures of the Middle East.* Prentice Hall, Englewood Hills, New Jersey.

Beaumont, P., Blake, G. H., and Wagstaff, J. M. (1976). *The Middle East: A Geographical Study.* John Wiley, New York.

Bell, J. B. (1969). *The Long War: Israel and the Arabs Since 1946.* Prentice-Hall, Englewood Cliffs, New Jersey.

Ben Gurion, D. (1967). *Meetings with Arab Leaders.* Am Oved, Tel-Aviv.

Benedick, R. E. (1979). The high dam and the transformation of the Nile. *Middle East Journal* 33:119–44.

Bennett, O., ed. (1991). *Environment and Conflict.* Panos Institute, London.

Beschorner, N. (1992). Water and instability in the Middle East. *Adelphi* 273, International Inst. for Strategic Studies, London.

Bill, J. A., and Leiden, C. (1984). *Politics in the Middle East.* Little, Brown, New York.

Biswas, A. K. (1970). *History of Hydrology.* North-Holland, Amsterdam.

Bonnifield, P. (1979). *The Dust Bowl: Men, Dirt, and Depression.* University of New Mexico Press, Albuquerque, New Mexico.

Borowski, O. (1987). *Agriculture in Iron Age Israel.* Eisenbrauns, Winona Lake, Indiana.

Bouwer, H. (1978). *Groundwater Hydrology.* McGraw-Hill, New York.

Bras, R. (1990). *Hydrology: An Introduction to Hydrologic Science.* Addison Wesley, Reading, Massachusetts.

Brice, W. C., ed. (1981). *An Historical Atlas of Islam.* E. J. Brill, Leiden.

Brinkman, R. (1976). *Geology of Turkey.* Elsevier Scientific, New York.

Briscoe, J. (1992). Poverty and water supply: how to move forward. *Finance and Development,* December: 16–19.

Brundtland, G. H. (1987). *Our Common Future: Report of the World Commission on Environment and Development.* Oxford University Press, Oxford, England.

Butzer, K. W. (1976). *Early Hydraulic Civilization in Egypt: A Study in Cultural Ecology.* University of Chicago Press, Chicago.

Butzer, K. W. (1982). *Archaeology as Human Ecology.* Cambridge University Press, Cambridge, England.

Cano, G. (1989). The development of the law in international water resources and the work of the International Law Commission. *Water International* 14:167–171.

Caponera, D. A. (1985). Patterns of cooperation in international water law: principles and institutions. *Natural Resources Journal* 25:563–588.

Caponera, D. C. (1973). *Water Law in Moslem Countries.* United Nations Food and Agriculture Organization, Rome.
Central Intelligence Agency. (1988). *The World Factbook.* U.S. Government Printing Office, Washington.
Clarke, R. (1991). *Water: The International Crisis.* Earthscan Publications, London.
Clawson, M., Landsberg, H. H., and Alexander, L. T. (1971). *The Agricultural Potential of the Middle East.* American Elsevier, New York.
Cluff, C. B., Dutt, G. R., Ogden, P. R., and Kuykendall, J. K. (1972). *Development of Economic Water Harvest Systems for Increasing Water Supply.* The University of Arizona, Tucson, Arizona.
Cobban, H. (1985). *The Making of Modern Lebanon.* Westview Press, Boulder, Colorado.
Cohen, N. M. (1977). *The Food Crisis in Prehistory: Overpopulation and the Origins of Agriculture.* Yale University Press, New Haven, Connecticut.
Cressey, G. B. (1958). Qanats, karez, and foggaras. *Geographical Review* 48:27–44.
Cuenca, R. H. (1989). *Irrigation System Design.* Prentice Hall, Englewood Cliffs, New Jersey.
De Candolle, A. (1967). *Origins of Cultivated Plants.* Harper, New York.
Dewdney, J. C. (1971). *Turkey: An Introductory Geography.* Praeger, New York.
Dieleman, P. J., ed. (1977). *Reclamation of Salt Affected Soils in Iraq: Soil Hydrological and Agricultural Studies.* International Institute for Land Reclamation and Improvement, Wageningen, The Netherlands.
Dorfman, R., Jacoby, H. D., and Thomas, H. A. (1972). *Models for Managing Water Quality.* Harvard University Press, Cambridge, Massachusetts.
Downey, T. J., and Mitchell, B. (1993). Middle East water: acute or chronic problem? *Water International* 18:1–4.
Dregne, H. E. (1983). *Desertification of Arid Lands.* Harwood Academic, Chur, Switzerland.
Drysdale, A., and Blake, G. (1985). *The Middle East and North Africa: A Political Geography.* Oxford University Press, New York.
Dunne, T., and Leopold, L. (1978). *Water in Environmental Planning.* Freeman, San Francisco.
Efrat, M. (1967). Syria's dam on the Euphrates. *New Outlook* 10:39–46.
Egyptian Ministry of Culture and Information. (1972). *The High Dam.* State Information Service, Egypt.
El-Ashry, M. T., and Gibbons, D. C. (1986). *Troubled Waters: New Policies for Managing Water in the American West.* World Resources Institute, Washington.
El-Katsha, S., and White, A. (1989). Women, water, and sanitation: household behavioral patterns in two Egyptian villages. *Water International* 14:103–111.
Falkenmark, M. (1990). Global water issues confronting humanity. *Journal of Peace Research* 27:177–191.
Falkenmark, M., and Lindh, G. (1976). *Water for a Starving World.* Westview Press, Boulder, Colorado.
Falloux, F., and Mukendi, A., eds. (1988). *Desertification Control and Renewable Resource Management in the Sahelian and Sudanian Zones of West Africa.* Technical Paper Number 70, The World Bank, Washington.

Farr, E., and Henderson, W. C. (1986). *Land Drainage.* Longman, Harlow, England.
Feliks, J. (1963). *Agriculture in Palestine in the Period of the Mishna and Talmud.* Magnes Press, Jerusalem.
Finkel, H. J. *Semiarid Soil and Water Conservation.* CRC Press, Boca Raton, Florida.
Fisher, S. N. (1978). *The Middle East: A History.* Knopf, New York.
Fisher, W. B. (1978). *The Middle East: A Physical, Social, and Regional Geography.* Methuen, London.
Frederick, K. D. (1993). *Balancing Water Demands with Supplies: The Role of Management in a World of Increasing Scarcity.* Technical Paper Number 189, The World Bank, Washington.
Frederick, K. D., and Gleick, P. H. (1989). Water resources and climate change. In: Rosenberg, N. J., Easterling, W. E., Crosson, P. R., and Darmstatdter, J., eds. *Greenhouse Warming: Abatement and Adaptation.* Resources for the Future, Washington.
Frey, F. W. (1993). The political context of conflict and cooperation over international river basins. *Water International* 18:54–68.
Friedman, T. L. (1989). *From Beirut to Jerusalem.* Doubleday, New York.
Fukuda, H. (1976). *Irrigation in the World: Comparative Developments.* University of Tokyo Press, Tokyo, Japan.
Garraty, J. A., and Gay, P. (1987). *The Columbia History of the World.* Harper & Row, New York.
Gischler, C. (1979). *Water Resources in the Arab Middle East and North Africa.* Middle East & North Africa Press, Cambridge.
Glueckstern, P. (1982). Preliminary considerations of combining a large reverse osmosis plant with the Mediterranean-Dead Sea project. *Desalination* 40:143–156.
Goldschmidt, A., Jr. (1988). *A Concise History of the Middle East.* Westview Press, Boulder, Colorado.
Gordon, D. C. (1983). *The Republic of Lebanon: Nation in Jeopardy.* Westview Press, Boulder, Colorado.
Gore, A. (1992). *Earth in the Balance: Ecology and the Human Spirit.* Houghton Mifflin, Boston.
Gorse, J. E., and Steeds, D. R. (1987). *Desertification in the Sahelian and Sudanian Zones of West Africa.* Technical Paper Number 61, The World Bank, Washington.
Goudie, A. (1982). *The Human Impact: Man's Role in Environmental Change.* MIT Press, Cambridge, Massachusetts.
Green, D. E. (1973). *Land of the Underground Rain: Irrigation on the Texas High Plains.* University of Texas Press, Austin, Texas.
Greenwood, N., ed. (1992). *Israel Yearbook and Almanac.* Israel Business, Research, and Technical Translation/Documentation, Tel Aviv.
Grinwald, Z. (1980). *Water in Israel.* Israel Water Workers Association, Tel Aviv.
Gruen, G. (1991). *The Water Crisis—The Next Middle East Crisis.* Simon Wiesenthal Center, Los Angeles.
Gubser, P. (1983). *Jordan: Crossroads of Middle Eastern Events.* Westview Press, Boulder, Colorado.
Haas, P. M. (1990). Towards management of environmental problems in Egypt. *Environmental Conservation* 17:45–50.

Haddadin, M., and Gur, S. (1992). *Water and the Peace Process: Two Perspectives.* Research Memorandum Number 20, The Washington Institute for Near East Policy, Washington.

Halabi, R. (1981). *The West Bank Story: An Israeli Arab's View of Both Sides of a Tangled Conflict.* Harcourt Brace Jovanovich, San Diego.

Hallo, W. W., and Simpson, W. K. (1971). *The Ancient Near East: A History.* Harcourt Brace Jovanovich, New York.

Hallsworth, E. G. (1987). *Anatomy, Physiology, and Psychology of Erosion.* Wiley, New York.

Hays, J. B. (1948). *T.V.A. on the Jordan: Proposals for Irrigation and Hydro-Electric Developments in Palestine.* Public Affairs Press, Washington.

Held, C. C. (1989). *Middle East Patterns: Places, Peoples, and Politics.* Westview Press, Boulder, Colorado.

Hillel, D. (1970). Artificial inducement of runoff as a potential source of water in arid lands. In: McGinnies, W. G., ed. (1979). *Food, Fiber, and the Arid Lands.* University of Arizona Press, Tucson, Arizona.

Hillel, D. (1971). *Soil and Water: Physical Principles and Processes.* Academic Press, New York.

Hillel, D., ed. (1972). *Optimizing the Soil Physical Environment Toward Greater Crop Yields.* Academic Press, New York.

Hillel, D. (1976). Soil management in arid regions. In: *Yearbook of Science and Technology.* McGraw-Hill, New York.

Hillel, D. (1976). A new method of water conservation in arid zone soils. In: Mundlak, Y., and Singer, S. G., eds. *Arid Zone Development.* Published for the American Academy of Arts and Sciences by Ballinger, Cambridge, Massachusetts.

Hillel, D. (1980). *Fundamentals of Soil Physics.* Academic Press, San Diego, California.

Hillel, D. (1980). *Applications of Soil Physics.* Academic Press, San Diego, California.

Hillel, D., ed. (1982–87). *Advances in Irrigation, Vols I–IV.* Academic Press, San Diego, California.

Hillel, D. (1982). *Negev: Land, Water, and Life in a Desert Environment.* Praeger Scientific, New York.

Hillel, D. (1987). *The Efficient Use of Water in Irrigation: Principles and Practices for Improving Irrigation in Arid and Semiarid Regions.* Technical Paper Number 64, The World Bank, Washington.

Hillel, D. (1990). Role of irrigation in agricultural systems. In: *Irrigation of Agricultural Crops.* Agronomy Monograph Number 30, American Society of Agronomy, Madison, Wisconsin.

Hillel, D. (1991). The problem of desertification: a critical reconsideration. Arid Lands Research Center, Tottori, Japan, *Annual Report* for 1990–91:33–35.

Hillel, D. (1992). *Out of the Earth: Civilization and the Life of the Soil.* University of California Press, Berkeley.

Hillel, D., and Rosenzweig, C. (1989). *The Greenhouse Effect and Its Implications Regarding Global Agriculture.* University of Massachusetts, Amherst, Massachusetts.

Holden, D., and Johns, R. (1982). *The House of Saud: The Rise and Rule of the Most Powerful Dynasty in the Arab World.* Holt, Rinehart and Winston, New York.

Hourani, A. (1991). *A History of the Arab Peoples.* Harvard University Press, Cambridge, Massachusetts.

Hourani, A. H. (1981). *The Emergence of the Modern Middle East.* University of California Press, Berkeley.

Howe, C. W., Schurmeier, D. R., and Shaw, W. D., Jr. (1986). Innovative approaches to water allocation: the potential for water markets. *Water Resources Research* 22:439–445.

Hudson, J. (1971). The Litani River of Lebanon. *Middle Eastern Journal* 25: 1–14.

Hudson, N. (1971). *Soil Conservation.* Batsford, London.

Hult, J. L., and Ostrander, N. C. (1973). *Antarctic Icebergs as a Global Fresh Water Resource.* A report prepared for the National Science Foundation by Rand Corporation, Los Angeles, California.

Hunt, R. C. (1988). The impact of the Aswan High Dam reconsidered. *Brandeis Review* 7:1–6.

Hunt, R. C., and Hunt, E. (1976). Canal irrigation and local social organization. *Current Anthropology* 17:389–411.

Hurewitz, J. C., ed. (1969). *Soviet-American Rivalry in the Middle East.* Proceedings of the Academy of Political Science, Vol. 29, Columbia University, New York.

Hwang, N. H. C. and Hita, C. E. (1987). *Fundamentals of Hydraulic Engineering Systems.* Prentice Hall, Englewood Cliffs, New Jersey.

Ibrahim, F. N. (1981). *The Aswan High Dam: Serious Human Interference with the Ecosystem.* Special publication of the Institute of Geo-Sciences, University of Bayreuth, Germany.

International Development Research Centre. (1988). *Fresh Water: The Human Imperative.* IDRC, Ottawa, Canada.

Israel Ministry of Agriculture. (1990). *Agriculture in Israel, 1990.* State of Israel Ministry of Agriculture, Tel Aviv.

Issar, A. (1985). Fossil water under the Sinai-Negev peninsula. *Scientific American* 253:82–89.

Issar, A. (1990). *Water Shall Flow From the Rock: Hydrogeology and Climate in the Lands of the Bible.* Springer Verlag, Berlin.

Jacobsen, T. (1982). *Salinity and Irrigated Agriculture in Antiquity.* Undena Publications, Malibu, California.

Jacobsen, T., and Adams, R. McC. (1958). Salt and silt in ancient Mesopotamian culture. *Science* 128:1251–1258.

James, L. D. (1974). *Man and Water: The Social Sciences in Management of Water Resources.* The University Press of Kentucky, Lexington, Kentucky.

James, T. G. H. (1979). *An Introduction to Ancient Egypt.* Farrar Straus Giroux, New York.

Jean, D. S., and Warne, A. G. (1993). Nile delta: recent geological evolution and human impact. *Science* 260:628–634.

Jordan, W. R., ed. (1987). *Water and Water Policy in World Food Supplies.* Texas A&M University Press, College Station, Texas.

Jovanovic, D. (1985). Ethiopian interest in the division of the Nile river waters. *Water International* 10:82–85.

Kally, E. (1989). *Water in Peace.* Sifriat Poalim, Tel Aviv.

Kedar, B. Z., ed. (1992). *The Horns of Hattin.* Variorum, London.

Khouri, R. G. (1981). *The Jordan Valley: Life and Society Below Sea Level.* Longman, London.

Klein, C. (1961). *On the Fluctuations of the Level of the Dead Sea Since the Beginning of the 19th Century.* Hydrological Service, State of Israel Ministry of Agriculture, Jerusalem.

Klein, E. (1987). *A Comprehensive Etymological Dictionary of the Hebrew Language.* Carta, Jerusalem.

Knapp, A. B. (1988). *The History and Culture of Ancient Western Asia.* Dorsey Press, Chicago.

Kolars, J. F., and Mitchell, W. A. (1991). *The Euphrates River and the Southeast Anatolia Development Project.* Southern Illinois University Press, Carbondale, Illinois.

Kuffner, U. (1993). Water transfer and distribution schemes. *Water International* 18:30–34.

Lamberg-Karlovsky, C. C., and Sabloff, J. A. (1979). *Ancient Civilizations: The Near East and Mesoamerica.* Benjamin/Cummings, Menlo Park, California.

La Riviere, J. W. M. (1989). Threats to the world's water. *Scientific American* 261: 80–107.

Lawrence, T. E. (1935). *Seven Pillars of Wisdom.* Doubleday Doran, Garden City, New York.

Lenczowski, G. (1980). *The Middle East in World Affairs.* Cornell University Press, Ithaca.

Leopold, L. B. (1974). *Water: A Primer.* Freeman, San Francisco.

Lewis, N. N. (1949). Malaria, irrigation, and soil erosion in central Syria. *Geographical Review* 39:278–290.

Lewis, N. N. (1953). Lebanon—the mountain and its terraces. *Geographical Review* 43:1–14.

Lonergan, S. C., and Brooks, D. B. (1992). *The Economic, Ecological, and Geopolitical Dimensions of Water in Israel.* Centre for Regional Development, University of British Columbia, Victoria, Canada.

Lowdermilk, W. C. (1944). *Palestine, Land of Promise.* Harper and Brothers, New York and London.

Lowdermilk, W. C. (1953). *Conquest of the Land Through 7,000 Years.* Bulletin 99, Soil Conservation Service, U.S. Department of Agriculture, Washington.

Lowi, M. R. (1993). *Water and Power: The Politics of a Scarce Resource in the Jordan River Basin.* Cambridge University Press, Cambridge.

Maass, A., and Anderson, R. L. (1978). *. . . and the Desert Shall Rejoice: Conflict, Growth, and Justice in Arid Environments.* MIT Press, Cambridge, Massachusetts.

Manners, I. R. (1974). Problems of water resource management in a semi-arid environment: the case of irrigation agriculture in the central Jordan rift valley. In: Hoyle, B. S., ed. *Spatial Aspects of Development.* John Wiley, Chichester, England.

Marchant, W. E., and Dennis, A. S. (1991). Weather modification as a response to variations in weather and climate. In: *Managing Water Resources in the West Under Conditions of Climate Uncertainty.* National Academy Press, Washington.

Marr, P. (1985). *The Modern History of Iraq.* Westview Press, Boulder, Colorado.

Mather, J. (1984). *Water Resources: Distribution, Use, and Management.* Wiley, New York.
Mazar, A. (1977). Water for Jerusalem: An historical perspective. *Kidma Israel Journal of Development* 3:20–24.
Meinardi, C. R., and Heij, G. J. (1991). *A Groundwater Primer.* IRC International Water and Sanitation Centre, The Hague, The Netherlands.
Mikesel, M. W. (1969). The deforestation of Mount Lebanon. *Geographical Review* 59:1–28.
Moore, D., and Seckler, D., eds. (1993). *Water Scarcity in Developing Countries: Reconciling Development and Environmental Protection.* Winrock International Institute for Agricultural Development, Arlington, Virginia.
Mostyn, T., and Hourani, A. (1988). *The Cambridge Encyclopedia of The Middle East and North Africa.* Cambridge University Press, New York.
Naff, T., and Matson, R. C. (1984). *Water in the Middle East: Conflict or Cooperation?* Westview Press, Boulder, Colorado.
Nagmoush, S. (1988). *History of Land Reclamation in Egypt.* Ministry of Land Reclamation, Cairo, Egypt.
National Academy of Sciences. (1974). *More Water for Arid Lands: Promising Technologies and Research Opportunities.* NAC, Washington.
Nativ, R. (1988). Problems of an over-developed water system: the Israeli case. *Water Quality Bulletin* 13:126–131.
Niblock, T. (1982). *Iraq: The Contemporary State.* St. Martin's Press, New York.
Nijim, B. K. (1990). Water resources in the history of the Palestinian conflict. *GeoJournal* 21:317–323.
O'Mara, G. T., ed. (1988). *Efficiency in Irrigation: The Conjunctive Use of Surface and Groundwater Resources.* The World Bank, Washington.
Oppenheimer, A., Kasher, A., and Rapoport, U. (1986). *Man and Earth in the Ancient Land of Israel.* Yad Ben Zvi, Jerusalem.
Orni, E., and Efrat, E. (1971). *Geography of Israel.* Israel Universities Press, Jerusalem.
Page, G. W. (1987). *Planning for Groundwater Protection.* Academic Press, San Diego, California.
Pearce, F. (1991). Wells of conflict in the Middle East. *New Scientist,* Spring 1991.
Pearce, F. (1992). *The Dammed: Rivers, Dams, and the Coming World Water Crisis.* The Bodley Head, London.
Peretz, D. (1986). *The West Bank: History, Politics, Society, and Economy.* Westview Press, Boulder, Colorado.
Polk, W. R. (1980). *The Arab World.* Harvard University Press, Cambridge, Massachusetts.
Porat, M. (1990). *Water Management in Israel.* Report of the State Comptroller, Jerusalem.
Postel, S. (1989). *Water for Agriculture: Facing the Limits.* WorldWatch Institute, Washington.
Postel, S. (1992). *Last Oasis: Facing Water Scarcity.* W. W. Norton, New York.
Postgate, J. N. (1992). *Early Mesopotamia: Society and Economy at the Dawn of History.* Routledge, London.
Powledge, F. (1982). *Water: The Nature, Uses, and Future of Our Most Precious Resource.* Farrar Straus Giroux, New York.
Pritchard, J. B., ed. (1958). *The Ancient Near East: An Anthology of Texts and Pictures.* Princeton University Press, Princeton, New Jersey.

Rabinovich, I. (1984). *The War for Lebanon 1970–1983.* Cornell University Press, Ithaca, New York.

Rabinovich, I., and Reinharz, J., eds. (1984). *Israel in the Middle East: Documents and Readings on Society, Politics, and Foreign Relations, 1948-Present.* Oxford University Press, New York.

Redman, C. L. (1978). *The Rise of Civilization: From Early Farmers to Urban Society in the Ancient Near East.* Freeman, San Francisco.

Reisner, M. (1986). *Cadillac Desert.* Penguin Books, New York.

Said, R. (1962). *The Geology of Egypt.* Elsevier, Amsterdam.

Said, R. (1981). *The Geological Evolution of the River Nile.* Springer-Verlag, New York.

Salama, R. B. (1984). *Groundwater Resources of Sudan.* Rural Water Corporation, Khartoum, Sudan.

Salameh, E. (1990). Jordan's water resources: development and future prospects. *American-Arab Affairs* 33:69–77.

Salem, O. M. (1992). The great manmade river project: a partial solution to Libya's future water supply. *Water Resources Development* 8:270–278.

Schiller, E. J., ed. (1992). *Sustainable Water Resources Management in Arid Countries: Middle East and North Africa.* Institute for International Development and Cooperation, University of Ottawa, Ottawa, Ontario, Canada.

Schioler, T. (1973). *Roman and Islamic Water Lifting Wheels.* Odense University Press, Copenhagen.

Schwartz, J. (1992). Management of the water resources in Israel. *Israel Journal of Earth Sciences* 39:57–65.

Shahin, M. (1989). Review and assessment of water resources in the Arab region. *Water International* 14:206–219.

Shuval, H. I. (1987). The development of water reuse in Israel. *Ambio* 16:186–190.

Shuval, H. I. (1992). Approaches to finding an equitable solution to the water resources shared by Israel and the Palestinians over use of the mountain aquifer. In: Baskin, G., ed. *Water: Conflict or Cooperation.* Israel/Palestine Center for Research and Information, Jerusalem.

Shuval, H. I., Adin, A., Fattal, B., Rawitz, E., and Yekutiel, P. (1986). *Wastewater Irrigation in Developing Countries: Health Effects and Technical Solutions.* Technical Paper Number 51, The World Bank, Washington.

Smith, R. A., and Birch, P. B. (1963). The East Ghor irrigation project in the Jordan valley. *Geography* 48:406–409.

Smith, S. E., and Al-Rahwahy, H. M. (1990). The Blue Nile: potential for conflict and alternatives for meeting future demands. *Water International* 15:217–222.

Sofer, A. (1992). *Rivers on Fire: The Conflict of Water in the Middle East.* Am Oved, Tel Aviv.

Stanhill, G. (1981). The Egyptian agro-ecosystem at the end of the eighteenth century—an analysis based on the 'Description de l'Egypte.' *Agro-Ecosystems* 6:305–314.

Starr, J. R. (1990). *Economic Development in the West Bank and Gaza Strip: Illusion, or Vision.* The Armand Hammer Fund for Economic Cooperation in the Middle East, Tel Aviv University, Tel Aviv.

Starr, J., and Stoll, D., eds. (1988). *The Politics of Scarcity: Water in the Middle East.* Westview Press, Boulder, Colorado.

Stein, L. J. (1983). *The Balfour Declaration.* Magnes Press of the Hebrew University, Jerusalem.
Stern, P. (1979). *Small Scale Irrigation.* International Irrigation Information Center, Bet Dagan, Israel.
Survey of Israel. (1985). *Atlas of Israel.* Government Press, Tel Aviv.
Susskind, L., and Cruikshank, J. (1987). *Breaking the Impasse: Consensual Approach to Resolving Disputes.* Basic Books, New York.
Taylor, A., ed. (1971). *Focus on the Middle East.* Praeger, New York.
Tekeli, S. (1990). Turkey seeks reconciliation for the water issue induced by the Southeastern Anatolia Project (GAP). *Water International* 15:206–216.
Teveth, S. (1985). *Ben Gurion and the Palestinian Arabs.* Shocken, Tel Aviv.
Thesiger, W. (1964). *The Marsh Arabs.* Dutton, New York.
Thirgood, J. V. (1981). *Man and the Mediterranean Forest: A History of Resource Depletion.* Academic Press, London.
Thomas, L., ed. (1956). *Man's Role in Changing the Face of the Earth.* University of Chicago Press, Chicago.
Tolba, M. (1990). Climate change and water management. *Water International* 15:56–59.
Turan, I. (1993). Turkey and the Middle East: problems and solutions. *Water International* 18:23–29.
Turner, B. L., Clark, W. C., Kates, R. W., Richards, J. F., Mathews, J. T., and Meyer, W. B., eds. (1990). *The Earth as Transformed by Human Action: Global and Regional Changes in the Biosphere Over the Past 300 Years.* Cambridge University Press, Cambridge, England.
Twain, M. (1966). *The Innocents Abroad.* Harper & Row, New York (first published in 1869).
Ubell, K. (1971). Iraq's water resources. *Nature and Resources* 7:3–9.
U. N. Food and Agriculture Organization. (1980). *The Law of International Water Resources.* Legislative Study No. 23, FAO, Rome.
United States Agency for International Development. (1993). *Water Resources Action Plan for the Near East.* USAID, Washington.
United States Army Corps of Engineers. (1991). *Water in the Sand: A Survey of Middle East Water Issues.* U.S. Defense Dept., Washington.
United States Bureau of Reclamation. (1964). *Land and Water Resources of the Blue Nile Basin: Ethiopia.* U.S. Interior Dept., Washington.
United States Departments of Interior and State. (1967). *Water for Peace.* Proceedings of International Conference, Washington.
Van der Leeden, F., Troise, F. L., and Todd, D. K. (1990). *The Water Encyclopedia.* Lewis Publishers, Chelsea, Michigan.
Van Tuijl, W. (1993). *Improving Water Use in Agriculture: Experiences in the Middle East and North Africa.* Technical Paper, The World Bank, Washington.
Vesilind, P. J. (1993). The Middle East's water: critical resource. *National Geographic Magazine* 183:38–70.
Vlachos, E. (1990). Prologue: water, peace, and conflict management. *Water International* 15:185–188.
Wakil, M. (1993). Analysis of future water needs for different sectors in Syria. *Water International* 18:18–22.
Waterbury, J. (1979). *Hydropolitics of the Nile Valley.* Syracuse University Press, Syracuse, New York.

Waterhouse, J. (1982). *Water Engineering for Agriculture.* Batsford, London.
Watson, A. (1983). *Agricultural Innovation in the Early Islamic World.* Cambridge University Press, Cambridge.
Weber, P. (1991). Desalination's appeal evaporates. *WorldWatch* 4:8–9.
Weisgall, M. W., ed. (1977). *The Letters and Papers of Chaim Weizmann.* Israel Universities Press, Jerusalem.
Westcoat, J. L., Jr., and Leichenko, R. M. (1992). *Complex River Basin Management in a Changing Global Climate: The Indus River Basin in Pakistan.* Department of Civil, Environmental, and Architectural Engineering, University of Colorado, Boulder.
Westing, A. H., ed. (1986). *Global Resources and International Conflict.* Oxford University Press, Oxford.
Whittington, D., and Guariso, G. (1983). *Water Management Models in Practice: A Case Study of the Aswan High Dam.* Elsevier Scientific, Amsterdam.
Whittington, D., and McLelland, E. (1991). *Opportunities for Regional and International Cooperation in the Nile Basin.* University of North Carolina, Chapel Hill.
Williams, W. D. (1987). Salinization of rivers and streams: an important environmental hazard. *Ambio* 16:180–185.
Wishart, D. (1989). Economic approach to understanding Jordan Valley water disputes. *Middle East Review* 21:45–53.
Wittfogel, K. A. (1956). *Oriental Despotism: A Comparative Study of Total Power.* Yale University Press, New Haven, Connecticut.
Wolf, A. (1992). *The Impact of Scarce Water Resources on the Arab-Israeli Conflict: An Interdisciplinary Study of Water Conflict Analysis and Proposals for Conflict Resolution.* Ph.D. Dissertation, University of Wisconsin, Madison, Wisconsin.
Wolf, A. (1993). The Jordan watershed: past attempts at cooperation and lessons for the future. *Water International* 18:5–17.
World Bank. (1992). *Development and the Environment: World Development Indicators.* Oxford University Press, New York.
World Resources Institute. (1991). *A Guide to the Global Environment.* Oxford University Press, New York.
Yakobovitz, M. (1971). *Water in Israel.* Shikmona Publishing Co., Haifa.
Yakobovitz, M. (1981). *The Water Lexicon.* Maariv, Tel Aviv.
Yergin, D. (1991). *The Prize: The Epic Quest for Oil, Money and Power.* Simon & Schuster, New York.
Zarour, H., and Isaac, J. (1993). Nature's apportionment and the open market: a promising solution to the Arab-Israeli water conflict. *Water International* 18: 40–53.

Glossary

Abbasid: A dynasty (descended from Abbas, Muhammad's uncle) that ruled the Muslim Empire from its capital in Baghdad, 750–1258 C.E. One of its rulers was Harun al-Rashid (786–809), famous for leading the Muslim holy war against the Byzantines and for the splendor of his court.

Ain (Ein): A spring of water.

Alluvium: River-deposited sediment, such as silt or clay laid down on riverine floodplains and deltas.

Amir (Emir): An Arab prince, or a ruler of a domain.

Amirate (Emirate): The domain ruled by an Amir.

Anatolia: Geographically, the peninsula between the the Black, Aegean, and Mediterranean seas, also known as "Asia Minor." The parts of modern Turkey extending eastward of that peninsula are now included under the designation "East Anatolia."

Aquifer: A water-bearing subterranean formation feeding springs or capable of supplying water to wells.

Arab Revolt: British-backed rebellion of Arabs (led by the Hashemite Clan of Hedjaz, together with Lawrence of Arabia) against the Ottomans during the First World War.

Aridity: A measure of the dryness of an area, generally characterized in terms of the ratio of water supply to demand, i.e., of precipitation to potential evapotranspiration. The lower this ratio is (below unity), the more arid the area.

Arid region: In agricultural terms, where rainfed farming is entirely marginal, as in most years precipitation is insufficient to sustain crop growth.

Baal: The Canaanite fertility and nature god, who was believed to be the provider of rain.

Balfour Declaration: A declaration issued in 1917 by British Foreign Secretary Arthur Balfour, stating that His Majesty's government favors the establishment in Palestine of a national home for the Jewish people. Because of the purposely vague wording, it permitted different interpretations and became a subject of controversy until the actual establishment of Israel in 1948.

Baluchistan: The desert region of western Pakistan.

Barlev Line: Israel's defense line on the eastern bank of the Suez Canal prior to the 1973 war with Egypt. Its eastern embankment was breached in October 1973 by the Egyptians using water cannons.

Basalt: A dark, dense, fine-grained volcanic rock; congealed lava. It typically weathers to form a darkish, expansive clay.

B.C.E.: Before the Common Era, equivalent numerically to the Christian designation B.C.

Bedouin: Desert dweller, usually a member of a nomadic pastoral tribe. (Arabic *badawi* is singular, the suffix *in* signifies plural.)

Bronze Age: A period of human history, between the Stone Age and the Iron Age, which in the Middle East included the third and most of the second millennia B.C.E. It was characterized by the use of tools made of bronze (an alloy of copper and tin, often with traces of other metals).

Caliph: Head of the Islamic community, successor to Muhammad (from the Arabic *khalifah,* meaning "replacer").

Canaan: The land conquered by the Israelites (circa 1200 B.C.E.) and called by them *Eretz Yisrael* (Land of Israel). The country was renamed Palestine by the Romans after they suppressed the Jewish rebellions in the first and second centuries C.E.

Canaanite: Member of the Semitic people inhabiting Canaan before the Israelites (Hebrews).

Catchment (watershed): An area from which runoff drains naturally into a river or lake. Natural catchments or watersheds are separated topographically by lines called watershed-divides.

C.E.: Of the Common Era, equivalent to the Christian designation A.D.

Century Storage Scheme: A comprehensive water development plan for the upper Nile, proposed in 1902 by William Garstin of the Egyptian Public Works Department (then under British administration).

Cistern: A rock-hewn reservoir generally filled with runoff water collected from a catchment or diverted from a stream. In highlands, towns, and desert regions devoid of perennial water sources (springs or rivers), cisterns have long served as the major water supply for humans and their animals.

Cloud seeding: A method of inducing or increasing rainfall from clouds by releasing into them substances (such as dry ice or crystals of silver iodide) capable of triggering the formation of raindrops.

Cropping intensity: The average number of crops grown on an area of land per year. Under rainfed farming or flood-irrigation, cropping intensity in the Middle East does not exceed unity, and is less when fallowing is practiced (to restore soil fertility). Under perennial irrigation, cropping intensity may be double or more.

Crusades: Series of European Christian expeditions to the Holy Land and neighboring territories (between the eleventh and fourteenth centuries) in an effort to return them to Christian control.

Cuneiform: The script invented in Mesopotamia. Writing was done by impressing a wedge-shaped reed stylus onto moist clay surfaces that were then dried or fired to form permanently hardened tablets.

Dead Sea: The landlocked lake that is the terminus of the Jordan and its tributaries. It occupies the lowest valley on the earth's continental surface (some 400 meters below sea level) and is the earth's saltiest body of water (some 27 percent salt by mass). Both Israel and Jordan extract minerals from its waters (chiefly potassium chloride, an important fertilizer).

Delta: A fan-shaped expanse near the outlet of sediment-laden rivers, where the river deposits its load and divides into several distributaries (so named because of its triangular shape, resembling the Greek letter delta).

Desalinization: The conversion of saline water to fresh water by removal of the salts, using any of several methods: distillation, reverse osmosis, freezing, or dialysis. The process is still very expensive, owing to energy costs, but may well become feasible in the future.

Desert pavement: The natural gravel bed that covers the surface of extensive areas in rocky (as distinct from sandy) desert regions.

Desert: In agricultural terms, an extremely arid region (a land "deserted" or uninhabited) where annual precipitation is insufficient to sustain agricultural crops. The natural vegetation is typically sparse and xerophytic—able to survive long periods of drought.

Diaspora: Dispersion. Generally refers to the scattering of the Jews after their displacement from Israel by the Romans in the first and second centuries C.E. Palestinians use the same term to describe their refugees and emigrants living outside Palestine.

Dolomite: A sedimentary rock composed of calcium-magnesium carbonate.

Drainage: The removal, by natural or artificial means, of excess water from a given area. Surface drainage is the removal of water ponding over a soil, whereas subsurface or groundwater drainage is the removal of water saturating the soil from below. The latter is accomplished by means of ditches or buried perforated tubes.

Drip irrigation: A modern method of high-frequency, low-volume irrigation, by which water is dripped directly into the root zone. Developed in Israel, it permits greatly improved efficiency of water use and water conservation.

Drought: A prolonged period of lower than usual precipitation and higher than usual potential evaporation.

Druze: Member of a secretive religious sect that broke off from the main body of Islam; named after the religious leader Isamil al-Darazi (died 1019 C.E.), who preached that the Fatimid Caliph al-Hakim was Muhammad's true successor and a latter day prophet. The Druzes reside in mountain districts in Syria, Lebanon, and Israel.

East Ghor Canal: The Kingdom of Jordan'a major water development project, diverting water from the Yarmouk River and conveying it along the eastern edge of the lower Jordan Valley for irrigation in that valley and for supplying towns and industries as well.

Enuma Elish: The ancient Babylonian creation-epic.

Evapotranspiration: The sum of evaporation from the soil and transpiration from plants in a particular area.

Fatah: An Arabic acronym, spelling (in reverse) the words *Harakat al-Tahrir al-Falastiniya* (Palestinian Liberation Movement).

Fatimid: The Muslim dynasty (claiming descent from Ali and Fatimah, Muhammad's daughter) that ruled Egypt and other parts of North Africa during the tenth to twelfth centuries C.E.

Fault: A large fracture in the bedrock resulting from the movement of the earth's crust.

Fellah: Traditional farmer, or peasant (Arabic for "tiller"). Plural: *fellahin*.

Fertile Crescent: Arc of territory extending from the eastern seacoast of the Mediterranean northeastward and eastward over the foothills of what is today northern Syria and southeast Turkey, and thence southeastward along the Zagros piedmont of northeastern Iraq and western Iran. It is believed to be where the advent of farming took place (the so-called Neolithic transformation).

GAP: Acronym for *Guneydogu Anadolu Projesi* (Southeast Anatolia Project), being Turkey's regional development plan designed to harness the headwaters of the Euphrates and Tigris rivers for hydroelectric power and irrigation.

Gaza Strip: The strip of territory along the southern coast of Palestine that was seized by Egypt in the 1948–49 war and occupied by Israel in 1967. It is densely populated by Palestinian refugees and is due to join with the West Bank in establishing a Palestinian national entity.

Ghor: Flat-bottomed river valley.

Gilgamesh: Mythological king of Erech in southern Mesopotamia, and hero of a great epic tale of adventure, including the story of a flood that covered the earth (somewhat similar to the biblical story of Noah).

GNP (Gross National Product): In economics, the total value of goods and services generated by a certain country (both from domestic and from foreign transactions) in a given year.

Golan Heights: A strategically important volcanic plateau that forms the eastern edge of the upper Jordan valley and sits astride some of the headwaters of the Jordan and of the Yarmouk. In the early 1960s the Syrians attempted to divert two of the upper Jordan's tributaries away from Israel through the Golan plateau. Consequently it was seized by Israel in June 1967.

Graben: A block of the earth's crust, bounded by fault lines, that descended below the surface of the adjacent formations. The Syrian-African Rift Valley exhibits a series of grabens.

Granite: An intrusive igneous rock of various colors, generally exhibiting visible crystals of such minerals as quartz, feldspar, and mica.

Great Man-Made River: A Libyan project tapping the waters of a fossil aquifer in the Sahara and piping it northward to the coast for irrigation as well as industry and domestic use.

Greenhouse effect: The effect of certain gases in the atmosphere (including water vapor and carbon dioxide) in trapping the heat radiated from the earth, thus causing the earth's surface to be warmer than it would be otherwise. It is a natural effect, but is being enhanced artificially by the accelerated emission of carbon dioxide and other trace gases (e.g., methane, nitrous oxide) due to human activity. It is predicted to cause global warming.

Haj: Muslim pilgrimage to Mecca. The *haj* is required of all faithful Muslims. A person who has performed the *haj* is given the honorific *Haji.*

Halophyte: A plant conditioned to growing in a saline medium.

Hamadah: A barren stony expanse, often called "desert pavement," characterized by a surface cover of (generally flinty) gravel.

Hammurabi: The Babylonian king (1792–1750 B.C.E.) who codified the laws of ancient Mesopotamia.

Harmon Doctrine: An anachronistic principle stating that an upstream country has the unrestricted right to use the waters of an international river irrespective of downstream consequences.

Hashemite: The ruling family of Hedjaz (1916–25), Syria (1918–20), Iraq (1921–58), and the Kingdom of Jordan; presumed to descend from Hashem, the Prophet Muhammad's clan of the Qureishi tribe.

Hawr: Any of several large marshes in southern Iraq, inundated by shallow water and grown with a thicket of reeds and other aquatic plants; the abode of the marsh Arabs.

Hectare: An internationally recognized unit of land area, measuring 10,000 square meters (2.47 acres).

Helsinki Rules: A set of principles setting the basis for international law regarding the equitable allocation and use of international rivers, promulgated by the International Law Commission.

Horst: An uplifted geological structure, bounded by faults. The opposite of a *graben.*

Humid region: In agricultural terms, a region in which precipitation is practically always sufficient to sustain crop plants, and may at times be excessive.

Hydraulic civilizations: The riverine civilizations that developed alongside the major rivers of the Middle East, including the Tigris-Euphrates, the Indus, and the Nile. Some scholars (e.g., K. A. Wittfogel) have suggested that, because of the high degree of coordination required for water management, such civilizations tended to become centrally controlled and even despotic.

Ibn Khaldun: Noted Arab historian and philosopher (d. 1406).

Igneous: Type of rock formed of the original molten magma, either extrusive (due to volcanic activity, like basalt) or intrusive (cooled gradually in underground masses, like granite).

Interfluve: Area lying between roughly parallel rivers.

Intermontane: A valley or plain lying between mountains.

Intifada: A popular rebellion of Palestinians against Israeli rule over the West Bank and Gaza, beginning in late 1987 (Arabic word meaning "uprising").

Iron Age: The period in human history that followed the Bronze Age. It started in the Middle East around 1200 B.C.E. and was characterized by the advent of iron metallurgy.

Irrigation: The artificial supply of water to crops to supplement insufficient rainfall in semiarid and arid areas.

Islam: The religion founded by Muhammad early in the seventh century C.E. (generally dated to the *hejira*—Muhammad's flight from Mecca to Medina in the year 622). The literal meaning of *Islam* is "submission" to, or "acceptance" of, Allah's will.

Isohyet: Line on a climatic map connecting points of equal precipitation.

Jabal (jebel): Arabic "mountain."

Jezreel (Esdraelon): Major intermontane valley of northern Israel, connecting the coast with the Jordan Valley. (Mythical site of the prophesied Battle of Armageddon between forces of good and evil.)

Johnston Plan: A plan promulgated in the 1950s by American Ambassador Eric Johnston (sent by the Eisenhower administration) for the equitable division of the waters of the Jordan basin among Israel and its Arab neighbors. His proposed formula was accepted by the technical experts of the countries concerned but was vetoed by the Arab League for political reasons.

Jonglei Canal: A projected canal in southern Sudan designed to convey the waters of the White Nile through the Sudd swamps and to reduce the huge evaporation losses occurring in those swamps. Implementation of this project has been thwarted by the rebellion in southern Sudan.

Judea: The biblical name (after the Tribe of Judah) of a mountainous province of ancient Israel that is now the southern part of the West Bank.

Karst: A phenomenon characteristic of limestone rocks, whereby rainwater dissolves the calcium carbonate to form caverns, which in turn may collapse to form sinkholes.

Khamsin: Dry, hot desert winds that are especially prevalent during the transition seasons of autumn and spring (from the Arabic word meaning "fifty"—presumed to be the length of the spring season).

Kilometer: An internationally recognized unit of distance measuring 1,000 meters.

Kinneret: The oval-shaped lake (hence its Hebrew name, meaning "violin" or "harp") in the upper Jordan valley, also called Sea of Galilee or Lake Tiberias. It now serves as Israel's principal surface-reservoir of fresh water.

Koran: The Muslim holy scripture (literally, "recitation"), held to be the message conveyed by Allah via the angel Gabriel to the Prophet Muhammad.

Kurd: Member of a central-Asian ethnic group that settled in the highland districts of northern Iraq, western Iran, northeastern Syria, and southeastern Turkey. Though they number many millions, the Kurds are a minority in each country, having being denied a unified sovereign domain (Kurdistan) by the Allied Powers who partitioned the region after World War I.

Levant: The territory along the eastern coast of the Mediterranean.

Limestone: A sedimentary rock composed of calcium carbonate. It tends to be dissolved by rainwater and to form caverns and sinkholes, features commonly designated under the term *karst.*

Loess: A type of soil or geological material formed of wind-deposited dust. Typical loess is a uniform, yellowish soil with a texture of fine sand or silt.

Lowdermilk Plan: A plan conceived by Walter Clay Lowdermilk, an American soil conservationist, who visited Palestine in 1939. The plan envisaged the diversion of the Jordan for irrigation and of seawater from the Mediterranean to the Dead Sea for generating electric power.

Mandate: Authority granted to Britain (over Iraq, Palestine, and Transjordan) and to France (over Syria and Lebanon) by the League of Nations, to govern and guide these countries toward independence.

Marduk: The chief god in the Babylonian pantheon, who smote the evil goddess Tiamat and formed the human creature out of mud and blood.

Marsh Arabs: The tribes that live in the marshes of southern Iraq. They practice fishing and a unique type of farming on floating reed islands, and also build their distinctive homes and vessels from reeds.

Mawali: Non-Arab converts to Islam during the early period of Muslim expansion. The *mawali* soon outnumbered their Arab conquerors, and eventually accepted Arabic culture and identity.

Mesophyte: A plant adapted to growing in a well-aerated soil. Most crop plants (with the exception of such crops as rice and sugarcane) are mesophytes.

Mesopotamia: Greek name for the land between the Tigris and Euphrates rivers.

Meter: The basic international unit of length, equivalent to 3.28 feet, or nearly 1.1 yard.

Microirrigation: A general term applied to modern methods of irrigation that consist of small-scale devices designed to apply water at high frequency to a fraction of the soil surface. Includes drip, microsprayer, and bubbler irrigation.

Mongol invasion: Destructive invasion of the Middle East (especially Iraq) by nomadic horsemen from east-central Asia in the thirteenth century C.E.

Monsoon: A term derived from the Arabic word for "season" and signifying the seasonal reversal of winds over Ethiopia and across southern and eastern Asia, generally bringing summer rains.

Mughani: The traditional qanat diggers in Iran.

Muslim: A member of the Islamic religion (literally a "submitter" to the will of Allah).

Nabatean: Of the ancient Arabian nation that controlled the spice trade in the late centuries B.C.E. and early centuries C.E. They established settlements along their caravan routes in the Edomean and Negev highlands, and these grew into towns (their chief town being Petra). The Nabateans excelled in techniques of land and water husbandry, including water harvesting and runoff farming.

National Water Carrier: Israel's main water supply system. It is an integrated complex combining water from several sources (lakes, springs, groundwater), but its principal component is an aqueduct that conveys water from Lake Kinneret southwestward to the coast and to the northern Negev.

Negev: The southern desert of Israel, the name denoting "dryness."

Nomad: Member of a tribe that traditionally migrates over a territory of desert or semidesert in search of pasture. Arab nomads (known as *Bedouin*) do not wander aimlessly but generally follow regular patterns of transhumance. They regularly trade with settled farmers, but in times of prolonged drought they tend to encroach upon agricultural lands.

Orographic: Of mountainous topography. Orographic rainfall, for example, results from air masses uplifted and cooled by the presence of mountains across their path. The orographic effect enhances the rainfall occurring in the Middle East. Lands downwind of mountain ranges are generally in the "rain-shadow" of the moisture-bearing winds and tend to be arid.

Ottoman Empire: Multinational Islamic state (based in Anatolia and dominated by Turks) that ruled for more than six centuries (1299–1918) over vast territories in the Middle East as well as parts of the Balkans in Europe.

Overgrazing: The destruction of an area's vegetation as a consequence of excessive grazing (mainly by goats and sheep), thus contributing to permanent land denudation and soil erosion.

Palestine: The name given to the Land of Israel (or Judea) by the Romans after they quelled the Jewish rebellions of the first and second centuries.

Palestinian: Arab inhabitant of Palestine or displaced from Palestine.

Pan-Arabism: The idea that all the Arab people in the Middle East belong to one great nation that should unite politically rather than be divided into separate states.

Paris Peace Conference: Meeting of the victorious Allies after World War I at which the fate of the Middle East was decided.

Peace Pipeline: A proposal made in 1987 by the late Turgut Ozal, then prime minister (later president) of Turkey, for the conveyance of water from two Turkish rivers (Seyhan and Ceyhan) southward to some of the water-short countries of the region. The plan was disavowed by Ozal's successors.

Persia: Ancient name for Iran.

Philistines: A seafaring tribe from the Mediterranean (or Aegean) islands who settled along the southern coast of Canaan around the twelfth century B.C.E. They fought with the Israelites, who entered the country from the east about the same time. Within a few centuries, the Israelites prevailed and the Philistines disappeared from the stage of history. Their name was metamorphosed into "Palestine" and "Palestinian" though they never controlled the entire country so named and were ethnically and culturally unrelated to the Arabs.

Phoenicians: A nation that occupied the coast of today's Lebanon during the latter part of the second and early part of the first millennium B.C.E. The Phoenicians were expert seafarers and traders, who manufactured glass and dyes and established far-flung colonies (including Carthage). They were named by the Greeks after the red dye they extracted from the murex sea snail.

Phreatophyte: A plant growing in areas with a high water-table.

PLO: Palestine Liberation Organization.

Potential evaporation: The total evaporation of water from an area the surface of which is kept constantly moist. As such, it represents a climatic index to characterize the evaporational demand (maximum possible evaporation, or evaporativity) of the atmosphere. In vegetated areas, the index is usually called "potential evapotranspiration."

Ptolemaic: The Hellenic dynasty that established itself over Egypt after that country's conquest by Alexander the Great (332 B.C.E.).

Qanat: An Arab term meaning "canal" and referring to lateral tunnels conveying groundwater to the surface. Qanats are especially prevalent in Persia, today's Iran, where they are called *karez;* as well as in North Africa, where they are called *foggaras.* In English, they are often described as "chain wells."

Rainfed farming: Farming practiced in the relatively humid areas of the Middle East, where rainfall is sufficient to sustain crop growth. Generally, the minimal seasonal rainfall required is 300 (or so) millimeters.

Rain-shadow: The lee side of a mountain range lying across the path of rainbearing winds. The windward side of such a range (where the air rises, cools, and condenses its moisture) is generally rainy; the leeward side (where the air descends and warms up) is generally arid.

Rift: A narrow, elongated valley bounded by cliffs resulting from parallel geological faults. The Jordan Rift Valley is a segment of the Great African-Syrian Rift, which includes the Red Sea.

Riparian: Pertaining to rivers. "Riparian rights" refer to privileges of access to a river, and "riparian nations" are nations through whose territories an international river flows.

Rival: This word, which in modern English implies competitor or adversary, originally (in Roman law) referred in a neutral sense to a neighbor along a *rivus* (stream) sharing its water with others.

Salah al-Din: Arabic name for the Kurdish military leader who defeated the crusaders (1187) and became ruler of Syria and Egypt (called Saladin by the Europeans).

Salinity of water: The total concentration of soluble salts, generally expressed in terms of milligrams per liter, equivalent to parts per million (ppm). Fresh water generally does not exceed 600 ppm. Brackish (slightly saline) water may contain as much as 3,000 ppm. Saline water contains an even higher concentration. Seawater salinity is typically 30,000–40,000 ppm. Dead Sea brine contains over 250,000 ppm.

Samaria: Biblical name for the mountainous district that was the domain of the ancient Israelites, then of the Samaritans, and is now the northern section of the West Bank.

Sanjak: An administrative district within the Ottoman Empire.

Saqia (waterwheel): An animal-powered water-lifting device by which a yoked ox, donkey, or camel (generally blindfolded) walking round and round activates a wheel with buckets or other containers that fill with water from a well and spill it in an irrigation ditch.

Sassanid: The Persian empire that existed from 227 to 651 C.E. or its ruling dynasty, named after its founder, Sassan.

Saudi: Of the Saudis, the dynastic family of Ibn Saud, founder of the kingdom known as Saudi Arabia.

Sea Peoples: A loose group of island seafarers who raided the southeastern coasts of the Mediterranean during the late thirteenth century B.C.E. They even attacked Egypt (1230), but were defeated in a sea battle by the forces of Rameses III. One of their tribes, the Philistines, then settled along the southern coast of Canaan.

Sedimentary: Geological deposits composed of sediments, materials laid down by water or wind, and usually stratified. The term includes rock formations such as limestone, sandstone, and shale, as well as unconsolidated deposits such as loess and sand dunes.

Semiarid region: In agricultural terms, a region in which precipitation is sufficient in most seasons for crops to grow, but in which droughts occur frequently enough to make the practice of rainfed farming a hazardous venture.

Semitic: Pertaining to an ethnic and linguistic group of western Asian peoples, including the ancient Akkadians, Canaanites, Arameans, Hebrews, and Arabians; as well as the modern Jews and Arabs. The Semitic languages use consonantal writing systems and have similarly inflected grammars based on three-consonant roots.

Shadouf: A human-powered water-lifting device consisting of a levered beam, one end of which is weighted with a stone or lump of clay while the other end has an attached vessel for drawing water from a river or a well.

Shatt al-Arab: Confluence of the Tigris and Euphrates rivers; the united channel through which they discharge into the Persian Gulf.

Sheikh: An Arab tribal leader or chieftain (literally, an "elder").

Shiite: A member of the Muslim division (literally "partisan") adhering to Ali, Muhammad's son-in-law, believed to have been the rightful successor to the Prophet.

Soil salinization: The process by which soil is infused with salts, which render it unfit for crop growth.

Steppe: A semiarid ecological zone of short grasses and scattered shrubs (with occasional trees), intermediate between the subhumid forested zone and the desert.

Sudd: A vast area inundated by the White Nile and its tributaries in southern Sudan (literally, "barrier").

Sumer: The first civilization of southern Mesopotamia, which began as early as the fifth millennium B.C.E. and was later absorbed by the Semitic states of Akkadia and Babylonia. The Sumerians are believed to have invented cuneiform writing.

Sunni: A Muslim who accepts the established succession of *Khalifs* who followed the death of Prophet Muhammad (from *sunnah,* the "correct" Muslim belief and behavior).

Sura (Surah): Any one of the 114 sections (or chapters) of the Koran. The word means "row" or "line."

Surface runoff: The amount of water that trickles off sloping surfaces during a rainstorm. It occurs when rainfall intensity exceeds the rate of water infiltration into the soil.

Sykes-Picot Agreement: Secret agreement among the British, French, and Russians during the early stages of the First World War over the future disposition of the territories of the Ottoman Empire.

Tambour (Archimedes' screw): A human-powered water-lifting device consisting of a spiral conveyor in a drum, one end of which was dipped in the river (or reservoir) while the other end was turned by hand to convey the water upward into a trough or channel.

Tel: An archaeological mound, the site and ruin of a succession of ancient settlements.

Terracing: An ancient technique to conserve water and soil by transforming hillslopes into a series of flat fields, with stone-built vertical retaining walls. Remnants of such terraces are prevalent in upland areas of Israel, Lebanon, and elsewhere.

Third River: The recently completed artificial channel in southern Iraq, between the Tigris and Euphrates, designed to drain that marshy area.

Transhumance: The traditional, regular pattern of seasonal migration practiced by pastoral tribes in semiarid and arid regions.

Umayyad: A Muslim Arab dynasty and the empire it ruled from Damascus from 661 to 750 C.E.; also the dynasty's continued rule in Spain until 1030.

Vilayet: An administrative province within the Ottoman Empire.

Wadi: An intermittent stream bed in a dry region (known in the American Southwest as an "arroyo"), or a valley within which such a stream runs.

Water harvesting: The practice of collecting surface runoff in arid areas from sloping lands, either for filling reservoirs or for irrigating crops planted in the bottomlands.

Waterlogging: The process by which a rising water-table saturates a soil, which in an arid zone is generally accompanied by soil salinization. It can be prevented by means of drainage.

Water-table: The level to which groundwater rises in a well, generally indicating the top of the fully saturated zone in a porous formation of soil or rock.

West Bank: The central mountainous parts of Palestine (biblical Samaria and Judea), annexed by Transjordan's King Abdallah in 1949 thus becoming the West Bank of his state, renamed Jordan. After its occupation in 1967 by Israel, King Hussein relinquished his claim to the territory, which is now the focus of Palestinian national aspirations to independence.

Xerophyte: A plant adapted to live in a dry zone. Special adaptations include long vertical or lateral roots, small and waxy leaves, and the ability to survive prolonged periods of drought.

Yarmouk River: Principal tributary of the Jordan and main water resource of the Kingdom of Jordan. In history, the site of a decisive Arab victory over the Byzantines (636 C.E.).

Zagros: Mountain range in western Iran, some of whose streams feed the Tigris.

Zionism: Jewish national movement supporting the creation and development of a Jewish state in *Eretz Yisrael* (Land of Israel).

Zor: A riverine floodplain, the lowest strip of valley floor over which a braided river meanders and occasionally floods.

Index

Abassid caliphs, period, 73, 98
Abdallah Ibn Hussein (emir, later king), 78, 81, 87, 88, 150
Abraham (Biblical patriarch), 24, 61
Absolute and relative water scarcity, 213
Absolute sovereignty over water resources, 270
Absorptive capacity, mandatory Palestine, 158, 159
Abu Ali stream (Lebanon), 181
Abu Muhammad Abu Zeid (Muslim sage), 268
Abu Muhammad Bin Mahsud (Muslim sage), 267
Abu Musa Bin Minas (Muslim sage), 267
Abu Said Bin Lub (Muslim sage), 268
Abu Simbel (Temple of Rameses II), 127
Adam and Eve, 44
Addasiye (Jordan), 171
Aerial wells (refuted hypothesis), 70
Agricultural self-sufficiency, 226, 227
Agricultural transformation (Neolithic), 44
Agriculture, beginning of, 43
Agro-chemicals, 35
Agro-industrial centers, 284
Agronomists, 285
Akkad (Mesopotamia), Akkadians, 51, 53, 58
Alami, Musa (Palestinian leader), 149
Albert Nile, 115
Alexander of Macedon (Greek general and king), 50, 179
Alexandretta, Iskenderun (Turkey), 189
Alexandria (Egypt), 15, 61
Alfasi, Yitzhak (Jewish sage), on water rights, 268
Allenby, Edmund (British general), 79
Allocation of Nile water, 127
Alluvial fans, 192
Alluvium, alluvial deposits, 13, 52, 93
Al Quds (Jerusalem), 84
American assistance, development of water resources, 162
Amman (Jordanian capital), 174
Anatolia (Turkey), 15, 37, 80, 88, 92
Ancient trade in spices, aromatics, medicinal herbs, 64
Ancient waterworks, 97
Ansaria district (Syria), 188
Anti-Lebanon range, 180, 188
Antioch, Antakya (Turkey), 188, 189
Anti-Semitism, 150
ANZAC forces, World War I, 79
Application efficiency, water, 215, 217
Apsu (Mesopotamian water god), 57
Aqaba (Jordanian seaport), 84, 260
Aquatic weeds, water hyacinths, 129, 220
Aquifer depletion, 166, 200
Aquifer degradation, 191
Aquifer recharge, 40, 200, 201, 262
Aquifers, 16, 31, 36, 39, 40, 42, 107, 190–209
Aquifers of Israel and the West Bank, 203
Arab-Israeli dispute, 292
Arab-Israeli war of 1973 (Yom Kippur War), 266
Arab Legion, 88, 150
Arab Monetary Fund, 318
Arab nationalism, 80, 81, 82
Arab refugees, 150
Arab Revolt (World War I), 78, 79
Arabia, 83, 192
Arabian Peninsula, 8, 63, 72, 81, 87, 194
Arabic culture, spread of, 76
Arabic language and words, 23, 26, 72, 73
Aral Sea (Central Asia), 55
Aramaic language, 72
Arava (Araba) Valley, 40, 64, 258, 260
Archaeological sites, 128, 166, 307
Archimedes' screw (tambour), 61, 62
Arid zones (regions), 21, 22, 35
Aristotle (Greek philosopher), 251
Armand Hammer Center, Tel Aviv University, 315
Armenian mountains (eastern Turkey), 95, 104
Armenians, 77, 88, 95
Artesian aquifers, 191
Artificial lakes, Arava Valley, 258
Asa (Judean king), 63
Ashur (ancient capital of Assyria), 93
Asia Minor, 15
Assad Dam, Tabqa Dam (Syria), 37
Assad Lake (Syria), 107, 109

Assyria, Assyrians, 46, 49, 53, 58, 93, 145
Aswan, 13, 112, 118–119, 134, 220, 266, 307–309
Aswan Dam, positive vs. negative impacts, 128–131
Aswan High Dam, 121, 122, 123, 125–131, 138, 220, 266, 307–309
Aswan Low Dam, 121
Asyut Dam (Egypt), 121
Atacama Desert (South America), 70
Ataturk (Mustafa Kemal, founder of modern Turkey), 88
Ataturk Dam (Turkey), 37, 104, 105, 106, 267, 305
Atbara River (Sudan), 113, 116, 117, 118, 122, 125
Atmospheric processes, 235, 236
Augmentation of Nile flow, 132
Augmenting water supplies, 232–263
Awali River (Lebanon), 83, 163, 181, 184, 185, 188, 249

Baal (Canaanite god of rain), 23, 179
Baalbek (Lebanon), 181, 188
Babylon, Babylonians, 24, 46, 49, 51, 53, 58, 64, 145
Badlands (Jordan Valley), 156
Baghdad (capital of Iraq), 56, 93, 96, 100, 101
Bahr el-Ghazal (tributary to Nile), 115, 131–132
Bahr el-Jebel (tributary to Nile), 115
Bahrain (Persian Gulf), 8, 194
Balfour, Arthur James (British foreign minister), 81, 303
Balfour Declaration, 81, 85, 149, 186, 303, 304
Balikh River (Syria), 93, 107
Baluchistan (desert province, Pakistan), 9, 16, 57
Banias (tributary to Jordan), 153, 157
Barid stream (Lebanon), 181
Bar Kochba (leader of Jewish rebellion against Romans), 146
Barlev Line (Suez Canal), 89
Barrages (diversion dams), 96, 97, 99, 121
Basin irrigation, 59, 121
Basra (Iraq), 83
Beaufort (Crusaders' fort in Lebanon), 181
Bedouin tale of mistrust, 296
Bedouins, 6, 14, 70, 72, 79, 87, 95, 107
Beer-Sheba (Negev), 24
Beirut (Lebanese capital), 83, 177, 180
Beit Netofa (valley in Galilee), 162
Beit Shean Valley, 261, 262
Bellagio Draft Treaty (international aquifers), 275
Ben Gurion, David (Israel's first premier), 79, 149, 296
Ben Gurion University of the Negev, 312
Ben Meir, Meir (Israeli water commissioner), 207
Benefits of coexistence, 293
Benghazi (Libya), 196, 198
Beqaa Valley (Lebanon), 83, 153, 180, 186, 188
Bethlehem (West Bank), 206
Bible, 5, 47, 61, 148
Bilharzia, schistosomiasis (water-borne disease), 129
Birthplace of three religions, 152, 160
Bitumen, 53
Black September, 187
Blue Nile, 11, 13, 38, 58, 112–142
Bolshevik Revolution (1917), 78
Borders (political boundaries), 75, 77, 82
Bosporus (straits), 78
Boutros-Ghali, Boutros (U.N. Secretary-General), 134, 266
Brackish groundwater, 230
Brackish water, 218, 224, 253, 283
Brackish water irrigation research, 312
Britain, British Empire, 76–79, 81–83, 88, 121, 122, 123, 125, 149
British colonialism, 123
British mandate over Palestine, 82, 84, 158, 303, 304
Bronze Age, 63, 144
Byzantium Empire and era, Byzantines, 61, 65, 69, 72, 146

Caesar, Julius (Roman emperor), 251
Cairo (Egyptian capital), 119, 120, 221
Caliphate of Baghdad, 73
Caliphate of Damascus, 73
Caliphs, 72
Camp David negotiations (1977), 314
Canaan, 5, 25, 61, 143, 144
Canals, 28, 52, 59, 95, 99
Capillary rise, 54, 56
Caravans, 64
Carrying capacity of water for silt, 301
Carter, Jimmy (U.S. president), 151
Carter Administration offer, Yarmouk development, 168
Carthage, 50, 178, 179
Casus pacis (occasion or reason for peace), 287
Casus belli (occasion or reason for war), 163, 287
Cataracts (of the Nile), 12, 112, 118
Catchment (watershed), 67, 138, 272
Cedars of Lebanon, 16, 50, 177, 178, 180
Cenomanian-Turonian formations, 204
Center for Strategic Studies, 288
Center-pivot irrigation, 14, 224
Chain wells (qanat), 16, 17
Champollion, Jean-François (French Egyptologist), 62
Chiang Kai-shek (Chinese leader), 266

Christian community of Lebanon, 80, 83, 181, 183, 187
Christian status in Lebanon, diminished, 187
Churchill, Winston (British prime minister), 304
Cisterns, 47, 65, 66, 68, 70, 240, 302
Cities (growth of), 34, 36
Clay, 223
Climate change, 35, 74, 134, 142, 264, 300
Climate fluctuations, 43
Climate variability, 38
Climatological zones, Middle East, 19
Clogging of emitters in drip irrigation, 224
Cloud seeding, 233–239
Cloud seeding, downwind effects, 239
Coastal plain of Israel, 159, 165
Coastal aquifer of Israel, 165, 200–207
Cold War, 125, 264, 281, 297
Colonial interests, 75
Columbia River (U.S.A.), 270
Communal identity in the Middle East, 282
Compensatory water transfers, 287
Competition over Nile waters, 134
Concepts regarding water rights, 270
Confidence-building measures, 284
Contamination (pollution) of water, 35, 194, 229, 265
Contradictory positions on water rights, 306
Controlled-environment greenhouses, 229
Convective clouds, 237
Cooperative ecological security, 264
Coptic religion (Egypt), 113
Cost vs. price of water supplies, 214, 217
Costs and benefits of weather modification, 234
Costs of desalinization, 252–254
Costs of water wastage, 217
Cotton production, 121, 131
Crop domestication, 44
Crop failure, 225
Cropping patterns, 218, 221, 230
Cropping intensity, 141, 213
Crops, wild progenitors of, 43
Crusaders, defeat of, 265
Crusades, 79, 146, 294, 303
Culture of waste, 214
Cumulus clouds, 236
Cuneiform writing, 53, 56
Customary international law, 273
Cyprus, 76
Cyrenaica (Libya), 72, 198
Cyrus (Persian king), 146

Dajani, Shihadeh (Palestinian educator), 149
Damascus (Syria), 79, 80, 81, 87, 107, 109
Damietta (Dumyat) branch of the Nile, 120
Damman aquifer, 39, 194
Dams, 34, 37, 40, 94, 97, 102, 105
Dams in Ethiopia, 138, 153
Dan spring (Upper Jordan), 86, 153
Dardanelles (straits), 78
Darfur (Sudan), 9, 120
David (Israelite king), 26, 144
David's Lamentation, 313
Day of Atonement (1973), 90
Dead Sea, 4, 40, 152, 153, 155, 156, 157, 158, 159, 169, 171, 256
Dead Sea, falling level of, 168, 257, 310
Dead Sea Works, Israel, 257, 315
Deborah (Biblical prophetess), 25
Debt-for-nature swaps, 133
Deforestation, 35, 50, 54, 138, 178
Degradation of soil and water, 36
Degraded farmland, 178
Degraded landscape, 147
Degrading use of water resources, 269
Delta (of the Nile), 13, 14, 15, 23, 58, 119, 120, 121, 218
Deluge, 57
Demand management, 210–231
Demilitarized zone (Syria-Israel), 162
Demirel, Suleiman (Turkish president), 246
Denudation of watersheds, 57, 74
Depletion of fossil aquifers, 231
Depletion of water supplies, 35
Desalinization, 40, 196, 204, 251–263, 286
Desert aquifers, 194, 195, 171
Desert Belt (N. Africa and S.W. Asia), 42, 63, 151
Desert dust, effect on rainfall, 238
Desert ecology, 5
Desert farming, 65, 67, 68, 72
Desert nomads, 63
Desert pavement, 69
Desert reclamation, 137, 138, 142
Desertification, 31, 32, 33
Deserts, 31, 42, 54, 58, 63
Development of irrigation methods in Israel, 221–225
Dew, 26, 27
Dikes, 52, 56, 59, 65, 73, 94
Dindar River (tributary to the Blue Nile), 116
Disi fossil aquifer (Jordan and Arabia), 310
Distillation, 253
Diversion of floodwaters, 241
Diversion of Jordan River and tributaries, 162–163
Diversion of river waters, 52
Diyala River (tributary to the Tigris), 93, 98
Diyarbakir (Turkey), 96
Downstream vs. upstream users, 102, 110, 267

Drainage, 46, 56, 93, 97, 99–102, 106, 129, 138, 171, 214–219, 231
Drainage basin (catchment), 272
Drainage discharge, Jordan River, 168
Drainage water reuse, 136
Drinking water quality, 254
Drip (trickle) irrigation, 18, 136, 213, 223, 225–220
Drought, 123, 128, 142, 193, 196, 232, 239, 9, 13, 31, 33, 38, 42, 51, 58, 109, 117, 123, 131, 167
Druze, 77, 80, 183
Dual-purpose wells, 206
Dulles, John Foster (U.S. secretary of state), 125, 160
Dye making (Phoenician), 178

East Africa Rift Valley, 115
East Ghor Canal, 170, 172, 173
Eastern Anatolia, 37, 88, 92, 95
Eban, Abba (Israeli leader), 287
Ecological damage of wasteful practices, 212
Ecological zones, Jordan, 171
Economic efficiency of water use, 218
Economists' report, Israel, 290
Ecosystem, 6, 44
Edomean mountains, 64
Efficiency of water use, 209, 221
Efficient irrigation methods, 165, 212
Egypt, 8, 38, 47, 72, 76, 83, 87, 88, 102, 111–142, 145
Egypt, early farmers, 59
Egypt, population of, 38, 63, 77, 130
Egyptian agriculture, 59
Egyptian Civilization, 58
Egyptian language, 72
Egyptian Pharaohs, 178
Egyptian Revolution (1952), 123
Egyptian water management, 218–221
Egypt-Israel peace treaty, 151
Egyptology, 62
Egypt's land master plan, 136
Egypt's loss of agricultural land, 130
Egypt's monuments endangered, 129–130
Egypt's new lands 136
Eilat, 64, 178, 260
Eisenhower, Dwight David (U.S. president), 125, 284
El Salaam Canal (proposed), 247–249
Electrodialysis (desalinization method), 253–255
Elisha (Biblical prophet), 71
Engineers, role of, 285
Entrapment of bitter history, 294
Enuma Elish (Mesopotamian myth), 23, 57
Environmental crime, 271
Environmental crisis, 36
Environmental degradation, 57, 73, 134, 138, 184, 265, 282
Environmental impacts of technology, 220
Environmental Modification Convention (1977), 271
Environmental refugees, 34
Environmental rehabilitation, 282
Eocene formations, 204
Equitable allocation of Jordan waters, 161
Equitable allocation of water, 84, 103, 133, 134, 273
Eratosthenes (Greek geographer), 112
Eretz Yisrael (Land of Israel), 84
Erosion, 31, 35, 45–51, 54, 57, 65, 67, 74, 138–139, 147, 184, 239
Esarhaddon (Assyrian king), 302
Eshkol, Levi (Israeli premier), 311
Estuaries, 265
Ethiopia, 11, 13, 38, 42, 58, 113, 116–142
Ethiopian highlands, 117, 120, 139
Ethiopia's hydrological position, 308
Euphrates River, 15, 23, 37, 51, 57, 83, 88, 92–110
Eutrophication, 163, 220, 310
Evaporation pans, Dead Sea, 257
Evaporation, 22, 34, 58, 97, 102, 103, 106, 109, 110, 113, 125, 127, 132, 140, 141, 166, 191, 215, 224
Evaporation losses, White Nile schemes, 141
Evaporative power of desert winds, 255
Evpotranspiration, 201
Exodus from Egypt (Israelites), 143
Exotic rivers, 37, 57
Expansion of irrigation in Egypt, 136–137, 220
Expansion of irrigation in Ethiopia, 140
Extraction of potash, Dead Sea, 258
Eytan, Raphael (Israeli minister of agriculture), 288

Fallow, fallowing, 56, 121
Famine, 31, 33, 58
Fariah stream, valley (West Bank), 155, 206
Fatimids (caliphs), 73
Fatwas (Islamic edicts), 267, 268
Feisal Ibn Hussein (emir, later king, of Iraq), 78, 81, 84
Feisal-Weizmann agreement, 84–85, 304
Fellahin (cultivators), 59, 87, 95
Fertile Crescent, 8, 15, 42, 43, 46, 57, 78
Fertilizer injection in irrigation, 223
Field capacity (soil), 222
Finger of Galilee, 86
Fires, 51
Fisheries, Egypt, 129
Flash floods, 5, 25, 65, 71, 191, 240
Flood control, 103
Flooding, 13, 46, 51, 52, 57, 58, 94, 97, 110, 120, 121, 122, 132
Flood irrigation, 18, 56, 99, 121, 213, 216, 242
Floodplain, 51, 54, 58, 113, 119, 156
Floodwater, 181, 229

Floodwater diversion, 71
Food security, policy of, 212, 218
Forest clearing, 147
Fossil groundwaters, aquifers, 192, 194, 196–200, 283
Fourth Geneva Convention, 317
France, French, 76, 77, 78, 79, 82, 83, 86, 88, 125, 181
Frankfurter, Felix (American Zionist leader, jurist), 304
Free trade zone for water, 290
Freezing method of desalinization, 253
French rule over Lebanon and Syria, 188
Fungal diseases (plants), 224
Furrow irrigation, 216

Galilee (Israel), 47, 79, 83, 85, 146, 150, 178
Gallipoli (Turkey), World War I battle of, 88
Ganges River (India), 23
Gaza, District, Strip, 77, 79, 89, 125, 151, 200, 202, 244, 247–249
Geneva Convention (1977), 271
Gezira (Sudan), 113, 122, 306, 307
Ghab Valley (Syria), 188
Ghor-Amman diversion, 173
Ghor of Jordan, 155
Gideon (Israelite judge) 25
Gihon (Siloam) spring, 26, 144–146
Gilead plateau, 155
Gilgamesh epic (Mesopotamia), 51, 110, 300, 301
Glass making (Phoenicia), 178
Global warming, 38, 300
Glorification of war, 293
Goats, 49, 50, 51, 300
Golan Heights, Plateau, 39, 40, 85, 86, 89, 151, 163, 226
Gravel, 223
Gravel mounds (tuleilat el einab), 69, 70
Grazing, 31, 49
Great Zab River (tributary to Tigris), 93, 96
Great Man-Made River (Libya), 198–200
Greek geographers, 112
Greek language, 72
Greeks, 50, 61, 77, 177
Greenhouse effect, warming, 264, 300
Greenhouses, 39
Groundwater, 28, 37, 39, 56, 109, 161, 170, 190–209, 239
Groundwater depletion, 138, 191
Groundwater development, 196
Groundwater extraction, restrictions of, 193, 202
Groundwater, overpumping of, 193
Groundwater recharge, 34, 191, 202, 204, 212, 250
Groundwater salinization, 205
Gulf of Aqaba, 163
Gulf States, 39
Gullies, erosion, 48, 65, 239
Gypsum in soil, 107–108

Habib, Philip (U.S. diplomat), 168–169
Hagar (mother of Ishmael), 25
Hai Gaon (Jewish sage), 268
Haifa, Haifa Bay, 80, 159, 262
Haj (Muslim pilgrimage), 87
Halab (Aleppo), 80, 107
Halley, Edmund (British scientist), 28
Hama (Syria), 79, 80
Hammer, Armand (American oilman), 197, 311, 315
Hammurabi (Babylonian king, lawmaker), 55
Hannibal (Carthaginian leader), 179
Hapi, Apis (Nile god), 111, 127, 128
Harmon Doctrine, 270, 273
Harod stream (Israel), 155
Harun al-Rashid (caliph), 73, 99
Hasbani (tributary to Jordan), 85, 153
Hashem al-Qirbah Dam (Sudan), 123
Hashemite Kingdom of Jordan, 81, 86, 89, 150
Hashemites, 78, 81, 87
Hasidic tale, 292
Hasmoneans (Judean priests), 146
Hatay District (ceded to Turkey by the French), 189, 305
Hattin (Holy Land), battle of, 316
Hawr al-Hammar (Iraq), 94, 96, 102
Hays, James (American engineer), 159
Hazar Golu lake (Turkey), 96
Hebrew Bible, 296
Hebrew language and words, 28, 72
Hebron (West Bank), 88, 206
Hedjaz (Arabia), 77, 79, 81
Helsinki Rules (1966), 271, 272
Herdsmen, 61
Hermon (Mount), 86
Herod (Judean king), 146
Herodotus (Greek geographer and historian), 52, 58, 112, 192
Herzl, Theodor (founder of modern Zionism), 310
Hezekiah (Judean king), 145
Hezekiah's tunnel (Jerusalem), 145–146
Hieroglyphics, 62
High-frequency, low-volume irrigation, 217, 221, 224
High-value crops, 220, 229
High water-table conditions, 191
Hillslope cultivation, 178
Hindayah Barrage (Iraq), 99
Hiram (Phoenician king), 50, 178
Historical water rights, 110, 122, 158, 270
Hittite Empire, 46
Holocene, 43
Holy sites, 85
Homs (Syria), 79, 80

Horns of Hattin, site of historic battle, 265, 316
Huboobs (wind storms), 120
Huleh, drainage of, 154
Huleh lake, basin (Upper Jordan Valley), 85, 154, 163, 249
Human body (water content), 21
Humid equatorial zone, 122
Hunter-gatherers, 43, 44
Hussein Ibn Ali (sharif of Mecca), 78, 81
Hussein Ibn Talal (king of Jordan), 81, 151, 267, 312
Hussein, Saddam (ruler of Iraq), 305
Hydraulic civilizations, 23, 52, 60, 299
Hydroelectric power, 135, 138, 139, 158, 162, 184, 185, 189, 262, 103, 104, 105, 106, 107, 122, 123, 128, 131
Hydrological cycle, 29, 30, 314
Hydrological imperative, 283
Hydrological security, 208
Hydrological units, 270
Hydrologists, 205, 285
Hydrology, 29, 45
Hyksos (ancient Semitic invaders of Egypt), 61

Ibn Saud (Arabian king), 81
Ice age, 43
Ideology of farming, Israel, 226, 227
Imperial Valley, California, 262
Importing water by sea, 250, 251
Improving water use efficiency in Israel, 229
India, 55, 64, 78
Indian subcontinent, 42
Inducing and collecting runoff, 240–243
Indus River Valley, 23, 51, 55, 277
Indus River Civilization, 57
Infiltrability, 239
Infiltration, 46, 47, 71
Inland lagoons, Egypt, 129
Instability of Nile flow, 306
Integrated development of Nile basin, 123, 124
Intensification of crop production, 123
Interface between saline and fresh water, 201, 202
Intermontane valleys, 46, 179
International Water Conference (1977), 272
International assistance agencies, 265
International Center for Agricultural Research in Dry Areas, 287
International Conference on Water for Peace, 284
International Court of Justice (Hague), 276, 277
International groundwater convention, draft, 275
International law, 271–278
International Law Association, 271
International Law Commission, 272, 273
International Monetary Fund, 277
International program of weather modification, 237
International rivers, basins, 269, 271
International supervision of shared waters, 288, 289
International watercourses, non-navigable uses, 273, 274
Interstate (transnational) water transfers, 175
Interventions in Lebanon's civil war, 183
Intifada (Palestinian uprising), 151
Investments in research, 229
Iran, 8, 16, 17, 37, 39, 80
Iraq, 8, 31, 39, 73, 77, 81, 82, 88, 92–110
Iraq-Iran War, 101
Iron Age, 63, 144
Irrigated farming, 51, 55, 57, 194, 226
Irrigated farming, earliest evidence, 152, 153
Irrigating land in Sinai, 248
Irrigation, 3, 18, 22, 31, 34, 36, 38, 39, 47, 52, 54, 56, 59, 73, 93, 97, 105, 103, 104, 105, 106, 107, 109, 120, 128, 135, 138, 256, 207
Irrigation-based civilization, 57
Irrigation canals, 60
Irrigation development, 181
Irrigation development, Jordan Valley, 172
Irrigation efficiency, 215, 219
Irrigation methods, modern, 206
Irrigation research, 222–224
Irrigation research in California, 221
Irrigation trends in Israel, 228
Isaac (son of Abraham, considered progenitor of Hebrews), 24, 25
Isa Bin Ali al-Alami (Muslim sage), 268
Ishmael (son of Abraham, considered progenitor of Arabs), 5, 25
Islam, 72, 75
Islamic architecture, 28, 29
Islamic fundamentalist movement, 183
Islamic gardens, 28
Islamic law, 267
Isna barrage (Egypt), 121
Israel, 4, 8, 38, 40, 43, 50, 90, 91, 143–176
Israel, establishment, 88
Israel/Palestine Center for Research and Information, 318
Israeli-Palestinian interaction, 206, 207
Israeli security zone in southern Lebanon, 183
Israeli settlements in the West Bank, 207, 208
Israeli technological innovations, 225–230
Israelites (ancient), 25, 47, 49, 63, 143, 144, 148

Israel's water policy, critique of, 229
Istanbul, 78
Italians in Libya, 196, 197

Jacob (Biblical patriarch), 24, 61
Jazirah (Iraq), 93
Jebel Aulia Dam (Sudan), 117, 122, 123, 140
Jehoshaphat (Judean king), 63
Jericho (West Bank), 149, 152, 153, 158, 206
Jerusalem, 79, 80, 84, 88, 144, 151, 178, 206
Jewish Brigade (World War I), 79
Jewish calendar, dual nature of, 309
Jewish immigration, 158
Jewish National Fund, 309
Jewish national home, 85, 303, 304
Jewish rebellions against Romans, 303
Jewish sages, edicts on water rights, 268, 269
Jewish survivors of the Holocaust, 150
Jezreel Valley, 3, 159, 165, 261, 262
Jihad (holy war), 72
Johnston, Eric (U.S. ambassador), Plan, 161, 162, 173, 249
Joint desalinization, Israel-Jordan, 255–263
Jonglei Canal, 38, 115, 131, 132, 133
Jordan, kingdom of, 8, 39, 40, 88, 194, 206, 208, 231, 249
Jordan River, 3, 23, 38, 39, 51, 75, 85, 86, 87, 143–176
Jordan River, diversion of, 89
Jordan tributaries, 170
Jordan Valley, 4
Jordan Valley Authority, 159
Joseph the Wise, 317
Judaism, 72
Judea, Judeans, 47, 70, 145
Judea and Samaria (Israeli terms), 312
Judean kingdom, destruction of, 64
Judean rebellion against the Romans, 146
Juha (legendary prankster), 279, 280

Kabalega Falls (Murchison Falls), Ugauda, 115, 131
Kagera River (tributary to White Nile), 113
Karakaya Dam (Turkey), 105
Kara Su River (upper tributary to Euphrates), 95
Karez (qanat), chain wells, 16
Karstic geological formations, 181
Karun River (Iran), 96
Keban Dam (Turkey), 95, 105
Khabur River (Syria), 93, 96, 107, 109
Khamsins (desert winds), 120
Khartoum (Sudan), 10, 12, 13, 112, 116, 117, 122, 128
Khashm al-Qirbah Dam (Sudan), 122
Khidr (Koranic prophet), 27
Khuzistan (Iran), 96
Kibbutz, kibbutzim (Israel), 148, 158
Kingdom of Jordan, 152, 155, 157
Kissinger, Henry (U.S. secretary of state), 284
Koran, 20, 26, 27, 28, 41, 55
Kordofan (Sudan), 9, 120
Kufra oasis (Libya), 197, 198, 200
Kurdish rebels, 267
Kurdistan, 37, 93, 104
Kurds, 8, 37, 77, 80, 83, 88, 95, 104
Kuwait, 8, 102, 170

Labor Party, Israel, 208
Lake Abu-Dibbis (Iraq), 99
Lake Albert (Mobutu), 113, 115, 131
Lake Assad (Syria), 107, 109, 385
Lake Edward (upper White Nile), 115
Lake Habbaniyah (Iraq), 99
Lake Kyoga (upper White Nile), 115
Lake Mobutu (Albert), 113, 115, 131
Lake Nasser (Egypt), 13, 38, 118, 123, 127, 128, 131, 196, 305
Lake Nasser, evaporation from, 138, 139
Lake Tana (Ethiopia), 116, 138
Lake Tharthar (Iraq), 99
Lake Van (Turkey), 95
Lake Victoria (source of White Nile), 113, 115, 123
Lakes for mariculture and recreation, 262
Land degradation, 51, 136, 214, 217
Land ownership, 220
Land for peace formula, 40
Land reclamation, 136
Land subsidence, 108
Lawrence, T. E. ("Lawrence of Arabia"), 4, 5, 79, 81, 84, 281, 304
Leaching of salts, 59, 99
League of Nations, 82, 84, 303
Lebanese economy, 182
Lebanese National Pact, 182
Lebanese Republic, 182
Lebanon, 8, 50, 75, 77, 82, 83, 88, 163, 177–189, 249–250
Lebanon, climate of, 181
Lebanon range, 180, 188
Lebanon's civil war, 183
Levant, 76
Levees (dikes), 52, 56, 94, 97
Libya, 8, 196–200
Likud Party (Israel), 208
Limestone (mountain) aquifer, Israel, 204–209
Lisan marl (Jordan Valley), 155, 156
Litani (Qasimiah) River, 40, 83, 85, 181, 183–188, 249, 250
Litani River, diversion of, 184, 186, 188, 250
Litani Development Authority, 184

Litani to Jordan water transfer scheme, 249–250
Little Zab River (tributary to Tigris), 96
Lloyd George, David (British prime minister), 186, 303
Loess (aeolian soil), 70
Lowdermilk, Walter Clay (American hydrologist), 159, 257, 261
Lower Egypt, 58, 120
Lower Jordan River, 157
Low frequency irrigation, 222
Low-flow household appliances, 212
Low-intensity sprinkler irrigation, 223
Low-value crops, 227

Machar marshes (Sudan), 132
Madan (marsh Arabs), 94–95
Main, Charles (American engineer), 161
Malakal (southern Sudan), 115
Malaria, 46, 147
Maqarin Dam, Unity Dam (Yarmouk River), 168, 170
Marduk (Mesopotamian god of creation), 23
Marginal water resources, 220, 229
Marinus of Tyre, 112
Market mechanisms, water prices, 290
Maronite Christians of Lebanon, 182
Marqaba power station (Lebanon), 185
Marsh Arabs (Madan), 94–95, 102
Marshes, 46, 94, 96
Materialism, 76
Matruh-Sidi Barrani area (Egypt), 195
Mecca, 81, 87
Med-Dead Canal, 259–263
Medieval Islamic fatwas, 267
Mediterranean, 64, 85, 119, 120, 128, 151, 159, 177, 181, 195, 196, 235, 237, 238, 251, 256
Mediterranean climate, 169
Mediterranean–Dead Sea Canal, 257–263
Medusa bags, 250
Melville, Herman (American writer), 147
Membrane filtration (desalinizatiom method), 254
Mesopotamia, 7, 15, 23, 45, 46, 47, 50, 51, 53, 55, 56, 57, 59, 61, 73, 78, 79, 80, 81, 92–110, 178
Metering devices, valves, 213, 223
Methods of desalinization, 252–254
Methods of water spreading, 241–242
Microcatchments (microwatersheds), 242–244
Microirrigation, 224
Microsprayer irrigation, 136, 223
Microwatersheds (microcatchments), 242–244
Middle East (definition), 7–8
Milking rain from clouds, 233–239
Mineral extraction from Dead Sea, 258
Mini-Peace Pipeline, 247
Mining non-renewable groundwater, 174
Mishnah (Jewish Rabbinical tract), 146
Modern vs. traditional irrigation methods, 217
Mongol invasion of Mesopotamia (Iraq), 73, 98, 99
Monsoons, monsoonal rains, 13, 58, 64, 113, 116, 117, 119
Moses, 5, 24, 27, 257, 296, 300
Mosul (Iraq), 79, 83
Mountain aquifer (Israel), 200–207
Mount Carmel, 200
Mount Hermon, 153, 154, 156, 180
Mount Lebanon, 83, 177, 180
Movement to Save the Waters of Israel, 291
Mughani (qanat diggers), 16
Muhammad (prophet of Islam), 72
Muhammad Ali (ruler of Egypt), 121, 306
Muheiba Dam (Yarmouk), 170
Multiple cropping, 97
Murat River (tributary to the Euphrates), 95
Murchison Falls (Kabalega Falls, Uganda), 115, 131
Murex sea snail (source of purple dye), 178
Murray River (Australia), 55
Muslim conquest, 65
Muslim conquest of Palestine, 146
Muslim-Jewish Golden Age, 268
Muslim rule, 72
Muslim Shiites, 302
Muslim Sunnis, 302
Mutual dependence of Israelis and Palestinians, 289

Nabateans, 64, 65
Nablus (West Bank), 88, 206
Nahrawan Canal (diversion from Tigris River), 98
Naj Hammadi Dam (Egypt), 121
Nakht, tomb of, 61
Napoleon (invasion of Egypt), 61
Nasiriyah (Iraq), 96, 101
Nasser, Gamal Abdel (Egyptian president), 123, 125, 163
National Water Carrier (Israel), 159, 164–165, 200–202, 310
Natufian culture, 43
Navigation, Nile, 121, 128, 132, 219
Nebuchadnezzar (Babylonian king), 146
Negev, 4, 5, 24, 47, 63, 65, 66, 68, 69, 72, 150, 159, 160, 192, 196, 226, 240
Negev highlands, 67
Nejd (Arabia), 77, 81
Neolithic age, 43
Nero (Roman Emperor), 112
Nessana papyri, 302
New Lands (Egypt), 14, 123, 219, 220, 307

New methods of water management, 148
New Testament, 55
Nile, origin of name, 306
Nile to Gaza water transfer, 247–249
Nile riparian states, 113
Nile River, 10, 12, 13, 14, 23, 37, 38, 42, 51, 59, 111–142
Nile River, origins of, 112
Nile Valley, Basin, 57, 60, 63, 102
Nile tributaries, 114
Nilometers, 112
Nineveh (ancient capital of Assyria), 93, 96
Noah and the flood, 51
Nomads, nomadic herders, 49, 72
Non-renewable (fossil) groundwater, 170
North Africa, 63, 72
Notion of honor, 293
Notions of security, 288
Nubia, 112
Nubian population resettlement, 125
Nubian sandstone aquifers, 196–200
Nuclear energy, 251
Nuclear reactors for desalinization, 284

Oases, 192, 196, 258
Occidental Petroleum, 197, 198
Ogallala aquifer (High Plains, U.S.A.), 194
Oil (petroleum), 197, 219
Oman, 8, 39, 194
Operational storage capacity, Sea of Galilee, 166, 167
Ophir (Biblical land of gold), 178
Optimal cropping patterns, 221
Optimal development, 270
Optimal water requirements, 225
Optimal water use, 230
Organization of African Unity, 134
Ornamental gardens, water use in, 285
Orographically enhanced clouds, 236, 237
Orographic effect (on rainfall), 151
Orontes (Asi) River, 51, 107, 109, 181, 188, 189
Osiris (Egyptian god of resurrection), 60
Ottoman Empire, Ottomans, 4, 7, 73, 75, 76, 77, 78, 79, 80, 82, 87, 149, 279, 280
Overexploitation of aquifers, 174
Overgrazing, 16, 31, 33, 35, 50, 54, 65, 139, 147
Overirrigation, 224
Overland flow (see also: runoff), 239
Overtilling, 31
Owens Falls Dam, 113, 123
Ozal, Turgut (Turkish president), 246, 247, 267, 286

Paganism, 72
Pakistan, 16, 55
Palestine, 3, 8, 72, 77, 79, 80, 82, 83, 84, 85, 88, 156
Palestine, borders of, 85–86
Palestine, environmental change, 147, 148
Palestine, origin of name, 146, 303
Palestine, partition of, 88
Palestine Liberation Organization (P.L.O.), 151
Palestinian guerillas, 163
Palestinian and Israeli experts, meetings, 291
Palestinian refugees, 183
Palestinians, 38–40, 84, 90, 91, 149, 150, 162, 170, 206–209, 244
Pan Arabism, 82, 87, 294
Papyrus reeds, 59
Parable of Sea of Galilee and Dead Sea, 176
Paris Peace Conference, 85, 304
Partial-area irrigation, 224
Past rights and wrongs, irrelevance of, 292, 293
Pastoral civilizations, 42
Peace agreement, Egypt-Israel (1977), 90
Peace arrangements and trade-offs, 287
Peace Pipeline, 314
Peace process, 151
Peat, 163
Per capita supply of fresh water, 175, 230
Perched water-table, 191
Percolation, 22, 54, 191, 215, 239
Perennial irrigation, 121
Perrault, Claude (French scientist), 28
Persia, Persian, 15, 16, 23, 57, 61, 192
Persian Gulf, 37, 81, 87, 92, 93, 94, 95, 96, 100
Persian Gulf states (emirates), 77, 171, 194
Persian Gulf War (1991), 39, 100, 170, 266, 271, 305
Petra, 64
Petroleum (oil), 15, 35, 53, 78, 79, 80, 83, 87, 102, 103, 107, 138, 192, 196, 197, 244, 252, 265
Pharaohs, 59, 60
Philistines, 24, 25, 26, 144, 146, 308
Phoenicia, Phoenicians, 15, 23, 50, 177–179
Phoenician colonies, 178, 179
Phreatophytes, 56
Pibor River (tributary to the White Nile), 115
Picot, Charles François (French diplomat), 80, 85
Plato (Greek philosopher), 300
Pleistocene, 43, 196
Pleistocence formations, 204
Plow, 45
Policy of water management, 218
Political Islam, 281
Pollution, 35, 74
Pollution of aquifers, 207
Population growth, 33, 34, 36, 44, 50, 74, 88, 107, 122, 130, 138, 142, 161, 169, 170, 203, 227, 264, 282, 283
Population pressure, 42

Potential benefits of peace, 290
Potential evaporation, 170
Prayers for rain, 313
Preferential allocation of water to efficient users, 175, 226, 286
Preserving biodiversity, 133
Pressurized irrigation systems, 136
Price and cost of water, 213, 214, 217, 227, 228
Promised Land, 47, 80, 296
Ptolemaic occupation of Egypt, 61
Ptolemy (Greek geographer), 112
Public education toward water conservation, 211
Pumps and pumping, 99, 192
Punjab (Indus River Valley), 18

Qaddafi, Muammar (leader of Libya), 197–200, 311
Qanat, karez (chain wells), 16, 28, 192, 193
Qatar, 8
Qir'aun Dam (Lebanon), 184–85
Quetta (Baluchistan), 16
Qurnah (Iraq), 96

Rabel II (Nabatean king), 64
Rachel (biblical matriarch), 24
Radius of influence of wells, 200
Rahad River (Sudan), 116
Rain, rainfall, 24, 25, 26, 27, 28, 31–33, 42, 46, 47, 58, 65, 68, 72, 97, 113, 120, 128, 133, 140, 152, 156, 170, 196, 231–239
Rainbands, 237
Rainfall inducement (enhancement), 233–239
Rainfed civilizations, 42
Rainfed farming, 16, 22, 31, 47, 108, 120, 169, 179, 206
Rain inducement in Israel, 237–239
Rain-shadow effect, 15, 152
Rajasthan (northwest India), 42
Ramallah (West Bank), 206
Reagan, Ronald (U.S. president), 199
Rebekah (biblical matriarch), 24
Recharge of aquifers (groundwater), 165
Reclaimed sewage, 218, 225
Reclamation of desert land, 131, 196
Reclamation of wastewater, 200
Recycled effluents, 229, 231
Recycled water, 34, 35
Red Sea, 40, 64, 153
Red Sea, navigation, 64
Red Sea-Dead Sea Canal, 256–263
Reforestation, 159, 179
Refugees, 265
Regional hydropolitics, 170
Regional water-sharing arrangements, 246
Religious fundamentalism, 88
Research-based guidance to irrigators, 225
Reservoirs, 102, 103, 138
Reusing wastewater, 283
Reverse osmosis, 253–255
Rhine River Commission, 317
Rift Valley, 4, 152, 159, 180, 256
Rift Valley, African, 115
Rill erosion, 48
Rio Grande (U.S.-Mexico), 270
Riparian disputes and regional instability, 266
Riparian rights, 270
Riparian states, 103, 157, 158, 269
Riparian vegetation, 215
Ripon Falls (Upper White Nile), 113
Rivalries over common aquifers, 194
River basin commission, 274, 275
River diversions, 98
River integrity, principle of, 270
Riverine civilizations, 42
Role of technical experts, 285
Roman destruction of Carthage, 179
Roman Empire, 63, 65, 146
Roman hegemony, 146
Roman waterworks, 180
Romans, 49, 61, 64
Root zone, 215, 217, 222, 224
Roseires Dam (Sudan), 123, 125, 139
Rosetta (Rashid) branch of the Nile, 120
Rosetta Stone, 62
Runoff, 28, 46, 47, 65–67, 71, 116, 117, 151, 169, 180, 201, 239
Runoff farming, 67, 68, 69, 242–243
Russian (Bolshevik) Revolution (1917), 78, 82

Sadat, Anwar (Egyptian president), 38, 90, 127, 140, 151, 247–248,266
Sahara Desert, 9, 12, 14, 33, 120, 140, 192, 196, 197, 200
Sahel (African semiarid savanna), 9, 11, 33
Saladin, Salah ed-Din (Muslim leader), 79, 265, 316
Saline marshes, 15
Saline soil, 70
Saline springs, 167
Salinity, 99, 129
Salinization, 35, 51, 54, 56, 57, 74, 98, 99, 100, 107, 120, 129, 138, 167, 168
Salinization of wells, 201
Salt-tolerant crops, 230
Saltwater intrusion into aquifers, 191, 218
Samaria, 47, 70, 93, 145, 206
Samarra (Iraq), 93
Samuel, Herbert (British high commissioner, Palestine), 84
San Remo Conference, 82
Sand, 223
Saqia (water wheel), 61, 62
Sargon (Akkadian king), 178
Sarir (Libya), 197, 198
Sassanid Empire (Persia), 72

Satellite imagery, 196
Saudi Arabia, 8, 77, 194
Saudi Arabia, desalinization in, 252
Savanna zone, 115
Schaefer, Vincent (scientist), 233
Schistosomiasis (bilharzia disease), 129
Scorpion King (Egypt), 59, 60
Scouring banks, Nile, 129
Sea of Galilee, 4, 40, 85, 153–155, 162–168, 201, 202, 206, 238
Sea of Galilee, water quality, 166
Sea Peoples, 47, 144
Sea of reeds, 144
Sea transport of fresh water, 250, 251
Seawater canal proposals, Israel, 256–263
Seawater conversion, 251–255
Seawater intrusion, 129, 196, 201, 203, 205
Second Temple period, 85
Secularism, 76, 281
Security zone, Israeli (Lebanon), 187
Sediment, 46, 54, 113, 173
Sediment, transport and deposition, 138, 147
Sedimentation, 240, 265
Seeder-plow, 45
Seepage, 108, 109, 110, 127, 215
Semiarid regions (zones), 22, 31
Semiramis (Mesopotamian queen), 53, 301
Semitic alphabet, 50, 177
Semitic languages, 293, 297
Semliki River (tributary to Lake Albert), 115
Sennacherib (Assyrian king), 145, 146
Sennar Dam (Sudan), 121, 122, 123, 125
Senusi (Libyan king), 197
Sewage disposal, 229
Shadouf (water lifting device), 61, 62
Shakespeare, William, 299
Shalom and salaam, profound meaning of, 297
Shared aquifer, Palestinan vs. Israeli claims, 312
Shared culture of the Middle East, 295
Sharett, Moshe (Israeli premier), 187, 310
Shatt al-Arab, 96, 101
Sheep, 50
Sheet erosion, 48
Shiite Muslims, 80, 83, 101
Shiite Muslims of Lebanon, 183, 184
Shmitah (biblical land sabbatical), 47
Sidon (Lebanon), 177, 180
Siloam spring, Gihon (Jerusalem), 144–146
Silt, silting, 34, 51, 54, 55, 57, 59, 70, 74, 93, 96, 113, 116, 120, 128, 220
Silting of reservoirs, 139
Silver iodide, 233–239
Sinai, 4, 6, 47, 63, 79, 89, 90, 151, 192, 196, 248
Sind (desert province, Pakistan), 42
Siwah depression (Egypt), 195
Six Day War (1967), 163, 266, 284
Smuts, Jan (South African statesman), 82
Sobat River (Sudan), 115, 132
Soil aeration, 222
Soil compaction, 224
Soil conservation, 47, 67, 139
Soil degradation, 53, 129, 130, 191
Soil erosion, 159
Soil fertility, 22
Soil moisture reservoir, 222
Soil reclamation, 159
Soil salinity and salinization, 14, 39, 55, 56, 58, 212, 215, 219, 222
Soils of uplands, 46
Solar ponds, 255, 260
Solar radiation, 254, 255
Solomon (Israelite king), 26, 178
Solomon's coronation, 144
South Lebanese Army, 187
Southeast Anatolia Project (GAP), 37, 104–110, 246, 267, 305
Sovereignty over water, 209
Soviet Union, 125, 163
Specialty crops, 230
Spiritual kinship, Israelis and Palestinians, 295
Springs, 24, 26, 27, 28, 65, 206, 207
Sprinkler irrigation, 213, 216, 219, 220
State comptroller's report, Israel, 202
Statistical evidence of rain enhancement, 236–239
Stone Age, 43
Storm clouds, seeding of, 234–235
Strabo (Greek geographer), 120, 306
Subirrigation, 223
Subsidies for water, 36, 218, 228
Sudan, 8, 10, 12, 38, 78, 112–142
Sudan, natural region of, 126
Sudanese civil war, 132
Sudanese plain, 117
Sudano-Egyptian agreements, 125, 126
Sudd swamps (wetlands), 10, 38, 112, 115, 123, 131–133
Suez Canal, 4, 78, 87, 89, 90, 163, 266
Suez war (1956), 125
Sumer, Sumerians, 23, 46, 51, 52, 53, 58, 95, 99
Sumerian mythology, 52, 57
Summary tables, water in the Middle East, 260, 261
Summer crops, 122
Sunni Muslims, 83
Supplementary irrigation, 66
Surface crust, 70, 71
Surface irrigation, 213, 217, 223
Surface treatments for runoff inducement, 242
Sustainable water management, 219, 220
Swamps, 101, 147
Sykes, Mark (British diplomat), 80, 85
Sykes-Picot agreement, 80, 85

Syria, 8, 37, 40, 72, 77, 80, 81, 82, 86, 88, 89, 92–110, 151, 154, 158, 193
Syrian attempt to divert Jordan headwaters, 162–163
Syrian appropriation of Yarmouk water, 231
Syrian Greeks, 146

"T.V.A. on the Jordan," 159
Tabqa Dam, 107, 108, 109
Tadmor, Naftali, 71
Talmud, Jerusalemite (Rabbinical treatise), 146
Tambour (Archimedes' screw), 61
Tazerbo (Libya), 198
Technical cooperation, Egypt-Israel, 221
Tel Aviv sewage reclamation project, 165
Tel Aviv University, 288, 290
Tel el-Qadi (site of Dan spring), 86
Tel Hai (Galilee settlement), 86
Tennessee Valley Authority, 159
Terraces and terracing, 47, 49, 50, 67, 68, 180
Terra Sancta (Holy Land), 84
Thar (desert), 42
Tharthar Depression (Iraq), 96
Thawra Dam (Syria), 267
Third River (Iraq), 100–102
Tiamat (Mesopotamian goddess of primordial waters), 23, 57
Tiberias (Israel), 155
Tigris-Euphrates Valley, 58, 75, 92–110
Tigris River, 15, 23, 37, 39, 51, 53, 57, 73, 83, 88, 92–110
Tillage, 46
Tiran Straits (Red Sea), 163
Tools of soil husbandry, 44
Tower of Babel, 53
Trade routes (ancient), 64
Traditional irrigation schemes, 217
Transboundary groundwaters, 317
Transhumance, 33
Transjordan, 77, 79, 81, 82, 83, 84, 86, 87, 144, 146
Transnational water transfers, 243–251
Transpiration, 21
Trickle (drip) irrigation, 223, 225
Tripoli (Lebanon), 83, 180
Tripoli (Libya), 196, 197, 198, 200
Tropical rainbelt of Africa, 42, 113
Turkey and Turks, 8, 37, 38, 75, 77, 79, 80,83,87,88,92,103,104,109
Turkey's Peace Pipeline, 245–247
Twain, Mark (American writer), 147–148, 233
Tyre (Lebanon), 50, 177, 180

U.N. International conference on water (1977), 140
U.N. Partition Plan (Palestine), 150
U.N. Peacekeeping units, 163
U.S. Agency for International Development, 90, 136, 198
U.S. Bureau of Reclamation (Ethiopian plan), 138, 139, 140
U.S. mediation in water disputes, 162
U.S. Soil Conservation Service, 159
Uganda, 78, 115
Ugaritic mythology, 23
Unconfined (phreatic) aquifers, 191, 201
Underground springs, 166
Underground water storage, 255
Underpricing for water, 227
Undugu (fraternity) group, 134
United Arab Emirates, 39, 194
United Nations, 88, 102, 274
United Nations charter, 271
United Nile, 116–142
United States Congress, 303
United States and Egypt, 124
Unity Dam (Yarmouk River), 38, 244–245, 250
Upper Egypt, 13, 58, 120
Upstream riparians, Nile, 126, 142
Upstream vs. downstream rights, 55, 270
Ur (Sumerian capital), 53
Urban society, earliest, 53
Urban water use, Jordan, 169
Usable storage, 166
Uzaym River (tributary to the Tigris), 93, 98
Uzziah (Judean king), 63–64

Vernadsky, Vladimir (scientist), 20
Victoria Nile, 115
Vonnegut, Bernard (scientist), 233

Wadi Halfa (Sudan), 118, 125, 127
Wadis and wadi beds, 5, 6, 68, 69, 155, 241
Wahat ad-Dakhilah (Egypt), 195
Wahat al-Bahriyah (Egypt), 195
Wahat al-Farafirah (Egypt), 195
Wahda Dam, Unity Dam (Yarmouk River), 173, 310
War of Basoos (legendary), 6, 7
War of 1948–49, 158, 160, 162
War of 1967 (Six Day War), 163, 170
Wars (Arab-Israeli), 89
Wasteful use of water, 210–231
Wastewater, treatment and reuse, 34, 175, 212, 219, 225, 229
Water-borne diseases, 129, 189
Water commissioner, Israel, 202
Water conservation, 34, 159, 210–231
Water as an export commodity, 246
Water-extravagant crops, 213, 219
Water harvesting, 65, 68, 240–243
Water lifting devices, 53, 61, 62, 97, 99, 121, 265, 305
Waterlogging, 51, 54, 98, 120, 129, 191, 217

Water as national resource, 269
Water for Peace program, 284
Water per unit of production, 222
Water Planning Authority (Tahal), Israel, 288
Water pollution, 33, 74, 109, 129, 138, 175, 212
Water quality, 36, 138, 162, 174, 175, 189, 209, 244
Water resources development, 19, 134, 184
Water resources and distribution, Jordan, 170, 173–175
Water retention by soil, 46
Water rights, 67, 209, 220, 267–278
Water-saving devices, technologies, 211, 286
Water spreading, 65, 67, 240
Water subsidies, 212, 213
Water supply (per capita), 34
Water-table, 54, 56, 59, 60, 191
Water-table fall, 174, 192, 196, 203
Water-table rise, 46, 99, 129, 215, 218, 219
Water-thrifty crops, 221
Water use efficiency, 34, 131, 135, 215, 221
Water as weapon and target of war, 265–267, 304
Waters of Strife, 296
Watersheds (catchments), 67, 68, 71, 84, 265, 269
Weather modification, 233–239
Weeds, 224
Weizmann, Chaim (Zionist leader), 84, 85, 186, 303
Well drilling, 192
Wells, 11, 16, 24, 25, 26, 28, 31, 39, 60, 65, 190–209, 206, 207
West Bank, 39, 40, 77, 88, 89, 150, 151, 152, 168, 173, 203, 204, 206–209, 244
West Ghor Canal (never built), 170, 173, 208, 250
Western Desert, 194
Wetlands, 265
White Nile, 10, 11, 12, 58, 112–142
White Nile pump schemes, 140–141
Wilderness of Zin, 4, 5, 296
Wilson, Woodrow (U.S. President), 82, 304
Wilting (plants), 22, 222
Winter crops, 58
Wittfogel, Karl (anthropologist), 23, 60, 299
Woolley, C. L. (archaeologist), 53
World Bank, 9, 274, 277
World Court, 276, 277
World War I, 4, 7, 74, 76, 77, 81, 88, 99, 149, 181, 186
World War II, 87, 88, 125, 149, 150, 197

Yabis stream (tributary to Lower Jordan), 155
Yarkon-Negev pipelines (Israel), 165
Yarkon springs (Israel), 165, 204
Yarkon-Taninim aquifer (mountain aquifer, Israel), 204–209
Yarmouk, historical battle of, 303
Yarmouk riparians, 169
Yarmouk River, 38, 39, 86, 107, 109, 155, 156, 158, 159, 168, 169, 171, 173
Yellow River, breaching dikes of, 266, 301
Yemen, 8, 77, 193
Yom Kippur War (1973), 89
Young Turks, 78, 79

Zagros mountains (Iran), 93, 96
Zaire, 115
Zamzam Spring (Mecca), 23
Zarqa stream (Jordan), 155
Zaslavsky, Dan (Israeli water commissioner), 315
Zifta barrage (Egypt), 121
Ziggurats (ancient Mesopotamia), 53, 102
Zionist movement and organization, 85, 86, 149, 303, 309